Working in Horticulture

CAREER PREPARATION FOR AGRICULTURE/AGRIBUSINESS

AGRIBUSINESS EDUCATION
Jasper S. Lee, Consulting Editor
Mississippi State University

CORE CURRICULUM
Max L. Amberson, Consulting Editor
Montana State University

Introduction to Agribusiness Management
Don L. Long J. Dale Oliver
Charles W. Coale

Human Relations in Agribusiness
John Hillison John Crunkilton

Agribusiness Procedures and Records
Delene W. Lee Jasper S. Lee

Advertising and Display in Agribusiness
James McGuire

Selling in Agribusiness
Larry E. Miller

Physical Distribution in Agribusiness
Ronald A. Brown John W. Oren, Jr.

Working With Animal Supplies and Services
Roy D. Dillon

Fertilizer, Chemicals and Seed
William S. Farrington Clyde F. Sartor
Paul R. Vaughn Ronald A. Brown

Working in Agricultural Industry
Jasper S. Lee

Working in Plant Science
Douglas D. Bishop Stephen R. Chapman
Lark P. Carter

Working in Animal Science
Paul Peterson Allen C. Christensen
Edward A. Nelson

Working in Agricultural Mechanics
Glen C. Shinn Curtis R. Weston

Leadership for Agricultural Industry
Bob R. Stewart

Learning Through Experience in Agricultural Industry
Max L. Amberson B. Harold Anderson

The Core Curriculum program includes student textbooks, student activity guides, audiovisual elements, and teacher aids.

The Agribusiness Education program includes student textbooks, student activity guides, and teacher aids.

Working in Horticulture

William B. Richardson
Purdue University
West Lafayette, Indiana

Gary E. Moore
Purdue University
West Lafayette, Indiana

GLENCOE

Macmillan/McGraw-Hill

Lake Forest, Illinois   Columbus, Ohio
Mission Hills, California   Peoria, Illinois

Library of Congress Cataloging in Publication Data

Richardson, William   (date)
  Working in horticulture.

  Includes index.
  SUMMARY: Text and related materials provide
instruction for skills basic to the field of horticulture and describe careers in ornamental horticulture, olericulture, pomology, and aboriculture.
  1. Horticulture—Vocational guidance. 2. Horticulture. [1. Horticulture. 2. Horticulture—
Vocational guidance. 3. Vocational guidance]
I. Moore, Gary E., (date) joint author.   II. Title.
SB317.6.R53   635'.023   79-17931
ISBN 0-07-052285-5

Working in Horticulture

Imprint 1991
Copyright © 1980 by the Glencoe Division of Macmillan/McGraw-Hill
School Publishing Company. All rights reserved.

Send all inquiries to:
Glencoe Division
Macmillan/McGraw-Hill
936 Eastwind Drive
Westerville, Ohio 43081

          10 11 12 13 14 15 VH 00 99 98 97 96 95 94 93 92 91

Sponsoring Editor/ Susan Horowitz
Editing Supervisor/ Katharine Glynn
Design Supervisor/ Caryl V. Spinka
Production Supervisor/ Kathleen Morrissey

Photo Research/ Mary Ann Drury
Cover Design and Text Illustrator/ F. Ronald Fowler
Technical Studio/ Burmar Technical Corp.

*This book was set in Optima by Monotype Composition Company, Inc.
It was printed and bound by Von Hoffman, Inc.*

CONTENTS

PICTURE CREDITS

PREFACE

The horticulture industry is an important part of our daily lives. Each day, we encounter the many aspects of horticultural products and services. For example, fruits and vegetables provide critical portions of our diet. Green plants and flowers are grown and sold for their beauty, for use in home decorating, for making places of work more attractive, and for use in recreational areas. Turfgrass is used on lawns and for sodded areas, such as golf courses. Ornamental trees and shrubs are used to landscape residential and commercial buildings, and to provide additional privacy.

As the demand for horticultural products and services has increased, the need for trained horticulturists to assume jobs and careers in horticulture has also increased. Horticulture industries and businesses have grown enormously and these firms need employees to produce horticultural products and to provide related services. For example, floriculturists grow and market flowers and foliage plants. These floral products are used for weddings and funerals as well as for floral displays in the home. In conjunction with providing these products, the floriculturists will often supervise arranging of flowers and floral displays on the premises. Another example of horticultural service jobs involves landscape maintenance firms that provide services to homeowners and businesses by installing and maintaining landscapes. These firms' employees assist in the landscape installation and maintenance.

The examples above are but two of many that exist in horticulture. Generally, the jobs and related careeer opportunities occur in the four branches of horticulture: (1) ornamental horticulture including landscaping, nursery, floriculture, and turfgrass; (2) olericulture (vegetables); (3) pomology (fruits); and (4) arboriculture (trees). The jobs and careers found in these four branches require employees who can perform specific horticultural tasks.

This text and related instructional material are designed to provide the basic information to develop these skills and tasks common to all four branches of horticulture.

Working in Horticulture: An Instructional Module

Working in Horticulture was developed as an instructional module to assist teachers in developing a systematic approach to teaching basic skills in vocational horticulture. The competencies developed in this module are the basis of a core curriculum in vocational horticulture.

The instructional module includes the following components:

- A student textbook that provides basic skills development in a career context.
- A student activity guide that includes application-type questions, problem solving, and laboratory activities to supplement the learning of basic skills.
- A teacher's manual and key that includes teaching suggestions and answers to the questions in both the text and activity guide.

STUDENT TEXTBOOK

Working in Horticulture provides students with the competencies needed for most entry-level horticultural jobs. Each unit and chapter is organized around a set of competencies that have been converted into behavioral goals. For example, Unit

Two is entitled "Growing Horticultural Plants" and Chapter 4 in Unit Two deals specifically with basic plant growing principles such as photosynthesis, respiration, and transpiration. One specific competency in Chapter 4 concerns the relationship between temperature and the rate of photosynthesis. As this area of content is discussed, the text attempts to illustrate why this knowledge is crucial to the horticulture employee. After reading Chapter 4 and completing all the exercises, a student should be able to perform tasks related to the basic plant growth principles. A similar pattern is followed with each unit and chapter in the text.

Behavioral goals are listed for each chapter. These goals were synthesized from regional and national research studies that identified the important skills needed by beginning-level horticulture employees. The student should review these goals prior to reading and/or discussing the content of the chapter. After completing the chapter, the student should again review the goals to see that each goal was mastered.

End-of-chapter questions are included to assist the student in internalizing the concepts taught in the chapter. Examples, case studies, and illustrations relate the general concepts of the world-of-work and daily living situations.

The text is divided into eight units, containing 23 chapters. Unit One (Chapters 1 and 2) provides an overview of the horticulture industry and the related careers. Unit Two (Chapters 3–5) discusses the competencies necessary to growing horticultural plants. Unit Three (Chapters 6–10) explores the techniques of controlling environmental factors that influence the growth of horticultural plants. Unit Four (Chapters 11–14) discusses the health of horticultural plants including a chapter on the safe use of pesticides. Unit Five (Chapters 15–17) describes the structures, tools, and equipment used in horticulture. Unit Six (Chapters 18–19) is designed to focus attention on plant products and services with specific emphasis on the care and maintenance of lawns and designing floral arrangements. Unit Seven (Chapters 20–21) explores horticultural experience and leadership programs that are unique to vocational horticulture. Unit Eight (Chapters 22–23) focuses on finding jobs and managing businesses in the horticulture industry.

Many people assisted in the preparation of the text. Special recognition should go to contributing authors Glen Shinn, Mississippi State University and John Wott, Purdue University; and to Richard Stinson, The Pennsylania State University, for review of the illustrative material. Appreciation also goes to the following members of our horticulture advisory committee for manuscript review: Donald Barrett, McHenry County Community College, Illinois; Sam Brashear, Seminole Community College, Florida; Jean Landeen, University of California, Davis; Kenneth Rhodes, Derry Area Senior High School, Pennsylvania; and Julian Smith, Chatham Central High School, North Carolina. We would also like to thank the following people for their reviews of the manuscript: Jay Disberger, Hutchinson Junior College, Kansas; Thomas Eltzroth, University of California, San Luis Obispo; Charles Griner, University of Georgia; Clifford Nelson, University of Maryland; Kenneth Parker, University of Massachusetts; D. Larry Spencer, South Mecklenberg High School, North Carolina. Grateful acknowledgment also goes to William H. Crouse and Donald L. Anglin for use of technical art found in Chapters 16 and 17.

Acknowledgment is also given to Katie Strain, Jackie Hertig, and Terry Robertson, horticulture education students at Purdue University, who contributed to the early development of portions of the manuscript, and to Bonnie Dobbins for typing the manuscript.

Working in Horticulture

ONE

The Modern Horticulture Industry

Each time you bite into an apple, bicycle through a park, buy a bouquet of flowers, or water a house plant, you're feeling the influence of the modern horticulture industry.

Fruits, vegetables, plants, shrubs, trees, grass—these are all part of horticulture, and all part of our daily lives—a part we often take for granted.

Stop and think a moment. Someone picked that apple from a tree long before it reached your hands. Someone planted and harvested the vegetable you ate for dinner last night. Someone potted and sold you the philodendron that's taking over your window sill. And someone is regularly caring for the shrubs and trees and grass that you see in parks and along highways. All these "someones" work in the horticulture industry.

The world of horticulture offers a wide variety of careers and jobs. There are outdoor jobs, such as those of the farmer and gardener, and indoor jobs, such as those of the horticulture salesperson and the horticulture teacher. There are jobs for people with limited experience, such as nursery workers, and jobs for highly trained specialists, such as greenhouse technicians. There are jobs for people who want to supervise other people or manage a business, and jobs for people who'd rather spend their time taking care of plants. In short, horticulture is a field that offers a wealth of opportunities to people of varying interests, goals, and skills.

In Chapter 1 of this unit, you'll explore the nature, scope, and importance of the horticulture industry. You'll learn about the four branches of horticulture, and about the kind of work involved in each branch.

In Chapter 2, you'll take a look at the range of specific career and job opportunities in horticulture, and you'll learn about the skills and background needed to succeed in this diverse field.

1

Overview of The Horticulture Industry

What comes to mind when you think of the word horticulture? Flowers, perhaps? Plants? Gardens and greenhouses? Actually, these and much more are all part of the world of horticulture.

Horticulture plays a major role in your everyday life. The fruits and vegetables you eat, the parks you walk through, and the plants you grow in your home and garden are all part of horticulture.

Today's horticulture industry is an important and dynamic segment of the American economy, as well as a significant factor in our daily environment. Horticulturists produce approximately 40 percent of our food supply. Their landscaping work adds beauty to homes, recreation areas, public buildings, highways, golf courses, and numerous other places.

The modern horticulture industry is a major influence on our lives, not only in terms of our daily diet, but also in its contribution to our surroundings.

CHAPTER GOALS

Your goals in this chapter are:

- To write a definition of horticulture
- To describe the relationship between horticulture and related plant sciences
- To explain the importance of horticulture as an industry in the United States
- To identify and describe the different branches of horticulture
- To describe the climatic regions and their influence on growing horticultural crops

What is Horticulture?

The next time you visit a park, notice the many different kinds of plants and how they are used. Grass, trees, shrubs, flowers—they all help to make the park an attractive and pleasant place, and also serve to divide it into recreation and picnic areas and separate it from nearby houses and businesses. Growing and caring for the plants used in parks and other public places is just one aspect of the modern horticulture industry. Let's see what else horticulture involves.

Horticulture as a Plant Science

The origins of the word horticulture are the Latin words for "garden," *hortus,* and "cultivation," *cultura.* The modern meaning of the word, however, is much broader.

Horticulture today refers to the science or art of growing fruits, vegetables, and ornamental plants. It is one of the four plant science areas in agriculture. The others are agronomy, botany, and forestry (see Figure 1-1). Examining the relation between horticulture and the other three areas will help you to understand exactly what horticulture is.

Agronomy is the study of field crops such as corn, wheat, soybeans, alfalfa, and rice. Normally these crops are field grown on very large

acreages. An individual farmer may have 600–1000 acres or more of a particular crop such as corn. Persons who study these crops are called *agronomists.* In addition to studying field crops, agronomists are also concerned with soils and soil conservation.

Botany is the study of the properties and life phenomena exhibited by a plant. In other words, *botanists* study plant life and the many things that affect plant life.

Forestry refers to the study and maintenance of forests. People who specialize in this area are *foresters.*

Horticultural crops are often distinguished from agronomy crops by the manner in which they are grown and the purpose for their growth. Normally, horticultural crops are grown in an *intensive* manner—that is, in small acreage, with much labor and at high cost. Agronomy crops are grown in larger acreages, and both the cost per acre and the labor required per acre are often less.

Horticultural crops are usually grown either as food or for their aesthetic value. To illustrate

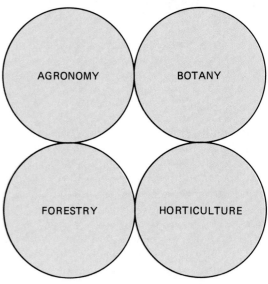

Figure 1-1. Agronomy, botany, forestry and horticulture are the four plant sciences in agriculture.

Figure 1-2. The products of orchards and farms provide 40 percent of our food supply. Vitamins, minerals, and other nutrients are supplied to us from horticultural products.

the difference, think of growing corn for grain (agronomy) and growing corn for human consumption (horticulture). The intensity of production is greater when corn is being produced for human food. Corn is an agronomic crop when it is grown in very large acreages (hundreds of thousands), whereas flower crops, such as the carnation, are horticultural crops that are grown intensively.

The relationship of botany to horticulture and agronomy to horticulture is similar. Botanists study plant life phenomena regardless of the plant's classification; agronomists study plants based on the amount to be grown.

Forestry contrasts with horticulture in the fact that trees grown in the forest are produced in large acreages. Much of this tree production is for wood and wood products, where ornamental trees used by horticulturists have food or aesthetic value.

Scope of Horticulture

In recent years people have become more diet and health conscious, and more aware of the part that fruits and vegetables play in well-balanced meals. As a result, the work of horticulturists involved in the production, processing, and storage of fresh fruit and vegetables has taken on increasing importance.

Figure 1-3. Horticulture beautifies our surroundings through the artistic use of shrubs, trees, grass, and flowers, and gives people an opportunity to express their creative skills.

The "green plant explosion" of the past few years has also focused attention on horticulture. People enjoy filling their homes, offices, and patios with green plants, and there has been a growing demand for horticultural information about the care of these plants. These are but two elements of modern horticulture. There are many more.

Food Value. The food-related importance of the horticulture industry comes not only from the fact that 40 percent of our food supply is horticulturally produced. In addition to providing a large portion of the food needed to feed our population, horticultural products supply many of the vitamins, minerals, and other food nutrients required for a balanced diet. In fact, horticultural products are the best source of

many of our vitamins and minerals. Many vegetable crops, such as broccoli and spinach, supply vitamin A and other essential vitamins. Minerals needed by your body, such as iron and phosphorus, are also available in many vegetables (see Figure 1-2).

Aesthetic Value. Another important aspect of horticulture is its value in beautifying our surroundings. Horticulture provides flowers, shrubs, ornamental trees, grass, landscape designs, and much more (see Figure 1-3).

The next time you see a well-groomed lawn with flowering shrubs; maybe a flower garden; tall, green hedges to provide privacy; and perhaps a few shade trees, remember that all these things are part of horticulture. Indeed, everything from the grass on tennis courts and golf courses to the colorful flowers that the florist sells is part of the horticulture industry.

Educational and Avocational Value.   A third aspect of modern horticulture is its use for educational and avocational purposes. Every year more and more people take up vegetable gardening or flower gardening as a hobby. While these people are not professional horticulturists, their gardens do provide them with an enjoyable and productive pastime.

At many retirement homes, mental health institutions, and similar agencies people learn to make flower arrangements, bouquets, and other items with horticultural plants. And, where space permits, they grow gardens to gain the benefits of both physical and mental activity.

Economic Value.   Horticulture is also an important part of our economy. It is becoming more and more dynamic, and the need for horticultural workers is increasing. Several support industries have developed as a result of the growth of horticulture. In addition to providing approximately 40 percent of our food supply (a major economic force), the modern horticulture industry has great economic impact.

Retail florists, garden centers, and nurseries handle millions of dollars of business in local communities each year. This business generates local taxes and jobs.

Landscape and turf businesses provide a community service and hire many employees.

Vegetable and fruit industries encompass jobs in production, processing, and marketing of food substances produced in horticulture.

Flower and plant production in the West, lower East Coast, and Florida generate major industries in these areas.

The need for businesses to process, store, and market horticultural products is expanding. With this expansion comes the need for trained workers. This type of economic growth has positive effects upon our economy.

Branches of Horticulture

Now that you have a clearer understanding of the importance and scope of horticulture, let's take a look at some of the different horticulture branches. In general, horticulture can be divided into four different areas: (1) *floriculture, turf, landscaping,* and *nursery production;* (2) fruit production, known as *pomology;* (3) vegetable production, called *olericulture,* and (4) tree and shrub cultivation, or *arboriculture.*

Ornamental Horticulture

The ornamental branch of horticulture deals with the cultivation and use of plants to beautify or improve the environment. Ornamental horticulture involves those activities connected with growing, arranging, selling, and maintaining flowers, grass, shrubs, and trees.

Floriculture.   Floriculture is the part of ornamental horticulture that deals with the cultivation of ornamental and flower plants. Floriculture can be further described as the business that grows and sells flowers, foliage, and related materials. This part of the modern horticulture industry offers employment opportunities with florists in planning, growing, designing, and selling floral products and services.

One of the important aspects of floriculture is the use of greenhouses. Since many of the flowers and ornamentals produced by horticulturists are very fragile, greenhouses are needed to provide a controlled environment. In this way, horticulturists can produce almost any type of plant, regardless of the outside weather.

Turfgrass.   The turfgrass portion of ornamental horticulture is becoming a valuable industry. Turfgrass is used in athletic fields, golf courses, parks, cemeteries, and numerous other places.

Workers in this industry must know how to establish a turf and how to maintain turfs once

Figure 1-4. The four work areas in horticulture provide many employment opportunities: (Top left) in the greenhouse; (Bottom left) in turfgrass; (Top right) in landscaping; (Bottom right) in nursery production.

they are established. Their work involves the operation of several pieces of equipment used in turf maintenance, such as lawn mowers, fertilizer spreaders, small tractors, trucks, and chemical applicators.

Turfgrass workers must also be able to deal with the various factors affecting the growth of good turf, such as fertilization, seeding, pest control, and proper watering.

Landscaping. The planning and construction of landscapes is an important part of ornamental horticulture. The landscape industry is con-

cerned with the establishment and maintenance of trees, shrubs, and other plants, and the planning and construction of landscape features such as walks, patios, rock gardens, and flower beds.

Landscape workers must be able to identify and care for a large variety of plants. They must know how to control insects, diseases, and weeds, and they need to be knowledgeable about pruning and fertilization techniques. In short, they must have the skills required to establish and maintain the many plants that make up a landscape.

Nursery Production. Another major segment of ornamental horticulture is the production, harvesting, and sale of ornamental trees and shrubs by nurseries.

Nurseries provide many of the plants used in landscapes. They may sell their trees and shrubs to garden centers or directly to the public. Nursery workers must know how to propagate, grow, and care for many kinds of plants.

Fruits—Pomology

The production, harvesting, and marketing of fruits, known as pomology, is a second main branch of horticulture. Fruit crops are more dependent on climate and geographical region than most other agricultural crops. Citrus crops,

for example, are grown in warm climates, such as Florida, Texas, and California, while cherries and blueberries grow best in cooler climates, such as Michigan.

People who work in fruit production must know how to propagate, plant, transplant, prune, fertilize fruit trees and vines, and protect them from insects and disease.

Vegetables—Olericulture

The third principal branch of horticulture, called olericulture, involves the production, processing, and storage of vegetables. Vegetables, like fruits, are fragile and grow best in certain geographical areas. For example, mass production of many vegetables takes place in the warmer

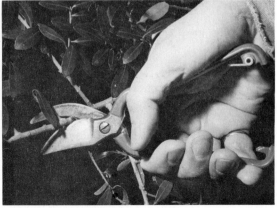

Figure 1-5. Transplanting, grafting, and pruning are some of the skills a student must learn to work in horticulture.

climates of the Southern, Southwestern, and Western portions of the United States.

Workers in the vegetable industry must know how to plant, grow, fertilize, and harvest vegetable crops. They must also be knowledgeable about methods of insect and disease control.

Arboriculture

Arboriculture, the fourth horticultural branch, concerns the cultivation of trees and shrubs. Specifically, it involves planting, transplanting, fertilization, disease control, pruning, bracing, and general care of trees and shrubs.

While there is some overlap between arboriculture and nursery operations, generally the trees and shrubs with which arboriculture is concerned are larger than those involved in nursery operations. However, this may vary in different regions of the United States.

Climate and Horticulture

The effect of horticulture on the economy is important, yet it varies throughout the United States. Also, the four branches of horticulture (ornamentals, fruits, vegetables, and arboriculture) are more important in some areas of the country than in other areas. The reason for the variation is the climate. Climate has a great influence on the growing of horticultural plants. A complete understanding of the modern horticulture industry includes a review of climate and its influence on horticulture.

Climatic Zones

The plant-growing regions of the world are classified into three very broad areas: tropical, subtropical, and temperate. These three regions normally are identified by a combination of their temperature and moisture characteristics.

The tropical regions occur near the equator. At the equator the sun's position is such that it shines most directly on the earth. This results in a very warm climate.

The areas around the equator with abundant rainfall are called *humid tropical regions,* whereas the tropical areas that do not have sufficient rainfall are called *arid tropical regions.* Most horticultural crops grown in tropical regions are grown where the rainfall is abundant. This is true of the banana, coconut, and papaya crops. Most of the houseplants you are familiar with have their origin in tropical areas. This is why these plants must be grown indoors, especially in cooler, nontropical areas.

The subtropical regions are in close proximity to the tropics. Whereas the tropics have an absence of winter, the subtropics may have a greater change in temperature and in some cases have frost and winter-type conditions. For example, some portions of Florida have subtropical conditions, yet winterlike temperatures can also occur. Citrus crops and avocados are often grown in subtropical climates.

The third climatic zone is the *temperate* one. These climatic zones do have winter temperatures and are often subdivided on the basis of the severity of the winter, whether it is mild or severe.

Regional Temperate Zones

Within the temperate climate region, the United States Department of Agriculture (USDA) has established *regional temperate zones.* The establishment of these zones is based on the approximate range of average minimum temperature (see Figure 1-6). Commonly these zones are called the *plant hardiness zones,* though plant hardiness zones are only indicators and may not correspond exactly to regional temperate zones.

Plants have different tolerances to cool and/ or warm temperatures. Some plants cannot grow in cold temperatures. Other plants grow well in

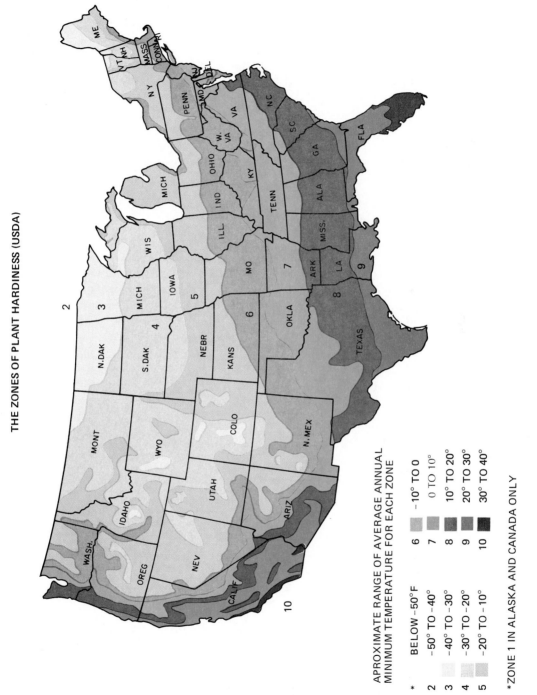

Figure 1-6. The zones of plant hardiness are based on the average minimum temperature. Since plants have different tolerances to temperature, these zones provide guidance for sensible planting.

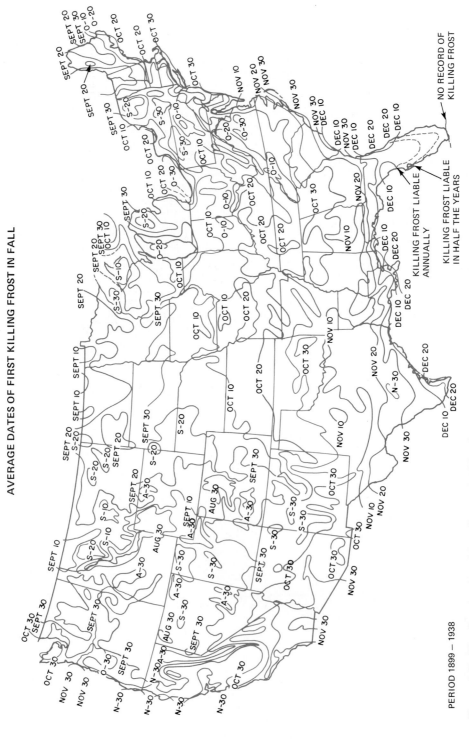

Figure 1-7. The first (top) and last (bottom) killing frost dates can help you determine the outdoor life of plants and crops.

AVERAGE DATES OF LAST KILLING FROST IN SPRING

KILLING FROST LIABLE ANNUALLY

NO RECORD OF KILLING FROST

KILLING FROST LIABLE IN HALF THE YEARS

PERIOD 1899 – 1938

cool temperatures but can also grow well in hot temperatures. The situation is often that the plant itself is not harmed by the cold or hot temperature, but its ability to produce fruit or to flower is damaged by the extreme temperatures. For example, cold temperatures will not necessarily kill the peach tree but the blooms can be destroyed by a frost at the wrong time of the year. The peach tree itself will continue to grow, but it will not produce fruit. And the peach tree must have a certain amount (700 to 1,000 hours) of cold temperature (below 45° F, or 7° C) before they will produce fruit.

Horticulturists are quite aware of the frost areas. In fact, most horticulturists and farmers can tell you immediately when the first frost is likely to occur in their area. Figure 1-7 provides a division of the United States based on the last killing frost as determined by USDA records. This information is very helpful to horticulturists in these areas, since it helps them to determine the times when certain plants can be moved either outdoors or indoors.

The Modern Horticultural Industry: A Review

Modern horticulture is indeed an important and dynamic industry. Its influence is felt each day in what we eat, where we live and work, and even how we spend our leisure time. In addition, the industry plays a major role in our country's economy.

The four main branches of horticulture—ornamental horticulture, fruit production, vegetable production, and tree and shrub cultivation—offer a variety of careers and jobs. Each has its own requirements for specialized skills and knowledge.

The modern horticulture industry offers challenge and excitement, as well as an opportunity to learn while making a significant contribution to our daily lives.

THINKING IT THROUGH

1. In your own words, define horticulture.
2. How does horticulture relate to the other areas of plant science such as agronomy, botany, and forestry?
3. Describe the importance of various aspects of modern horticulture in terms of its: (1) food value, (2) aesthetic value, (3) educational and avocational value, and (4) economic value.
4. What effect does climate have on the growth of horticultural plants in certain geographic areas?
5. What is meant by regional temperature zones?
6. Define and describe each of the major branches of horticulture.

2

Careers in the Horticulture Industry

The horticulture industry offers a wide variety of career opportunities to those who want to work with plants, flowers, and trees, or who have an interest in growing fruits and vegetables. All over the country, horticulturists are working to beautify the environment, develop new varieties of plants, and improve methods of cultivation. Nurseries, gardens, laboratories, and floral shops are just a few of the many places where someone trained in horticulture can work.

Whether you'd like to harvest crops on a vegetable farm, design or care for home landscapes, sell plants in a small shop, or work your way up to manager of a large nursery operation, the world of horticulture has a place for you in production, sales, or service, if you have the necessary skills and knowledge.

CHAPTER GOALS

Your goals in this chapter are:

- To identify career opportunities in ornamental horticulture, vegetable production, fruit production, and arboriculture
- To describe a typical career ladder in the horticulture industry
- To list personal and technical skills required for entry-level positions in horticulture
- To outline a model vocational horticulture education program

Career Opportunities in Horticulture

Each of the four principal branches of horticulture—ornamental horticulture (including floriculture, nursery, landscaping, and turf), vegetable production, fruit production, and arboriculture—offers a range of careers and jobs from which to choose. Some require a minimum of training, while others require the kind of skills that come with years of experience.

Functions of Horticulture Jobs

In horticulture, as in most other industries, the kind of work you do can be classified according to the job functions performed. Horticulture work is generally divided into production, sales, and service jobs.

Production jobs relate to the actual growing and harvesting of plants. A nursery worker, a greenhouse employee, and a floral designer all have jobs in which a product is produced and made ready for sale.

Sales jobs, of course, involve the actual selling of products to customers. A garden center employee, a nursery salesperson, and a landscape maintenance salesperson all work in sales.

Many horticultural businesses provide services, either to individuals or to businesses.

People who work for landscaping firms or floral shops, for instance, have service jobs.

Let's take a closer look at the variety of career opportunities in horticulture.

Floriculture

Careers in floriculture involve growing, distributing, and selling flowers and foliage plants. Floriculture businesses thrive in large metropolitan centers, as well as in areas that have long growing seasons and warm climates.

Career Profile. Gary is a former vocational horticulture student now working as a grower in a local greenhouse business. This business specializes in the production of floral crops. Gary chose the floriculture industry because he was interested in the growing and selling of flowers and foliage plants. Gary finds his greatest enjoyment in producing a quality potted plant. He takes pride in the growing of good quality

Fig. 2-1. Career opportunities in floriculture include employment in retail floral shops, greenhouses, flower farms, and garden centers.

plants as he propagates, prunes, and cares for the young plants. Finally, Gary's pride is best fulfilled when he sells a potted plant that he personally produced. He found that the career opportunities in floriculture included employment with retail florists, greenhouses, flower growers, flower farms, and garden centers.

Gary enjoys his job as a greenhouse worker. He hopes that job experience and additional horticulture education will enable him to become a greenhouse manager someday. He knows that the manager position will require years of work and experience, but he is willing to put in the time to reach his long-term career goal.

Career Opportunities. Opportunities in the floriculture industry revolve around the growing and selling of flowers and foliage crops. Plants may be grown either indoors, in the controlled environment of greenhouses, or outdoors on farms. The plants are then sold through floral shops, garden centers, and plant stores.

Some businesses combine the growing and selling of flower and foliage plants. For example, a florist shop might sell chrysanthemums that were grown in the shop's own greenhouse. Such businesses may require employees who are skilled both in growing the product and in selling it.

Jobs Related to Growing Plants. Many jobs related to growing flower and foliage plants involve the use of the greenhouse, in addition to or in place of an outdoor farm. In the states that produce major floral crops, you will find acres and acres of greenhouses. California, Florida, Ohio, Pennsylvania, New York, Colorado, and North Carolina, for example, all produce a large volume of chrysanthemums, roses, orchids, and similar plants, and have many acres of greenhouses, as well as acres of flower farms.

Greenhouses are especially important in areas where cool weather and short growing seasons limit outdoor production. But, a state such as

Figure 2-2. Some jobs in horticulture require a minimum of training while others demand an expertise that comes from years of experience.

Hawaii, with warmer weather and a longer growing season, can rely primarily on farms.

Careers and jobs related to growing plants span a wide range, and the level of skill and amount of experience they require vary.

Occupational opportunities, in general order of increasing responsibility, include: greenhouse grower, greenhouse technician, physical plant manager, production manager, and greenhouse manager. Although not every business includes all these specific job titles, all of the *skills* normally associated with these titles *will* be required. For example, in a small business the same person may be both manager and technician. However, in a large operation, there may be several supervisors and technicians.

Floriculture production jobs also exist in nongreenhouse settings, such as the flower farms in the South, southern East Coast, and California. The jobs closely parallel the job titles listed above. Thus, on a flower farm you would probably find a grower, a farm manager, and technicians.

Growers and greenhouse workers, or grower assistants, are responsible for the actual growing of the flower or foliage crops. The grower directs such processes as soil planning, crop propagation, fertilization, watering, pruning, pest control, chemical application, and harvesting. One or more assistants help the grower to carry out these functions. While the actual number of workers involved will vary with the size and scope of the business, proper performance of growing tasks is always crucial. A single mistake can mean thousands of dollars in lost revenue for the business.

The job of greenhouse technician often requires formal education beyond high school, as well as experience in flower and foliage crop production. Technicians must be knowledgeable in all the technical aspects of crop growth, from propagation to recognition and treatment of plant diseases and disorders. In addition, technicians must keep abreast of the latest growing techiques and any new developments in their field.

In the position of supervising technicians, growers, and other workers is the greenhouse or farm manager. The manager is responsible for the overall operation and coordination of the business. In many instances the manager may actually own the business.

Managers are required to make numerous decisions regarding the best use of the physical plant, personnel, and money. Such decisions require a high degree of skill in managing all three resources, if the business is to show a reasonable profit.

The manager's job requires a solid background in the business. Often managers work their way up through all the intermediate positions in order to gain the necessary experience and knowledge. In addition, many managers acquire formal training through college or through a vocational horticulture program.

Jobs Related to Selling Plants. The growing of plants accounts for only a portion of the floriculture industry. The plants must be sold. There are basically three kinds of businesses that sell plants: retail shops, garden centers, and wholesale houses.

One of the principal ways in which flowers and plants are sold is through retail shops. Floral shops and plant stores carry a wide variety of products, from bouquets of flowers to potted evergreens. A retail florist shop might sell thousands of dollars of plants and flowers on a weekend or holiday. In fact, before the major holidays such as Mother's Day, Christmas, Easter, and Valentine's Day, a shop might sell a large percentage of its annual sales. Catering to special activities such as weddings and funerals also comprises a good percentage of a floral shop's business.

Jobs in these retail businesses usually require people with some skill in designing and producing arrangements and constructing displays. In addition, employees must be able to relate well with customers and must have enough knowledge to be able to offer advice about plant care.

The most common job title is retail florist. The retail florist may operate and manage a floral shop and supervise one or more salespersons. In addition to having a working knowledge of plants and floral arrangements, the floral shop manager must possess both business management and personnel management skills.

Garden centers have become very popular in suburban and rural areas. The range of products sold by the garden center is quite varied. In addition to flower and foliage crops, a garden center may also sell fertilizers, seeds, chemicals, shrubs, trees, lawn supplies and equipment, bedding plants, and many other products.

The garden center may be privately owned, or it may be part of a chain. Jobs associated with garden centers include those of the manager, assistant manager, and garden center employees. The size and scope of the center determines the number and kinds of jobs performed by personnel.

Garden center managers and employees alike must know about flowers and foliage plants, must understand the care and maintenance of many plants, and must have selling skills.

Many floriculturists operate businesses that are called "wholesale houses." A wholesale house actually grows no plants. It serves as an intermediary between the wholesale producer and the retailer. For example, a floral shop may need a large quantity of Easter lilies at Easter to sell along with its locally grown lilies, and may rely upon a wholesale house for the additional quantities.

The jobs associated with the wholesale business generally concern growing the plants and selling the crop. A wholesale florist is a person who produces the plants that are sold by the wholesale house. Wholesale florists must be knowledgeable about flowers and foliage plants. In addition, they must know the seasons and geographic markets for flowers. A wholesale florist in Arkansas may well sell roses to a wholesale house that delivers flowers in Chicago, Dallas, or Denver. Thus the wholesale florist must keep up with the demands of markets in many areas of the country.

Nursery Operations

Jobs in nursery operations involve growing, harvesting, and selling a wide variety of woody plants. Nursery businesses may be either wholesale or retail.

Career Profile. Dianne is a former vocational horticulture student who is now employed by a nursery operation near her home. This nursery operation grows, harvests, and sells ornamental shrubs and trees. The main function of Dianne's

job as a nursery worker is to grow the nursery plants under the direction of a supervisor. This job requires a great deal of work with hand tools and some machinery. Dianne drives small tractors, uses pruning shears, digs shrubs with shovels, and uses a tiller to loosen soil in the nursery area.

Dianne plans to gain experience in the nursery so that she can eventually become a supervisor or an assistant manager in the business.

Career Opportunities. Like Dianne, many students who are trained in vocational horticulture begin as nursery workers. For people who enjoy working with plants and who want to gain some valuable experience in the industry, this job is an excellent starting point.

The next step up the ladder is the job of supervisor. The supervisor supervises crews of nursery workers as they grow the nursery plants. Supervisors oversee such important tasks as planting, pruning, fertilizing, irrigating, spraying, weed control, and digging. They may also help train newly hired and inexperienced workers. Supervisors usually report to an assistant manager.

The assistant manager is responsible for making decisions concerning plant production. The supervisor and work crews carry out their work on the basis of the decisions made by the assistant manager.

The nursery manager is the person responsible for the overall operation and coordination of the nursery. This means managing the personnel, facilities, and money that are available to the business.

In addition to these basic jobs, large nursery operations may also hire someone who is trained in the propagation of nursery plants. This job is critical to a nursery operation if the business is to produce quality plants. In smaller firms, the job is usually performed by the manager, assistant manager, or supervisor.

A large firm may also hire a quality control technician who has had special training in plant

Figure 2-3. Landscape (top), turfgrass (top right), and nursery (bottom right) employees apply their basic skills working on the job.

nutrition and pest control. Quality control technicians may also hold a college degree.

A nursery must have someone who is responsible for selling the nursery plants. This is the job of the sales manager. Large nursery operations may sell plants within a very wide area. In such cases the company may employ representatives or salespersons to handle customers and sales in specific areas called "territories."

Retail Nurseries.   Retail nurseries grow and sell trees, shrubs, and flowers to homeowners. In many ways the retail nursery is identical to the garden centers described earlier, except that a retail nursery may grow many of the plants that it sells.

Landscaping

Opportunities in the landscaping business revolve around construction and maintenance of home landscapes, including sodding, planting, and cultivation of trees and shrubs. Schools and business firms may also need the services of a landscape firm. In general, there are two types of landscape businesses: (1) landscape design and construction, and (2) landscape maintenance firms. Many firms do both. Garden stores and retail nurseries may provide these services.

Career Profile.   Don is employed by a landscape maintenance and installation firm. He wanted to work outdoors and to work with his hands, and his job gives him this opportunity.

Don is one of a group of workers who install and maintain landscapes under the watchful eye of a crew supervisor. Don's job requires him to plant trees and shrubs, water and fertilize plants, lay sod, weed flower gardens, mix and apply pesticides, and operate small hand and power equipment. His favorite task is laying sod, even though it is the hardest work. He enjoys seeing a barren spot covered with grass.

Don plans to gain experience with the landscape firm, and hopes to become a crew supervisor himself eventually.

Career Opportunities. Landscape firms offer careers ranging from landscape worker and landscape designer to construction supervisor-superintendent, salesperson, and manager.

As a person with high school training, you are most likely to be hired as a worker. The job of Don and other workers includes helping prepare a site for landscaping, planting trees and shrubs, laying sod, and constructing landscape sites such as gardens. Workers may also operate, adjust, and maintain machinery. The ability to drive a truck and operate a snowplow (in Northern states) is important to ensure year-round employment.

The landscape worker also reports to a supervisor. The supervisor is in charge of preparation of landscape sites; staking out where plants are to go; overseeing construction of walks, patios, and walls; planning, fertilizing, weeding, mowing, and general maintenance; and the supervision of the workers. In large firms a supervisor may also serve as superintendent. In such cases two people coordinate the work of several landscape crews.

Landscape supervisors and superintendents advance by gaining skills and knowledge through experience. In many cases they also received additional training in landscape principles through technical schools and/or colleges.

Landscape managers oversee the firm's entire operation. The manager is responsible for managing the equipment, personnel, and money of the firm and ensuring quality work. The manager is often the owner and secures the various landscape jobs the firm is to do.

The landscape designer prepares the plans for each landscape site. Designers are highly trained; they must be able to draw a landscape plan, recommend the correct plants and structures to be used, and in many cases supervise the construction of the landscape. Landscape designers must have a thorough knowledge of hundreds of kinds of landscape plants, and must know exactly where they will grow well.

Turfgrass

The turfgrass industry handles the production, installation, and maintenance of turf. Turf is used in many ways—around homes; as landscaping for commercial or industrial sites; in parks, golf courses, and athletic fields; and along highways.

Career Profile. Jack is employed year round as an assistant greenkeeper on a golf course. (In Jack's area, the climate is such that the golf course is open all year, which is not true in many states.)

The tasks Jack performs include mowing, fertilizing, pest control, irrigation, and general maintenance of the golf course. Jack has already learned a great deal about grass, and he hopes to eventually be promoted to greenkeeper.

Turfgrass Opportunities. Careers in the turfgrass industry involve various levels of responsibility with regard to turf production, installation, and maintenance. Entry-level positions include groundskeeper, greenworker, and landscape gardener. All three jobs require a knowledge of how turf is grown and how it is installed or established. In addition, workers must be able to carry out such maintenance functions as mowing, fertilizing, pest control, and irrigation. An ability to use hand tools and operate engine-powered machinery is also important.

Beginning workers in the turfgrass industry

may be supervised either by a supervisor or by a superintendent. For example, golf course workers like Jack report to a golf course superintendent or an assistant superintendent. On the other hand, if you are employed as a groundskeeper or a landscape gardener, you would report to a supervisor. Once you have gained experience as a greenkeeper, groundskeeper, or gardener, you may move up to assistant superintendent or crew supervisor. The assistant superintendent and supervisor positions require people who are experienced in working with turf, and also have the abilities needed to oversee the work of other employees. In some cases these jobs require additional training in a 2-year post-high school course.

The person who coordinates the work of all phases of the business is the superintendent or the manager. This high-level job includes the responsibility for managing personnel, keeping records, and ordering supplies and parts.

Fruit Production

The fruit production branch of horticulture involves the production of apples, cherries, oranges, pears, peaches, and other fruits. In addition, there are several fruit product industries that are related to fruit production, such as the orange juice industry.

Career Profile.   Beth is employed as a production worker in a vineyard, a job that has taught her a lot about how grapes are grown. Her work consists of pruning, fertilizing, applying pesticides, and harvesting and storing the grapes. Beth has to operate farm equipment, drive tractors and trucks, and use a variety of basic hand tools.

Beth reports to a production supervisor who is responsible for a crew of workers. She plans to gain experience as a production worker and become a supervisor herself someday.

Career Opportunities.   Careers in the fruit pro-

duction field involve two basic functions: producing the fruit and marketing it.

Fruit farms are quite large. The orchards and vineyards where fruits are grown need a great deal of maintenance. Producing high-quality fruit requires much experience and skill.

Marketing fruits is a large part of the food industry in the United States. Oranges consumed on the East Coast after being picked in Florida, Texas, or California have been handled by several agencies and many different people. Careers in marketing exist at various stages throughout this process.

Producing Fruit Crops.   Jobs in fruit production call for workers who are interested in growing and caring for the crops. The entry-level job is often that of production worker. On a fruit farm, production workers will primarily be taking care of trees or vines (such as grape vines). This involves pruning, pest control, disease control, and harvesting. Workers must be able to use hand tools and operate machinery.

Production workers are supervised by a production supervisor who is responsible for seeing that quality fruit is produced. The supervisor has experience in all phases of the fruit production process and works for the farm manager.

The farm manager is responsible for the entire fruit production farm. The manager is a decision maker who must have experience in the farm operation. In many cases the farm manager is also the owner of the farm.

Marketing Fruit Crops.   Jobs related to marketing fruit crops are complex. Many careers in fruit marketing require college training and/or considerable experience in fruit crop processing and marketing. These careers involve skills in such areas as packaging, pricing, storage, transportation, and selling.

Vegetable Production

The vegetable production branch of horticulture is concerned with the production of food and

Figure 2-4. Many fruits and vegetables are produced on large commercial farms.

food products. The majority of vegetables are produced by large commercial firms owned by corporations or by large farms owned by individuals.

Career Profile. Jack works with his father on their vegetable production farm. They raise a variety of crops suited to their location. Jack's responsibilities consist of preparing the fields for planting, doing the planting, applying pesticides, and harvesting and storing the crops.

Jack's immediate supervisor is his father, who assumes the dual role of production supervisor and farm manager. Jack's favorite job is transporting the crops to the nearby city. He drives the truck, helps load and unload the crops, and signs the weight receipts.

Jack hopes someday to become a full partner with his father, and when his father retires, Jack hopes to take over the farm.

Career Opportunities. The vegetable business, like the fruit business, involves two main functions: production and marketing.

The production phase deals with the actual growing of the crops. Since there are many different types of vegetable crops, the way in which they are grown will vary greatly. Today, most vegetable operations are large.

The marketing of vegetables has developed into a sizable industry since the 1960s. Specialized packaging, storage, and sales agencies have evolved which handle most of the vegetable marketing.

Producing Vegetable Crops. If you seek employment with a large vegetable farm, you are most likely to begin work as a production worker. Production workers assist with the growing of the crops. On most vegetable farms, this job involves soil preparation, planting the crops, fertilization, weed control, and harvesting. Workers must know how to use hand tools and operate engine-powered machinery.

Vegetable farm workers are supervised by a production supervisor. In addition to supervising the workers, the supervisor is responsible for

overseeing the field production process, and has often had considerable experience on the farm.

The farm manager, who may also be the owner, coordinates the entire operation and oversees the supervisor. The manager is responsible for all major decisions made on the farm, from selection of the varieties grown to assessing the proper harvest time for the crop. The manager's decisions have a tremendous impact on the ultimate profit or loss of the farm business. A good manager must develop good business sense as well as production skills.

Marketing Vegetable Crops. Those who seek marketing-related positions must have a basic understanding of the entire crop production process. The marketing sites require work with grading, packing, storage, pricing, and selling.

Arboriculture

Those who work in the field of arboriculture are engaged to repair, protect, and maintain trees and shrubs.

Career Profile. Jerry is employed by the city in which he lives. His job is to remove unwanted trees and shrubs, control tree and brush growth on city property, and prune damaged or diseased limbs from trees in the city park and on other city property.

Jerry enjoys this job because it allows him to work outdoors. He has always been challenged by climbing, and finds that climbing a large tree to prune a damaged limb is both exciting and interesting work.

Career Opportunities. You may have heard of tree surgeons and wondered just what they do. Jerry is a tree surgeon. He removes damaged and diseased limbs from trees, which is the kind of work that requires a great deal of skill. Career opportunities for those who are trained in the cultivation of trees and shrubs exist in three types of businesses or agencies: (1) private tree services, (2) utility companies, and (3) city and town governments. The kind of work in these three agencies is quite similar with regard to tree maintenance. In general, there are opportunities for ground workers and climbers, supervisors, and managers.

Ground workers and climbers must be physically strong and have little fear of heights, since their work involves climbing trees, removing damaged or diseased limbs, and performing any other services to the tree.

The supervisor oversees and directs the work of the ground workers and climbers. This job requires experience in the business as well as the ability to get along with people.

The manager is the decision maker for the business. It is the manager who secures jobs for the business and assigns the jobs to be done. It is also the manager's responsibility to supervise the repair and maintenance of equipment, and to hire personnel.

Career Ladder in Horticulture

As we have seen in examining the various career opportunities available in the four branches of horticulture, the more knowledge and experience a person has, the more responsibility he or she is able to take on. For example, a high-level position such as nursery manager requires much more background than an entry-level job such as nursery worker. And between manager and worker there are of course a number of other positions. A supervisor, for instance, would have more training than a worker, but generally not as much experience as the manager. Some of the basic characteristics of the jobs on this "ladder" are shown in Figure 2-5.

Most entry-level horticulture jobs require an ability to get along with people and a willingness to work. Entry-level jobs often involve a lot of manual labor. Many jobs are mechanized and technical, and employers are looking for em-

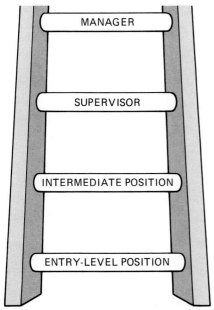

Figure 2-5. The career ladder shows the steps associated with many horticulture businesses.

ployees who have skills in using equipment and tools.

On the first rung of the horticulture career ladder, you will have to know or be willing to learn about plants, about the methods for using basic hand tools, and about the methods for operating small machinery.

Intermediate Positions

Once you have gained some experience in a horticulture business, your employer or supervisor may give you increased responsibility. Experience is very important, because most employers are reluctant to take a chance on inexperienced workers when business income is at stake. Once a worker has demonstrated skills and abilities, however, the supervisor will more readily assign new duties.

Intermediate-level jobs in horticulture generally require some business knowledge and experience. Workers may need to know about the various plants being produced and sold, the

operations being performed, and the methods of operating equipment and machinery. In addition, workers should show initiative in getting a job done properly.

Supervisory Positions

Supervisors direct the work of employees in one or more phases of a business operation. A supervisor of a landscape maintenance crew, for instance, might be responsible for overseeing a group of experienced and inexperienced workers in the maintenance of a school, business, or home landscape.

Supervisors must have a firm knowledge of their business, as well as an ability to direct and get along well with people, if they are to obtain quality work.

Managerial Positions

Managerial jobs often require specialized training and experience in the horticulture business.

Professional and Technical Occupations

The career ladder in Figure 2-5 presents the sequential career order most normally followed by horticulture students. However, there are other types of jobs that are not directly related to the career ladder. These are the professional and technical occupations.

Professional and technical jobs require experience and a great deal of formal education. For example, a greenhouse technician usually has a master's degree or a Ph.D., and a research scientist in a university or a private company will hold a Ph.D.

Another professional-level occupation is the job of extension agent. Extension agents, who are employed by the federal government and the land grant university in your state, perform an educational function by providing up-to-date information to horticulturists in local areas. One other example of a professional-level occupation is that of the horticulture teacher.

Technical job holders include plant propagators, who propagate plants and work closely with plant breeders to ensure good quality plants; and research scientists, who conduct research on the growing, processing, and distribution of horticultural products.

Skills Necessary for Horticulture Occupations

In addition to having a basic interest in and appreciation of plants, anyone planning to pursue a career in vocational horticulture needs two sets of skills: technical and personal-social.

Technical Skills

Technical skills are needed not only to enter many horticulture jobs, but also to advance up the career ladder. Technical skills refer to knowledge of horticultural plants, their production, and their uses. The need for skilled horticulture workers is one of the reasons why horticulture programs such as the one in your school were established.

Figure 2-6. The ability to get along with others is important in selling horticultural products.

Personal-Social Skills

Personal-social skills relate to an individual's personal qualities and social being. In horticulture work, as in most occupational fields, skills are very important; in some jobs, such as selling horticultural plants, they are critical. Have you ever encountered a rude salesperson? Did you return to that store? Most people would not return if they had encountered a rude salesperson. And, of course, whether or not your job involves dealing with customers, getting along well with your coworkers and supervisor is essential to the smooth operation of the business.

Working with People. Horticultural plants are grown to serve a purpose. What all their purposes have in common is that horticultural plants serve people. If you are to succeed in a horticultural career, you must be people oriented as well as plant oriented.

For instance, if you are interested in floral design, you must be able to design floral arrangements that people will like and want to purchase. And you must be able to communicate in a clear and tactful manner with all types of customers who may come into the floral shop.

If your interest is in turfgrass, this may lead to a job with a turf farm. In this case, you need to be able to work as part of a team with supervisors and coworkers in producing quality turf.

In these and many other cases, your ability to relate to and communicate with a wide variety of people will be a principal factor in determining your success on the job.

Willingness to Assume Responsibility. Almost any career that you choose will require you to assume some responsibility. The higher the level of the position, the greater this responsibility will be.

For example, as a golf course worker, an important part of your responsibility would be mowing the greens. Many golfers become upset if the greens are not properly mowed. You must

Figure 2-7. (Left) Awards and contests encourage students to compare their skills with others in their field. (Right) Supervised occupational experience (SOE) programs can help you develop the skills needed on your first job.

be willing to accept responsibility for learning how to mow the greens correctly and then carrying through with this task.

Personal responsibility is also a significant aspect of floral work. When the floral shop manager asks you to design centerpieces for weddings and other important functions, the reputation and financial status of the shop will be affected by your skill and conscientiousness in making each design a piece of living three-dimensional art. The manager assigns each job and then assumes it will be done properly—each arrangement, for example, must be the right size for the setting. The combinations and positions of flowers in the arrangements must be pleasing. And most importantly, the flowers must be fresh and free of blemishes. You must be willing to accept responsibility for all these elements if you are to be successful at your job.

Imagination and Initiative. An employer operates a business to make money. Often the amount of money one business makes compared to the amount another business makes is a function of the imagination and initiative of the employer and employees.

In a job such as floral designer, for example, imagination is a crucial skill. If you have imagination and can create attractive floral arrangements that customers cannot buy elsewhere, your employer's business will increase.

You must also have initiative. Initiative means doing things without having to be told to do them. You use your initiative in working with customers, displaying flowers and plants, maintaining a landscape, or pruning a tree. Employers want to assign you a task and have you show initiative in getting the task done.

Developing Skills

The development of skills in horticulture is enhanced by practicing those skills in situations resembling actual working conditions.

Future Farmers of America (FFA)

FFA is a student organization that is aimed at developing the personal-social and leadership skills necessary for employment in agriculture. Participation in FFA gives you an opportunity to learn by actually doing.

Some FFA activities, such as public speaking, learning parliamentary procedure, and participation in chapter activities, allow you to express and develop yourself as a person through fellowship and cooperation. Other activities specifically related to horticulture, such as contests and awards programs, encourage the development of technical skills and knowledge of floriculture, fruit and vegetable production, nursery operations, and turf and landscape maintenance.

There are several activities that your local organization can become involved in. In one state the horticulture students organized a green plant fair, where members of the school community came in and purchased plants for use in their classroom, at home, or as gifts. There are many other types of activities that can give you a chance to practice your horticulture skills.

Supervised Occupational Experience (SOE) Program

Your vocational horticulture program will also allow you to gain experience in the supervised occupational experience (SOE) program. In your senior year you might obtain release time from school to work in a horticulture business. This experience allows you to apply what you have learned in school to an actual working situation,

and provides you with a good transition from school to work.

Your vocational horticulture instructor will assist you in developing a program that matches your occupational interests. This program normally involves working part-time in a local horticultural business. You will be paid for this work, and you will be supervised by an employer who will act as your supervisor. In addition, your vocational horticulture instructor will assist and advise you in your job. Many students are offered permanent jobs if they do well in the SOE program. And since employers generally prefer to hire people who have had some actual work experience, participation in the SOE program increases your chances of getting the job you want once you finish school.

Careers in Horticulture: A Review

The modern horticulture industry is very large and includes many kinds of jobs. A variety of opportunities in production, sales, and service is open to people who have an interest in horticulture and are willing to get the necessary education and training.

As you begin your vocational horticulture program, you should aim to develop personal-social skills as well as technical skills. Both are necessary if you are to be successful.

If you choose your career path carefully, develop the necessary skills, and work hard, your future in horticulture will be rewarding.

THINKING IT THROUGH

1. List some of the career opportunities available in ornamental horticulture, vegetable production, fruit production, and arboriculture.
2. Describe a typical career ladder in one of the four branches of horticulture.
3. Name the three basic kinds of functions in horticulture, and give examples of each.
4. What are technical and personal-social skills, and why are they important?
5. Describe two opportunities for getting actual work experience while still in school.

TWO
Growing Horticulture Plants

There are three areas of importance to the modern horticulture industry, areas that encompass the special skills needed by most horticulture workers—the ability to grow plants while understanding why the plants grow, the ability to know and understand the basic plant parts and their functions, and the ability to identify horticultural plants. Very little goes on in horticulture that does not touch on these three major areas, and as a potential horticulture worker you must be able to perform these skills. Unit Two provides the basic skills and knowledge necessary to master the understanding of these three areas.

In Unit One you learned about the nature of the modern horticulture industry and the career opportunities available in horticulture. If you review the two chapters in that unit, you will recall that jobs exist in the following areas: ornamental horticulture, olericulture, pomology, and arboriculture. In this unit it will become clear that all these areas require employees who are able to handle the three areas of importance to grow plants, recognize and understand the functions of plant parts, and be able to identify plants. Consider the following examples.

A greenhouse employee who is to propagate geraniums must be able to make cuttings from stalk plants. This asexual method of propagation requires that the worker understand the functions of the stems and roots, as well as the mechanisms of plant growth. This requires learning the basic plant parts and functions. How can you make a stem cutting unless you know what a stem is and can identify its function?

An employee on a vegetable farm must understand the influence of weather on vegetable plants. The basic plant growth processes—photosynthesis, respiration, and transpiration—must

be understood by the employee to ensure that the plants are protected from the adverse effects of their environment. For example, tomato plants can be covered to reduce the harmful effect of the sun's energy during periods of high intensity. This requires that the worker understand how the sun influences the tomato plant.

Plants grown in greenhouses are also grown in controlled environments to maximize plant growth. By allowing the basic plant growth processes to function at their optimum level, greenhouse workers scientifically control the environment, light, temperature, moisture, and air, thus allowing the plants to grow well.

A landscape maintenance employee must know the names and identity of many shrubs, trees, and ornamental plants. A crew of landscape employees who are installing plants according to a landscape plan must be able to determine which plants are yews, which are junipers, and which are ivy. Without knowledge of the plant's identity, how can the worker place plants in the proper location? Also, it is important that landscape maintenance workers be able to identify these plants on sight, since time does not always permit the textbook approach to plant identification.

As you can see from these few examples, employees in the horticulture industry must be able to understand the growing of horticultural plants—specifically, plant parts and functions, how plants grow, and how to identify horticultural plants.

As you read and study Chapters 3, 4, and 5 in Unit Two, you will see how specific skills are applied to many other horticulture jobs. As you study further you can begin to see the necessity for horticulture employees to perform specific plant-growing tasks as a major part of their jobs.

3

Basic Plant Parts and Their Functions

Lonnie was visiting his uncle's nursery business and helping a couple of the employees to prune some of the plants. On this particular day they were removing old blooms from rose bushes. One worker explained to Lonnie that to remove these blooms correctly you must make the cut ¼ in (inch) [0.6 cm (centimeters)] above the axillary bud that accompanies the topmost five-leaflet leaf on the stem. Lonnie was impressed that the employees knew so much about the plants. One employee explained to Lonnie that a good horticulture employee had to learn a great deal about plant parts and their functions in order to do the job correctly. The employee illustrated the point by listing the number of plant parts and their functions that had to be understood to prune the rose bushes correctly. Particularly important were the identity of the axillary bud, the five-leaflet leaf, and the stem. Additionally, an employee had to understand why a cut was made at a particular point, and had to know the functions of each plant part.

Potential employers who operate a horticulture business expect you to have a basic understanding of plant parts. Plant growers who propagate their own plants must know the parts of the plant that are capable of regenerating new roots or stems, in order to provide whole plants; the tissues of the plants, in order to prepare cuttings; the root systems of plants, in order to prepare growing media; the parts of a flower, in order to produce flowering plants in time for a specific holiday; and the way in which turfgrass leaf blades elongate, in order to understand how mowing practices produce quality lawns. Plant growers must know all about plants, their parts, and their functions.

CHAPTER GOALS

Your goals in this chapter are:

- To recognize the types of stems and describe their functions
- To describe the structure and function of leaves
- To identify buds and describe their functions
- To recognize the basic parts of a flower and describe the function of each
- To distinguish between simple and compound fruit
- To identify the different types of root systems
- To describe the structure of the typical plant cell

After Lonnie's experience with the worker, he talked with his uncle and indicated how much the worker knew about plants. To illustrate the importance of understanding plant parts, Lonnie's uncle led him into the greenhouse, where he removed one of the plants from a bench. The plant was a geranium. Lonnie was asked to point out whatever parts of the plant he already knew. Lonnie was able to point to the leaves and to a stem, but his uncle explained that in addition to these parts, the geranium had roots, would soon develop buds, was able to produce flowers, and had several internal parts.

In order to acquaint Lonnie with the different parts of plants, he and his uncle strolled through the greenhouse and the other areas of the nursery where some plants were being grown. Lonnie's uncle pointed out some basic differences in plants based on plant characteristics.

Plant Classification

The plant classification system is important to the horticulturist not only in terms of plant identification and technical names (as discussed in Chapter 5) but also in terms of understanding plant parts and their functions.

A very detailed system has been developed to classify *all* plants. The category within the classification system of plants that the horticulturists are concerned with is called *Tracheophyta*. This category of plants includes almost all the higher plants—those with roots, stems, and leaves. The Tracheophyta category is further divided into several classes. The classes important to the horticulturists are the *Filicineae*, *Gymnospermae*, and *Angiospermae* classes. Ferns are examples of the Filicineae class of plants. The Gymnospermae class includes plants that bear seeds in scalelike holders with no covering (naked seeds). Examples of the gymnosperms are pine, fir, and spruce trees. The angiosperms bear seeds in a fleshy or enclosed holder. The apple, orange, geranium, and lily are examples of angiosperms.

Angiosperms are the most important class of plants to the horticulturists. Within the Angiospermae class of plants an important distinction

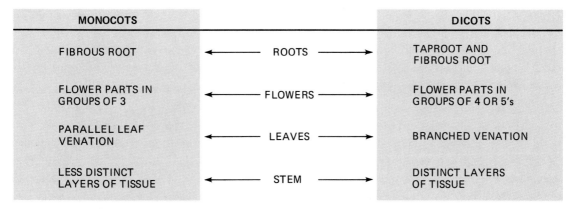

MONOCOTS		DICOTS
FIBROUS ROOT	← ROOTS →	TAPROOT AND FIBROUS ROOT
FLOWER PARTS IN GROUPS OF 3	← FLOWERS →	FLOWER PARTS IN GROUPS OF 4 OR 5's
PARALLEL LEAF VENATION	← LEAVES →	BRANCHED VENATION
LESS DISTINCT LAYERS OF TISSUE	← STEM →	DISTINCT LAYERS OF TISSUE

Figure 3-1. The differences between monocots and dicots.

exists. Plants in this class are in two subclasses—*monocots* and *dicots.*

When you examine a plant and look at the various plant parts, differences between monocots and dicots can be found. Normally, monocots have leaves with parallel leaf venation (where the veins in the leaves are parallel to each other) and flower parts in groups of three. The corn plant, bamboo, and lily are examples of monocots. There are about 50,000 different monocot plants.

The dicots have leaves that exhibit branched venation and flower parts in groups of four or five. Examples of dicots are tomatoes, apples, and geraniums. There are about 200,000 different dicot plants. Figure 3-1 provides a complete analysis of the differences between monocots and dicots. These differences will be further explained as the plant parts are discussed.

The importance of the differences between monocots and dicots relates to several factors. Asexual propagation (growing a plant from a part of an existing plant) discussed in Chapter 9 states that monocots are very difficult to propagate asexually. In fact, many monocots *cannot* be asexually propagated. A second factor relates to the use of pesticides. Chemicals have been developed that will kill broadleaf plants and not narrow leaf. Knowledge of the differences between monocots and dicots is very important to the horticulturist.

Stems

One of the major parts of each plant is the stem. The stem of the plant is like a backbone, because it supports the plant. The stem is also the main part for carrying water and minerals upward in the plant.

Types of Stems

Plant stems are called either herbaceous or woody, depending on how hard or soft the stems are. A *herbaceous* plant has a fleshy, soft stem that is easily broken. Peppermint plants, tulips, geraniums, and petunias are examples of herbaceous plants.

A *woody* stem is a stem that continues to grow from year to year. The stem is hard and woodlike. Examples of woody stemmed plants are maple trees, magnolia trees, and palms.

Monocot and dicot stems differ in the arrangement of the tissues within the stem. A dicot stem has distinct layers of tissues, while the monocot tissues are not so distinct (see Figure 3-1). When grafting and propagation are discussed in a later chapter, the importance of these differences will become apparent.

Nodes and Internodes

While Lonnie and his uncle were examining some of the plants that had just recently been

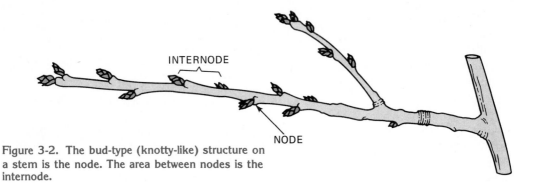

Figure 3-2. The bud-type (knotty-like) structure on a stem is the node. The area between nodes is the internode.

transplanted into containers, his uncle pointed out a forsythia stem and asked Lonnie whether he could identify the little knotty-looking structure on the stem. Lonnie was unable to name it, so his uncle mentioned that it was called the *node*. This enlarged portion of the plant is the place where leaves develop. The area between nodes is called the *internode* (see Figure 3-2).

Stem Variations

Plants have many variations of stems, one of which is called *modified stems*. Modified stems are divided into two groups, the above-ground stems such as crowns, stolons, runners, spurs, tendrils, and thorns; and the below-ground stems such as bulbs, corms, rhizomes, and tuber roots.

The knowledge of stem variations is important because these plants parts are helpful in identifying many plants, are used in propagating new plants, and are crucial to the growth of existing plants.

Above-Ground Stem Variations. The leaves of a strawberry plant appear to occur on a mass of almost white-colored stems compressed together. This fleshy lump of compressed stems is called a *crown*. Leaves and flowers come from the crown just as they do from any other stem.

Spurs are little stubby side stems that shoot out from the main stems on some woody plants.

Spurs are common on fruit trees such as pears, apples, and cherries, and it is the spurs that are responsible for most of the flower buds on these trees. The spur may after several years start to grow and become just like an ordinary stem. However, this does not occur very often.

A *stolon* is defined as a horizontal stem that lies along the top of the ground. A stolon is capable of forming roots from any node that touches the ground. Strawberry runners are stolons that have a relatively long distance between nodes. Stolons, like many other stems, have nodes that produce buds and flowers. A striped spider plant has stolons that can produce an entirely new plant.

Many grasses, such as Bermuda and St. Augustine, produce stolons. Golf course workers need to know how to grow grass from pieces of stolons. This method of growing grass is called *sprigging*.

The *tendril* is a stem modification that occurs on grape vines. Another stem modification that you have probably encountered is the thorn. If you have ever been pricked while handling a rose bush, then you know exactly where thorns occur.

Below-Ground Stem Variations. A potato is actually a stem variation called a *tuber*, an underground stem that stores food for the potato plant. The tuber, like every other stem, has nodes that produce buds. The eyes of a potato are actually the nodes on the stem.

Rhizomes are another kind of stem variation. They are very similar to a stolon, except that they grow horizontally underground rather than on top of the ground. They are also able to develop roots and shoots at any node. Plants that grow from rhizomes are Canadian thistle, iris, and Bent grass.

Tulips, lilies, daffodils, and onions are plants that produce underground stems known as *bulbs.* Bulbs are shortened stems with thick, fleshy scales. Buds develop at the base of the scales. Following low-temperature periods, these stems elongate to produce flowers. Flower growers must know the correct low temperature to use and the amount of time necessary for growing these bulbs. In October or November if you actually cut through the center of a tulip, daffodil, or assysius bulb, you can see all the flower parts in miniature within the bulb. A small hand lens that magnifies these parts can help in distinguishing the parts.

Although they look alike, bulbs and corms are not the same. The *corm* may have the same shape as the bulb, but it does not contain fleshy parts. A corm is a solid, swollen stem base that is enclosed in dry, scalelike leaves. A corm has small *cormels,* which are miniature corms de-veloping at the base of the parent corm that may be used to produce entirely new plants. Gladi-olus and crocus are examples of plants that have corms.

Workers in horticulture businesses have to be able to tell the difference between the top and bottom of the corm. This is so that they will not plant the corm upside down. If it is planted upside down, it will not grow.

It is a difficult process to distinguish between a below-ground stem and a root. One way to tell the difference is to look for nodes. A below-ground stem has nodes, whereas a root does not.

Stems are used to propagate plants. Thus, to the horticulturist a knowledge and identification of stems is important. Roots are also used in some form of propagation, yet you need to be able to distinguish between a root and stem in order to properly propagate a plant.

Leaves

The *leaf* is the major food-producing part of the plant. It is a function of the leaf to absorb enough sunlight so that the plant will produce enough

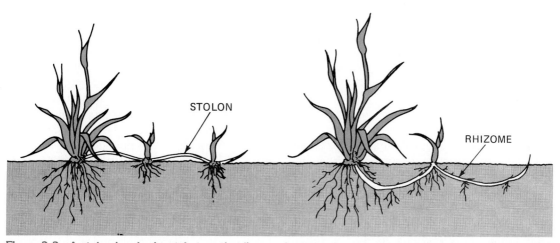

Figure 3-3. A stolen is a horizontal stem that lies on the ground; a rhizome is a horizontal stem that grows underground.

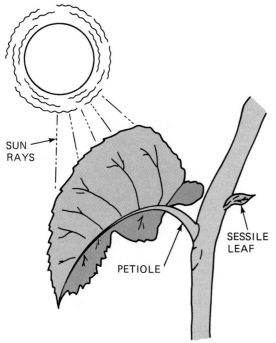

SUN RAYS

SESSILE LEAF

PETIOLE

Figure 3-4. The function of the leaf is to absorb sufficient sunlight to permit the plant to produce enough food to grow.

One of the parts of the leaf is the *epidermis,* which is a continuous cell layer that covers the entire plant, including the leaf. The epidermis helps to protect the leaf from dry air, sudden temperature changes, and wind, just as our skin protects us. The way in which the cells in the epidermis are arranged determines the texture of the leaf's surface. If you have ever felt a chrysanthemum leaf, you know that it is soft and velvety. These different textures result from differences in the ways the cells' epidermises are arranged. Some leaves have hairs that are an extension of certain cells of the epidermis. The African violet has so many that the leaf feels like velvet.

The Function of Cutin

The epidermis cell wall has an important structure called a *cuticle.* The cuticle is a waxy layer that is present on the exposed surface of the epidermal layer. The cuticle produces *cutin,* which is a waxy substance that protects the inner tissues from disease and also helps to keep the inner tissue from drying out. The amount of cutin that a particular plant has on its leaf surface is a response to the amount of sunlight it has received. A philodendron that has been in your living room with no sun will have very little cutin developed on its leaf surface, and if you move the plant into direct sunlight it is more than likely that the plant leaves will turn brown—at least on the margins. The reason for this is that the leaf had very little cutin on it to protect it from drying out, so when it was placed in the sunlight it could not protect itself. Any move to sunlight should have been gradual. In the case described, the philodendron has in a sense become sunburnt, just as you would if you were suddenly exposed to a great deal of sunlight in the spring when your skin is sensitive because it is not used to strong sun. When you are sunburnt, your skin dries out and you say that you are blistered. The philodendron must make the move to sun gradually, so that cutin

food to grow. Figure 3-4 shows a drawing of the leaf.

Horticulturists need to understand the leaf and its parts. For example, leaves are one part of the plant that tells the grower if there is a problem with the plant. Leaves and their parts are also crucial in the identification of plants. The leaf, as discussed in Chapter 5, is an important clue to plant identification.

Parts of a Leaf

The *petiole* is a slender stalk that connects the leaf blade to the stem. A leaf that attaches directly to the stem without a petiole is called a *sessile* leaf. Many grasses have sessile leaves. A leaf is attached to a stem at a node, and in some cases there is more than one leaf at a node. In the axil of each leaf there is a bud or several buds.

will have a chance to develop and protect the plant from drying out.

In sunlight the cutin layer is well developed, and there is no problem when a plant is moved to shade. The cutin simply begins to thin out. Even fruit has cutin on it. Have you ever polished an apple on your shirt in order to see the glossy shine? It is the large amount of cutin that produces this shine on fruit when it is polished.

The waxy cutin cuticle repels water just like the wax on a car. Put a drop of water on a leaf and you will see it bead up just like a car after a good wax job. This is the reason for using spreader-stickers or detergents in spraying a plant for insects or disease. Unless we use a spreader-sticker or detergent, the material we spray on will not stick to the leaf surface but will run off that cuticle layer on the leaf surface, and we will not get good results. Putting either spreader-stickers or detergents into our spray will do the same thing to the leaf of that plant that a detergent will do to a greasy plate when it is being washed. They simply get the water or pesticide down to the leaf surface in order to control the pest that is on the surface.

Guard Cells

There are little holes in the lower epidermis of the leaves in nearly all plants. These holes are called *stomates,* and they allow gas and moisture to flow between the inside of the leaf and the surrounding air. On each side of the stomata are cells that control the opening and closing of the stomata. These banana-shaped cells are called *guard* cells.

The amount of gas and moisture exchange that occurs is controlled by the shrinking and swelling of the guard cells. The action of the guard cells is controlled by light, with the stomata open during daylight hours and closed at night in most plants. A greenhouse employee should not water plants a few hours before sunset. The worker should know that the sto-mates will close and the plant will lose no more

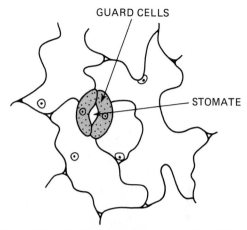

Figure 3-5. The stomata allows gas and moisture to flow between the inside of the leaf and surrounding air. The guard cells control the opening and closing of the stomate.

moisture during the night. Watering late in the day may present a risk of disease through infection of wet leaves. Water plants first thing in the morning.

Buds

Buds are plant structures that are capable of producing leaves or flowers, or both leaves *and* flowers. In most areas of the United States, many of the trees and shrubs lose their leaves in the winter. It is often difficult to identify trees in winter with their leaves gone. Bud characteristics are often used to determine the species of a plant, but a hand lens is often needed to detect the minute differences between buds.

Bud Locations

If you were planting trees in a landscape during the winter, bud identification would be useful in making certain the correct species of tree was planted.

Buds can be named by their locations on the plant. The most common position for buds is at

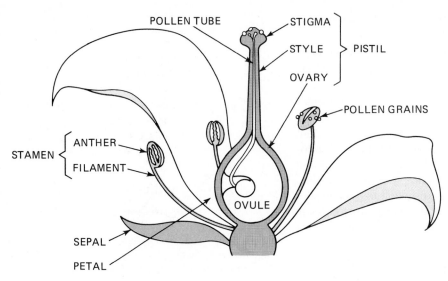

POLLEN TUBE

STIGMA

STYLE } PISTIL

OVARY

POLLEN GRAINS

ANTHER

STAMEN {

FILAMENT

OVULE

SEPAL

PETAL

Figure 3-6. The major parts of a flower.

the point where the leaf joins a stem. This point is called the leaf *axil* and is usually a node. Buds that are developed in the leaf axil are called *lateral buds.* The other position is at the tip of the twig or branch. The tip of the twig or branch is called the *stem apex,* and buds that develop at the stem apex are called *terminal buds.*

Although there are two normal positions for buds, a plant will often develop what is called *adventitous buds.* Plants that are injured or pruned excessively will develop adventitious buds. Sometimes called water sprouts, these buds grow quickly and are often removed from the plant by pruning.

Bud Dormancy

Buds do not normally grow in cold weather. Just like bears that hibernate during the winter when there is not much food available, buds of deciduous plants go dormant because it is too cold to grow. Thus it is said to be a *dormant bud.*

Buds of many plants require a certain number of days of low temperatures before they will start growing again. This is called a dormant rest period, and varies with different plants. For

example, the flower buds of the forsythia require a very short rest period, and will start growing in late winter if there is about a week of warm weather. A peach tree, on the other hand, requires a longer rest period of 700 to 1,000 hours of temperature below 45° F [7° C], and will not grow until it gets all the cold days it needs. But flowers of peaches can be killed by frost, and are frequently damaged this way because they bloom earlier than many other fruit trees.

Flowers

One of the plant parts that is of major importance to horticulture is the flower. Many horticulture plants are sold because of the attractiveness of the flower. Knowledge of flower parts and their functions is critical because: (1) workers who take jobs on flower seed farms, where good flower production is essential for producing seeds, need to know the many intricate aspects of the flower; and (2) knowing the flower parts and their function may help you to avoid accidently pollinating a plant such as the snap-dragon. Pollination of this plant will cause it to collapse overnight, rendering it useless to sell as a plant. Figure 3-6 provides a diagram of the major parts of a flower.

Sepals

Sepals are small green leaflike structures on the base of the flower. A sepal acts to protect the flower bud in the early stages of growth. When sepals are considered as a unit, they are called a *calyx*. While roses and carnations have ca-lyxes, tulips and lilies do not.

Petals

As you are aware, flowers come in many colors. The colored portion of the flower is called the *petal*. The bright colors help to attract insects and birds to the flower. Many flower petals have glands that secrete a liquid (nectar), which also attracts insects and humming birds.

Most flowers have more than one petal. In fact, the number of petals is often used in identifying a particular plant. For instance, roses have five petals on each flower. The petals as a group are collectively referred to as a *corolla*.

Pistils

In the center of the flower, there is a structure that is often shaped like a bowling pin and is called a *pistil*. It is the female reproductive organ of the plant. The pistil is composed of a stigma, ovary, and style. A *stigma* is the uppermost end of the pistil. It usually has a sticky substance to which pollen will cling. The *ovary* is the large end of the pistil, which contains one or more of the *ovules*. The *ovules* are the female eggs which, when fertilized, develop into seeds. The *style* is a long tube that connects the stigma and the ovary. The *pollen tube* produced by the pollen grain grows down the side of the style reaching the ovules in the ovary.

Stamen

The male reproductive organ is called the *sta-men*. It consists of two parts, a stemlike piece called a filament, and on its tip, the pollen-bearing part, called the *anther*. When a new anther dries up, it opens up and releases the pollen grains that contain the male reproductive units, the *sperm*. Both seed plants and higher animals reproduce by means of eggs and sperm.

Reproductive Functions

Why is it important for the brightly colored petals, fragrance, and sweet nectar to attract insects such as honey bees, moths, flies, and butterflies? Somehow the sperm and pollen grains on the anther must reach the ovules in the ovary if fertile seeds are to be produced. It is insects like the honey bee that help to move the pollen grain to the stigma. The transfer of pollen from the anther to the stigma is called *pollination*. Why is pollination so important? It provides the necessary male cell for union with the female egg (ovule) to produce a seed. As has already been mentioned, the sperm from the pollen grains must unite with the ovule inside the ovary to produce seed, which in turn will develop into the plant. The uniting of the sperm with the ovule is called *fertilization*.

Fruit

The fruit is actually the ripened ovule, together with any closely associated parts. In most plants, after the ovule or egg is fertilized by the sperm, the ovary matures into a fruit. As the maturing process occurs, many other changes occur as well. Carbohydrates are stored in the ovary tissue, and the tissue may hold increasing amounts of water. There are also changes in the pigment (color) of the fruit as it becomes fully developed. For example, a peach changes colors as it matures. It starts out green and by the time the peach is ready to eat, it may be a golden color with a pinkish tint to it.

Simple Fruit

Fruits are classified by their structure and number of ovules. *Simple* fruit develops from a simple ovary. Examples are coconuts, walnuts, peaches, and cherries.

Simple fruits can be fleshy or they can be dry. Berries are an example of simple fruits that have a fleshy tissue. Examples of berries are the tomato, banana, cranberry, and honeysuckle. The name "berry" when used in this manner does not apply to the berry fruits such as the strawberry, blueberry, and raspberry.

A simple fruit with fleshy tissue is further classified as *drupe* or *pome*. A drupe has a seed such as a stone fruit while pomes do not have a stone, but several seeds usually within a chamber. Examples of the drupe fruits are the cherry, peach, plum, and holly, as well as many woody ornamental plants. The inside portion of the fruit is fleshy. Examples of the pome fruits are the apple and pear.

Simple fruit can also have an ovary that is dry and brittle. There are two categories of the dry and simple fruits. The first category is called *indehiscent* fruits, which do not split open and release their seeds when ripe. Some examples of plants that have indehiscent fruits are buckwheat, corn, wheat, maple trees, and oak trees.

The second category of simple dry fruits are *dehiscent,* which split open when ripe and release their seeds. Examples of dehiscent fruits are the redbud, honeysuckle, coffee tree, peony, spirea, and forsythia.

Aggregate and Multiple Fruit

Fruits can also be classified as *aggregate* and as *multiple* fruits. Aggregate fruits are developed from one single flower that has many ovaries. Examples are the strawberry, raspberry, tulip, and magnolia trees.

Multiple fruit is a classification of fruit with flowers that are separated but closely clustered.

Examples of multiple fruits would be the fig, pineapple, and mulberry.

Roots

Another part of the plant that is crucial to almost all horticulturists is the *root*. Plants will not grow without roots. Interestingly enough, this part of the plant in most cases is not visible when one examines the plant. The roots are a major part of the plant and they often weigh over half of the dry weight of the plant.

The root system branches out and gives the plant contact with the soil. Why do you think this soil contact is so important? What functions do you think the roots serve for the plant?

Major functions of the roots are to serve as an anchor for the plant and to conduct nutrients and water from the soil to the upper parts of the plant. It is essential that the roots have good contact with the soil so that they will be able to absorb sufficient water and nutrients. However, it is also important that the soil have plenty of air space, because the roots need oxygen for respiration. If they do not get enough air through the roots they suffocate. How much the entire plant is able to grow depends upon how much the roots are allowed to expand.

Although the main function of the roots is to support the plant and absorb food, oxygen, and water for it, the roots may also serve as a storage house for food. Sweet potatoes and carrots are examples of roots used as storage houses.

How Roots Develop

When a seedling first develops, it sends into the soil its first root, which is called a *primary root*. If the primary root becomes the main big root of the plant, the system is called a *taproot system*. Carrots, beets, turnips, and some oak trees are examples of plants with a taproot system. If the primary root stops growing early in the growth of the young plant and starts growing new roots out of the stem, the primary root begins to develop a complex root system called a *fibrous root system*. Look at Figure 3-1

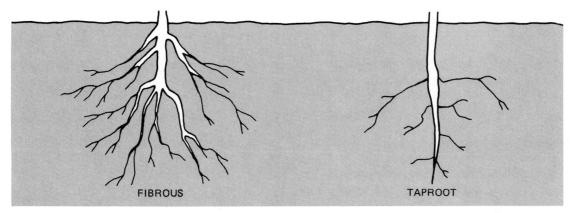

Figure 3-7. Compare a fibrous root system to a taproot system.

and note the difference between monocots and dicots based on type of root system.

Figure 3-7 compares fibrous and taproot systems. A fibrous root system allows the roots to make more contact with the soil. It is common practice for a nursery worker to cut off a taproot in order to force the plant to develop a fibrous root system. Since a fibrous root system makes better contact with the soil, it is often easier to transplant a fibrous rooted plant because it has a better chance of survival. In general, a taproot system grows much deeper into the soil than a fibrous root system, which is shallow rooted and takes food from the upper levels of soil.

One important structure found on the roots is root hairs. *Root hairs* grow from the epidermal cells of the growing roots. They occur only on the first ¼ in (.6 cm) of the root tips, and each hair lives only about 10 days. New root hairs are constantly being grown as the root tip grows through the soil. Root hairs are very delicate and quickly die if they become too dry or lack oxygen. If for any reason the root tip stops growing within 10 days, the root hairs will all die and the plant will be in serious trouble. Greenhouse growers of potted plants frequently check the health of the roots by knocking plants out of their pots to look at the color of the roots. White roots are healthy, black roots are dead.

Root hairs are important for the absorption of water and nutrients from tiny soil particles.

The Plant Cell

After Lonnie's uncle had given him an overview of the basic plant parts, he paused to tell Lonnie about the part basic to all plant life—the plant cell. The basic unit of plant and animal life is the cell. Your body is made up of millions of cells. Plants, like our bodies, consist of cells.

A typical cell is found in Figure 3-8. This cell contains many of the bodies that will be found in most cells.

The *cell wall* is the outer part of a plant cell. It encloses the cell and functions as a protective barrier around the cell. The cell wall of plants can be compared to our skin, which protects the internal structures of our body.

The *middle lamella* is a layer just inside the cell wall. It serves to maintain the structure of the plant cell, just as our bone structure maintains the shape of our bodies. When fruits begin to soften, the middle lamella breaks down.

The *cell membrane* serves to aid in giving support to the shape of the cell. It also functions as a barrier, similar to a screen door. Just as a screen lets fresh air in and manages to keep out pests, the cell membrane allows certain materials to enter the cell and keeps others outside the inner portions of the cell.

Although a plant cannot think, it does have a structure within each cell that controls the actions of the cell the way our actions are

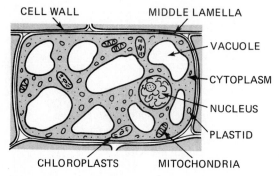

Figure 3-8. A typical plant cell.

controlled by our brains. This control center is the *nucleus*. Notice in Figure 3-8 that the nucleus is a dense, almost spherical-shaped body.

The *cytoplasm* is a very complex substance of the cell. It is complex because so many plant functions depend on the cytoplasm and the materials within it. The cytoplasm is composed of 85 to 90 percent water; the remaining 10 to 15 percent consists of substances such as salts, carbohydrates, proteins, and fats.

The plant cell also has its own power house. This power structure is called the *mitochondria*. The mitochondria contains large amounts of protein and is located in several areas throughout the cytoplasm of the cell. The mitochondria act along with various enzymes to function in the oxidation processes within the plant.

Vacuoles are located within the cytoplasm and are lined with a thin membrane. As shown in Figure 3-8 the vacuoles appear to be large, empty spaces. Actually these empty spaces are filled with a watery substance called *cell sap*. This cell sap contains many dissolved materials such as salts, pigment, and various necessary nutrients. Cells reproduce by dividing. In cells that are actively dividing, the vacuoles are very small; in older, more mature cells the vacuoles expand, become very large, and may occupy the entire center portion of the cell.

Chloroplasts are specialized rounded bodies that contain chlorophyll. Chloroplasts are essential to the plant cell because they are the complete unit necessary for photosynthesis, the process by which the plants make the food energy they need. Although chloroplasts are controlled by the nucleus, it often appears that they have a "mind of their own." They are capable of reproducing themselves.

Basic Plant Parts and Their Functions: A Review

Lonnie and his uncle examined and discussed the basic parts of most plants: stems, leaves, buds, flowers, fruit, roots, and the plant cell.

Each job Lonnie observed was related to a knowledge of plant parts. Fertilizing, pesticide applications, propagation, transplanting, pruning, and watering depended on a knowledge of flowers and roots.

THINKING IT THROUGH

1. What is the function of the stem?
2. Describe the nature of nodes and internodes.
3. List and explain the types of stem variations.
4. Explain how guard cells work to create openings in leaves, to allow water vapor and carbon dioxide to move into and out of plants. What are these openings called?
5. Describe the relationship between buds and dormant periods.
6. Identify the basic parts of a flower and describe the function of each part.
7. What are pollination and fertilization?
8. What is the difference between simple and multiple fruit?
9. What are the functions of roots?
10. List and describe the basic parts of a typical plant cell.

4

Plant Growth and Development

How many times have you or a member of your family either given or received a poinsettia at Christmas time? Many people do not realize that it isn't accidental that the poinsettia happens to bloom just prior to the holiday season. Did you know that the person who produces these plants—the greenhouse grower—carefully controls the plant's growth to produce the red bracts and green foliage at exactly the right time?

The greenhouse grower must know the precise time to begin growing the plant; the correct time to pinch it to induce branching; and the correct procedures for controlling sunlight, moisture, temperature, and plant nutrients.

For example, cuttings for poinsettias should be taken in morning hours. In order for the cutting to develop into the right size for a quality flowering plant for Christmas, the cutting should be taken during the first 2 weeks in September. The cutting will root in 3 weeks. The plant should be grown at nighttime temperature of 63° F (17° C). The grower uses light to control the "timing" of the flower bud formation. In central Ohio, flower bud formation naturally occurs about September 25, which results in flowering about 2 weeks before Christmas. The grower turns on lights each night at 10:00 p.m. for about 1 hour to prevent flower bud formation during the period from September 15 to October 10. In southern California, the grower must cover the plants from late afternoon until morning from October 10 to October 30, to be sure that the bud is set. The plants then bloom in time to be sold at Christmas.

The knowledge of this and other information about growing poinsettias is important to the greenhouse grower if the poinsettias are to be produced at just the right time.

The above example is just one of many that describes the everyday problems of workers in the modern horticulture industry. Similar concern for proper plant growth is characteristic of golf course employees, fruit and vegetable growers, retail florists, nursery operators, greenhouse growers, landscape maintenance workers, garden center employees, and employees of parks and recreational facilities. Can you think of other horticulturists in your community who grow plants for horticultural purposes?

Why do these horticulturists carefully monitor and control the plants' growth? One reason is that many of the plants they produce have seasonal usage, such as the poinsettia. If the poinsettia is not properly cared for during its growth, it may bloom too early or too late, or be of poor quality, which would cost the horticulturist money. Also, many of the plants are very delicate in nature and must be protected from adverse environmental conditions. Excessive heat will cause many flowers to bleach and become faded. This drastically reduces their economic value.

As a student who is interested in studying about horticulture and the horticulture industry, you too must learn the way in which plants grow and develop. A person who seeks a horticulture job may be asked to perform many tasks related to growing plants. You may secure a job in a nursery business where you are given the responsibility for watering nursery stock. Proper watering procedures must be followed if plants are to develop as they should. The tasks may vary with the type of plant being grown, but one thing is certain—the employer will insist on careful attention to the plants' growth.

CHAPTER GOALS

Your goals in this chapter are:

- To describe photosynthesis, respiration, transpiration, and translocation
- To identify the environmental factors that affect plant growth and development
- To describe the process by which plants grow to maturity

The proper care of plants in a horticulture business is of great economic importance, as is the care of a plant in your home. However, the economic impact of losing a crop of plants in the greenhouse is greater than the impact of losing a plant in your home. If you fail to water a plant or if you care for it improperly in your home, the economic loss may involve only a simple plant. However, the failure to properly care for the plants being grown by a horticulture business may result in thousands of dollars of lost income. In a similar manner, economic loss from the improper care of plants is just as critical for other horticultural businesses such as those of the nursery operator, the turfgrass specialist, the landscape gardener, the orchard owner, and the retail florist.

In this chapter you will be asked to assume the role of an employee in a greenhouse operation. You will be asked to care for a crop of

Figure 4-1. Working in a greenhouse, tending thousands of plants, is demanding yet rewarding work.

plants in order to test your knowledge of plant growing principles.

To take this role, assume that you have been employed by a greenhouse grower and assigned the job of assisting with the production of the vegetable plants. One of your jobs is to care for a crop of 10,000 tomato plants that are being produced in a greenhouse for a group of garden centers in your community. Your employer has given you the challenge of producing a healthy and vigorous crop of tomato plants.

There are several plant growth principles that affect your production of a high percentage of saleable plants. The environment in which the tomatoes are being produced must be properly controlled. The amount of sunlight, water, and temperature must be monitored because they have a strong effect on growing plants. The first plant-growing principle involves the way in which plants produce their food and the important life process they use to produce this food.

Photosynthesis

The production of food by a plant requires sunlight. Sunlight is important to plants, such as the tomato, because it it necessary to a plant growth process called *photosynthesis*.

Photosynthesis is the process by which plants produce their food. The word photosynthesis can be divided into two parts to determine its meaning: *photo*, which means "light," and *synthesis*, which means "to put together." Essentially, photosynthesis is a manufacturing process that uses sunlight energy to produce plant food.

However, this is only one part of the story. The photosynthesis process occurs inside living plant cells. As you may recall, the plant cell is the basic unit of plant life. Only certain cells of each plant are able to conduct the photosynthesis process. The leaves and stems generally contain the largest number of cells in which photosynthesis takes place.

TABLE 4-1
PHOTOSYNTHESIS EQUATIONS

General Equation

$$\text{Water} + \text{Carbon Dioxide} + \begin{array}{c}\text{Sunlight}\\ \text{and}\\ \text{Chlorophyll}\end{array} = \text{Simple Sugar} + \text{Oxygen}$$

Chemical Equation

$$12H_2O + 6CO_2 + \begin{array}{c}\text{Sunlight}\\ \text{and}\\ \text{Chlorophyll}\end{array} = C_6H_{12}O_6 + 6O_2 + 6H_2O$$

Inside the cells are bodies called chloroplasts. It is within green chloroplasts that photosynthesis takes place. Green chloroplasts can be viewed as the plant's food factory. Inside this living factory is a green substance called *chlorophyll*. (Incidentally, it is chlorophyll that gives the plant its green color.) This green coloring matter absorbs the sunlight (photo) and manufactures (synthesis) substances to produce plant food. The substances used in this process are water and carbon dioxide. Thus, inside living cells that contain green chloropolasts, sunlight and chlorophyll put together water and carbon dioxide to produce sugar, a plant food.

The photosynthesis process also requires that the plant environment have an adequate temperature. Many plants require a daytime temperature of 65 to 80° F [18 to 27° C], and a nighttime temperature of 45 to 70° F [6 to 21° C].

The plant food produced by photosynthesis is a simple sugar that is commonly referred to as a carbohydrate. After the simple sugar is formed, it can be transformed into other food substances for use by the plant for growth. Table 4-1 provides the chemical and general formulas for the photosynthesis process.

This process is important to the greenhouse grower. To produce strong, healthy, and vigorous plants, the environment in which the plants are growing must adequately allow for sunlight, water, carbon dioxide, and approximate temperatures. You, as the employee, must be aware of these environmental factors if you are to produce a high-quality crop of tomato plants. The control of these environmental factors is a skill that is often directly related to your job.

Photosynthesis and Sunlight

Have you ever picked up a board that has been lying on the grass in your yard? Did you notice the pale color of the grass under the board? This color results from the lack of sunlight received by the grass. Without the sunlight, the grass could not produce its food.

Photosynthesis is dependent on the availability of sunlight. Sunlight provides energy to combine water and carbon dioxide when a plant manufactures its food. Generally, the greater the sunlight intensity the greater the photosynthesis rate—which means greater food production by the plant.

The relationship between photosynthesis and sunlight is critical for the horticulturist. Many crops, such as tomatoes, respond best to maximum sunlight. The yield of tomatoes is reduced drastically as the amount of sunlight available to the plant is decreased. Only two or three tomato varieties will produce any fruit at all in greenhouses in late fall and early spring months.

Photosynthesis and Water

Water enters the plant through the root system and is transported throughout the plant. The importance of water to the plant is threefold:

1. Water is used in the photosynthesis process.
2. Water helps plants to maintain their turgor. *Turgor* is the firmness and tension given to living cells by the pressure of the fluids they contain. This pressure helps provide support through rigidity for the leaves and soft new stems.
3. Water transports dissolved substances throughout the plant.

The plant's water supply is in the soil. The plant grower must be aware of the amount of water in the soil. As a general rule of thumb, soil should be moist—neither too wet or too dry. As you gain experience as a horticulturist, you will be able to "feel" the soil and determine whether or not the plants require watering.

Water is removed from the top of the plant by *transpiration* (the process by which water vapor escapes from the plant). Yet the water is in the soil. This gives rise to two questions: (1) How is the water actually absorbed by the plant? and (2) How does the water move from the roots to other parts of the plant?

The process by which water is absorbed by the root system is called *osmosis*. The fine hairs on the roots are the plant part through which the water (together with dissolved substances) passes into the plant. These root hairs act as tiny filters that allow the water to pass while screening out other substances. This is similar to what happens when you strain the juice of fresh oranges—the juice is permitted to flow through while the pulp and seeds are not.

When the water is needed by the plant, a form of negative pressure called *osmotic pressure deficit* forms, which draws the water through the root hairs into the plant—an action similar to sucking liquid through a straw. Once the water is inside the root system, it moves to various parts of the plant. This movement through the plant is called *translocation*.

Water absorption and translocation are important to the horticulturist because without water plants become pale, weak, and flaccid (they wilt), reducing their economic value. Without water, the tomatoes you are growing wilt and reduce your chances of producing a quality crop. It has been found that only one severe wilting damages a plant so severely that it may not fully recover for a year or more.

Photosynthesis and Carbon Dioxide

Normally, air is about .03 percent carbon dioxide. Carbon dioxide enters the plant through the plant leaves. You remember that plant leaves have sets of specialized cells called guard cells, which open and close to create an opening in the leaf called a stomata. It is through these openings that carbon dioxide enters the plant.

Carbon dioxide in the air is usually not a limiting factor in the growth of plants, since the quantity of carbon dioxide in the air is adequate for proper plant growth. During midwinter months the carbon dioxide level in a tightly closed greenhouse may drop seriously on a bright day. Under these conditions better crops of roses, carnations, tomatoes, and certain other crops are produced if the carbon dioxide level is artificially raised to .12 percent.

Photosynthesis and Temperature

Photosynthesis is affected by temperature. Generally the rate of photosynthesis is greatest within a range of 65 to 80° F [18 to 27° C], and decreases with temperatures below or above this range. The best range, of course, varies with individual kinds of plants.

Normally, horticulturists list plants based on precise nighttime temperatures. These nighttime temperatures are used because photosynthesis (food manufacturing) only occurs in sunlight, while the other plant growth processes occur around the clock. Two of these processes are assimilation and respiration. If the nighttime temperature is too high, excessive energy from

Figure 4-2. There are many factors you can control in a greenhouse to produce healthy, saleable plants.

carbohydrates will be lost through respiration. If it is too low, the building of plant structure (assimilation) will be reduced. So precise nighttime temperatures provide a way to get maximum growth of the plant with the least loss of plant energy.

Tomatoes are a warm-season crop. They grow best in a nighttime temperature of 60 to 70° F [16 to 21° C]. The temperature also varies as a plant develops.

The greenhouse grower must always give careful attention to the temperature in the greenhouse. Many growers use both heating and cooling systems that are controlled by a thermostat to ensure proper temperature control.

Temperature is also of concern to horticulturists who grow plants outside greenhouses. In fact, temperature affects the location in the United States where many of our crops are grown. For example, oranges, which are a warm-season crop, grow best in warm areas such as Florida or California; apples, which require temperatures near freezing for proper flower bud development, are a cool-season crop, and grow best in the Northern areas.

Photosynthesis: A Summary

Now you have a basic knowledge of how a plant produces its food. You know that carbon

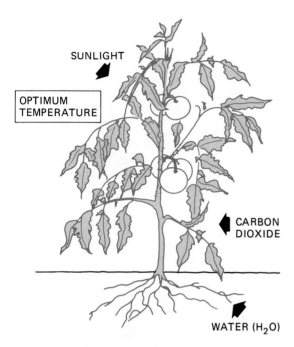

SUNLIGHT

OPTIMUM
TEMPERATURE

CARBON
DIOXIDE

WATER (H_2O)

Figure 4-3. The factors influencing photosynthesis.

temperature, and a fresh air supply are crucial to growing quality horticultural crops.

Respiration

The production of food by the plant is only the first step in the plant's growth process. The food supply produced by the photosynthesis process must be converted into energy by the plant. If food is allowed to accumulate it is of little value unless energy is released.

Plants need energy for growth. Humans produce their energy by burning foodstuffs stored in the body. In a similar manner, plants produce energy. In plants this energy production comes about as the carbohydrates produced by photosynthesis (see Table 4-2) are converted into energy. This process in plants (as well as in animals) is called *respiration*.

Respiration requires the use of oxygen. When oxygen is used to break down a substance it is called *oxidation*. Without oxygen, respiration stops. A visible example of oxidation is the process that occurs when you burn a match or a leaf pile, or build a campfire. The fire uses oxygen to burn the materials. The burning process releases energy as heat. If you remove or temporarily halt the oxygen supply, the burning process stops and energy is no longer being released.

The respiration process occurs inside living cells of plants. Unlike the match or campfire, it does not produce a large amount of heat. The materials it burns are the carbohydrates pro-

dioxide and water are needed by the plant in order to manufacture carbohydrates, and that control of environmental factors such as sunlight and temperature play an important role in producing good crops (see Figure 4-2).

Horticulturists must understand photosynthesis and the factors that affect it if they are to produce high-quality horticultural crops. And if you are to meet your challenge of producing strong, healthy, and vigorous tomato plants, you must know that appropriate sunlight, water,

TABLE 4-2
RESPIRATION

Chemical Formula
$C_6H_{12}O_6 + 6O_2 = 6CO_2 + 6H_2O + \text{Energy}$
General Formula
Carbohydrate + Oxygen = Carbon Dioxide + Water + Energy

duced by photosynthesis. This burning process releases energy. Thus respiration is a process that occurs in living plants and that uses oxygen to oxidize or burn carbohydrates. The resulting release of energy creates movement of materials within the plant and the conversion of carbohydrates into substances that make up the parts of cells. Table 4-2 provides the chemical and general formulas for respiration.

A question that arises is "Why is a knowledge of the respiration process important to a person producing horticultural plants?" To answer this question, we must examine the respiration process more closely, and we must determine what environmental factors influence the rate of respiration. Currently we know that respiration requires oxygen. There is a second environmental factor that is crucial to the respiration process—temperature. Without an adequate temperature level, respiration is drastically reduced. The horticulturists must be able to control both oxygen and temperature if quality plants are to be produced.

Respiration and Oxygen

A plant placed in an airtight room will have its respiration rate drastically reduced. The plant will eventually die. Plants get their oxygen from the air that surrounds the plant. This air contains about 21 percent oxygen. The amount is adequate for growing plants. Respiration occurs in all living cells, thus it does not occur only in the leaves. It takes place in stems, flowers, fruits, and roots. Each of these parts of the plant therefore requires a supply of oxygen. This is most critical in the root area. If the soil does not allow oxygen to be present in the root zone, the respiratory activity of the roots is drastically reduced. Without oxygen, the roots of most plants simply die, which results very quickly in the death of the above-ground portion of the plant as well. If you are growing plants you must ensure that there is an adequate supply of oxygen to all portions of the plant.

Respiration and Temperature

Respiration is greatly influenced by the temperature of the environment in which the plant is being grown, and not all plants grow best within the same temperature range. In fact, specific temperature range requirements are the primary reason why certain horticultural crops grow best in certain geographic locations in the United States. For example, artichokes and Brussels sprouts enjoy the moderate climate of a specific region of the California coast.

The temperature needed for respiration in plants is equally important to the person who produces plants in the greenhouse. The greenhouse operator can mechanically control the temperature in the greenhouse. Some plants grow better at higher or lower temperatures than

TABLE 4-3
COMPARISON OF PHOTOSYNTHESIS AND RESPIRATION

Photosynthesis	Respiration
1. Produces food	1. Uses food for plant energy
2. Energy is stored	2. Energy is released
3. Occurs in cells that contain chloroplasts	3. Occurs in all cells
4. Oxygen is released	4. Oxygen is used
5. Water is used	5. Water is produced
6. Carbon dioxide is used	6. Carbon dioxide is produced

others. For example, snapdragons grow best at 55° F [12° C] and the poinsettia at 62° F [17° C].

If you are to grow a high-quality crop of tomatoes, you must ensure that oxygen is available to all parts of the plant, including the roots. And you must monitor the temperature in the area where the tomatoes are being grown. Since all plants respire, the same principles apply to other crops. Turf grasses, trees, shrubs, and other horticultural crops require unpolluted air for good growth and will grow best in their optimum temperature. In unusually hot weather, Bent grasses on Northern golf greens must be lightly sprinkled several times a day to reduce the temperature of the leaf blades and prevent severe "burn" damage.

Respiration and Photosynthesis

By now you should begin to notice some relationships between respiration and photosynthesis. In fact, if you will carefully review the formulas in Tables 4-1 and 4-2, you will discover that respiration is the reverse of the photosynthesis process. Unlike photosynthesis, which occurs only in sunlight, plants respire day and night. Respiration occurs in all cells, not just those that contain chloroplasts. These two processes also differ in their use of oxygen, water, and carbon dioxide. Table 4-3 provides a descriptive comparison of these two processes.

Respiration occurs in all life forms. The release of accumulated carbon dioxide and intake of oxygen occurs at the cell level. In animals, blood carries both to the atmosphere inside of lungs and gills, where breathing allows exchange with the outside atmosphere (or water for fish). In plants there is simply diffusion into the open spaces within the leaf and exchange through stomata when they are open.

Respiration—A Summary

Now you know that for plants to grow they must produce energy. The way in which plants ac-

complish this is called respiration. This process in both plants and animals consumes food to release energy and requires adequate oxygen and optimum temperature. Thus, as a horticulturist, you must ensure that the plants you are producing have an adequate supply of fresh air and are grown in an environment that provides optimum temperature.

Transpiration

If you refer again to Table 4-3, you will note that photosynthesis produces carbohydrates and oxygen, while respiration produces energy, water, and carbon dioxide. Thus, inside the cells of plants these substances are constantly being produced. During daylight hours the plant actually uses more carbon dioxide and water in photosynthesis than it releases in respiration. At night the plant releases carbon dioxide and water in respiration.

The process by which the plant releases water is referred to as *transpiration*. Actually, about 90 percent of the water that enters the roots is released through the leaves in transpiration, and only 10 percent becomes involved in chemical processes or is tied up in plant structure.

The lower surface of leaves is dotted with special structures called stomata which were discussed in Chapter 3. By the action of guard cells, tiny openings are created during daylight hours and are closed at night. While they are open, there is free movement of water vapor from inside the leaf. This release of water is called transpiration. Gases such as oxygen and carbon dioxide are also freely exchanged when the stomata are open.

Environmental Factors Affecting Transpiration

The amount of water loss by the plant is governed by several environmental factors, namely, temperature, humidity, and air movement (see

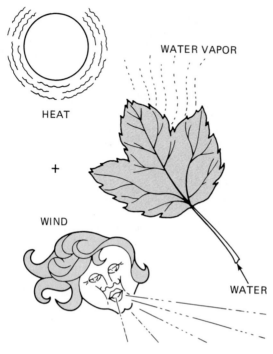

Figure 4-4. Temperature, humidity, and air movement are factors affecting transpiration.

Figure 4-4). Like photosynthesis and respiration, transpiration is influenced by temperature. As temperatures rise, the amount of water vapor that is expelled by the plant increases. With rising temperatures in the environment surrounding the plant, the vapor pressure becomes lower outside the plant, which tends to pull more water vapor from inside the plant into the atmosphere surrounding the plant. Think of what happens to a puddle of water in your driveway on a very hot day. The water will evaporate much more quickly than it will when the temperature is cooler.

Now, what happens to the water puddle if the air is dry and the wind is blowing? The evaporation is increased. This is why golf courses and lawns "dry out" on hot, dry, and windy summer days. The effect is the same on plants when temperature, humidity, and air movement are not controlled. The plants will dry out.

You understanding of transpiration is important as you grow the tomato plants. Abnormal transpiration losses will cause the plants to dry out. In tomatoes, if too much water is expelled the plants will wilt. The wilting is caused by the plant cells' losing their firmness or turgor. Thus, as you grow your tomato plants, you must take precautions to ensure that transpiration losses are held to a minimum.

Common practices used in greenhouses to reduce transpiration are (1) control of temperature, (2) control of humidity, and (3) avoidance of excessive air movement around the plants.

Humidity is defined as the amount of water vapor being held in the air at a given temperature. On humid days, when the water vapor is in larger concentrations in the air, the difference in vapor pressure inside the plant and the outside air changes are such that the pressure outside is greater, which does not allow excessive water vapor to escape. On dry or less humid days the opposite effect occurs—the vapor pressure is greater, which results in the escape of more water from the plant.

Wind or air movement has a similar effect. Air movement will cause the vapor pressure outside the plant to be less than inside the plant, and excessive vapor can escape.

Plant Growth Processes—A Summary

As you begin your task of caring for the tomato crop, you have become familiar with how plants produce their food. Also you are aware of the environmental factors that influence the products of plant growth processes.

You are aware of the need of sunlight, water, and carbon dioxide for photosynthesis, and of oxygen for respiration. And you are aware of how too much or too little water and low temperatures affect the way the plant will grow.

The greenhouse grower constantly checks the environment in which the plants are growing. Automatic controls allow the grower to control temperature, air circulation, carbon dioxide levels, and relative humidity. Similarly, you too must control the environment in which the tomato crop you are caring for is growing. One lapse in watering or temperature control could result in a ruined crop.

How Plants Grow

Plants grow and become larger in a way that is similar to the growth of animals and humans. Plants, like other forms of life, increase in size and weight until they reach maturity.

If you have provided oxygen, water, and carbon dioxide, and have controlled the temperature, the tomato plants you are growing should be getting larger. To fully understand how these plants actually grow, you must know the principles of plant growth discussed earlier in this chapter: photosynthesis, respiration, transpiration, and translocation. The question that arises is "How do these plant-growing processes actually increase the size and weight of the plant?"

Plants produce two types of growth: vegetative growth and reproductive growth. The *vegetative growth* phase consists of the development of leaves, stems, and roots—the vegetative parts of the plant. The *reproductive phase* is characterized by a maturing of the vegetative parts of the plant and the development of flowers, fruits, and seeds.

The person who grows horticulture plants must understand how to control vegetative and reproductive growth. For example, the economic part of the tomato plants you are growing is the tomato fruit. The tomato plant itself is not saleable. Thus, you want to understand how the tomato plant produces vegetative and reproductive growth so that you can maximize the production of tomato fruits (reproductive growth).

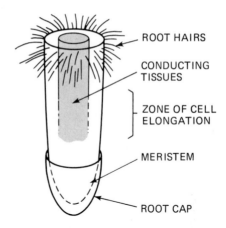

Figure 4-5. A longitudinal section of a root tip illustrates the parts involved in the growth process.

In contrast, if you were producing a green shrub, your concern would be to develop the vegetative parts of the plant, since this is the economic value for which the plant is grown.

In both cases, you as the horticulturist must carefully monitor the growth of the plant to ensure that the purpose for which the plant is being grown is fulfilled.

Vegetative Growth

You will remember that the cell is the basic unit of the plant. As your tomatoes grow, they are constantly developing new cells, enlarging those cells that are present, and grouping cells to form tissues. Vegetative growth is a three-step process. The first step in the process of making new cells is called cell division. The second step is the enlargement of existing cells or new cells and is called cell elongation. The third step is differentiation, in which groups of cells become different from other groups. Like groups are called tissues, such as pith, zylem, phloem, cambium, etc. The actual physiological process of how new cells are formed is very complex and not completely understood, thus the discussion will be concerned with the basic cell division and elongation process.

Cell division means making new cells. New cells require large amounts of carbohydrates. New cells are made by a tissue called the *meristem,* which uses plant foods to make various parts of the cells and to produce energy. The energy released from respiration is used by the new cells that are being formed. As these cells divide they must have this source of energy. A longitudinal section of a root tip is illustrated in Figure 4-5, which illustrates the growth process.

Normally, the rate of cell division is directly related to an adequate supply of carbohydrates. You remember that photosynthesis produces carbohydrates. The stems, roots, and other tissues of the plant where growth takes place must have the carbohydrates as new cells develop.

Cell elongation is the enlargement of new cells. This enlargement process requires water, certain plant hormones, and carbohydrates. Water is needed, since enlarging cells contain large amounts of water. *Hormones* are chemical substances secreted by the plant. Some hormones assist in helping cells become larger. The carbohydrates are needed, since the cell walls become thicker from additional layers of cellulose (a substance made from carbohydrates) around the cell wall.

If you examine your tomato plants you will note that the more mature parts of the plant are more firm. This results in part from elongating cells with thicker walls. The new growth will be somewhat softer. This is because some of the new cells' walls have not yet thickened.

The third phase of the vegetative growth is called cell differentiation and consists of the development of tissues. Examples of cell differentiation are the development of roots and stem tissues.

Reproductive Growth

The reproductive phase of plant growth consists of the formation of flowers, fruits, and seeds. Thus, in this phase of plant growth some plants

Figure 4-6. Plants need fertilizer which contains the nutrients they require for optimum health.

shift from production of new vegetative growth to the production of reproductive growth.

When the plant begins its reproductive growth, the production of leaves, stems, and roots is slowed. This means that an excess of carbohydrates is not being used for growth, but is being stored by the plant. In your tomato plants this is the tomato fruit.

Plant Growth and Nutrition

Plants also need nutrients to grow properly. Plant nutrients are substances that plants use to

Figure 4-7. Careful handling may determine the plant's fruit production and saleability.

produce leaves, stems, roots, fruit, and other plant parts. Just as your body needs vitamins, minerals, proteins, and other substances to ensure that you grow properly, plants need nutrients for growth. Plants produce those items that are needed for proper growth.

Most likely your greenhouse supervisor had you fertilize your tomatoes. Your supervisor may have explained that fertilizer is needed for vigorous growth of the plant. But what does the fertilizer contain that influences this vigorous growth? Do you wonder why the fertilizer was primarily applied in the early growth stage of the plant?

An important ingredient in most fertilizers is nitrogen. Almost all plants require nitrogen for growth and development. But how does the nitrogen interact with carbohydrates for cell division and elongation?

Nitrogen is needed to make proteins. Proteins are necessary for protoplasm, which is often referred to as the "living substance." If you recall from Chapter 3, protoplasm is a substance in cells. When a cell divides to produce a new cell, protoplasm must be formed. As more and more cells divide, more protoplasm is needed. Thus as a plant is developing, nitrogen is needed to form proteins, which in turn are necessary for making protoplasm.

Your tomato plants require large amounts of

nitrogen. This is especially true while the plant is young. The amount of nitrogen needed by the plant decreases when the vegetative growth of the tomato plant slows and the fruit is forming and ripening. Thus you fertilize the plant heavily during early stages of vegetative growth and do not fertilize the plant as much during later growth.

If you overfertilize the tomato plants you encourage soft, leafy growth that will produce relatively few and smaller fruits.

Control of nutrition is also important in the production of flowering plants. The production of chrysanthemums requires similar nutrition and fertilizer application. Mums are sold for their flowers and not primarily for their leaves and stems. Improper control of the growth process could result in large vegetative plants with only a few small flowers. If this occurs, the economic value of the plant is decreased.

Plant nutrients such as potassium and phosphorus are also essential to the plant's growth process. In addition, there are several minor and trace elements needed by many plants for certain plant growth processes to take place. Each species of plant requires different amounts of these essential nutrients. As a plant grower you must become knowledgeable about the nutritional needs of each kind of plant that you grow, and be prepared to treat each plant accordingly.

Plant Growth and Development: A Review

Plants are truly unique to our universe. The growth process of photosynthesis is unique to plant life. Horticulturists use knowledge about plant growth processes in the growing of high-quality horticultural plants.

As a greenhouse employee you have had to monitor the plant-growing processes to produce a quality crop of tomatoes. The importance of sunlight, water, carbon dioxide, temperature, and plant nutrients have been related to plant-growing processes. You have also learned that the control of environmental factors is essential in the production of a plant product. As you gain experience in the horticulture industry you will increase your understanding of the factors that influence plant growth and development, and increase your skill in manipulating plant growth.

THINKING IT THROUGH

1. Define photosynthesis.
2. What influence do sunlight, temperature, moisture, and air have on the phososynthesis process?
3. What is translocation?
4. What is respiration?
5. Contrast the photosynthesis and respiration processes.
6. What is transpiration and what environmental factors influence its rate?
7. What are the functions of water in the growth and development of horticultural plants?
8. Explain the relationships between vegetative and reproductive growth?
9. Describe the three processes involved in making new cells.
10. Why are nutrients needed for proper plant growth?

5

Identifying Horticultural Plants

Susan is a horticulture student in an area vocational school. While enrolled in the vocational horticulture program she became involved in the supervised occupational experience (SOE) program.

Susan was placed in a landscape installation and maintenance firm as part of her work experience. The owner of the firm also directed the work that was done on each site.

On Susan's first afternoon of work she was helping to unload a truck filled with plants that were balled and wrapped in burlap. The owner, Mr. McQuary, was telling Susan and the other workers where to place the plants. He said, "Set the six dark-green needle-leaved plants on the ground." As they finished this task Mr. McQuary pointed out that all six plants looked alike, but the two labeled T. baccata repandeus were spreading English yews, which grow low and have a graceful, slightly drooping appearance. Mr. McQuary said that these would be used by the entryway.

The two labeled *T. cuspidata capitata* would grow to be Christmas tree shaped, and nearly 12 ft (feet) [3.66 m (meters)] tall, and were intended for a back corner of the lot. These two plants are commonly called upright Japanese yews. The two marked *T. cuspidata densiformis* were dense Japanese yews, which would have a nearly rounded compact shape and were to be used at the entry to a paved terrace.

Susan was impressed by the knowledge Mr. McQuary had in identifying horticultural plants. However, she was puzzled, so she asked, "Why use such technical names?" Mr. McQuary said that common names were different in various parts of the country, yet technical names remained the same regardless of the region.

Mr. McQuary told Susan that she would gain experience in identifying plants, and that she would have to learn to identify the plant on sight. A person who was paying for a landscape to be installed would expect plants to be put in the right place. Hence, the spreading English yew must not be mistaken for the upright Japanese yew. The landscaper's idea would be destroyed if the plants were not installed as planned.

Other horticulturists, such as turf specialists, greenhouse operators, and fruit and vegetable growers, also have to be able to recognize the plants with which they work. After working for a short time with the landscape firm, Susan was convinced of the necessity of being able to identify horticultural plants.

CHAPTER GOALS

Your goals in this chapter are:

- To describe the importance of identifying horticultural plants
- To examine a plant and provide an oral analysis of the plant's identifying characteristics
- To describe the three life cycles of plants

How to Identify Plants

Susan's work experience provided her with insights into the need for a good working knowledge of plant identification. After a few months on the job she was able to identify most of the plants commonly used by her boss, Mr. McQuary.

Yet Susan realized that she must constantly work on being able to identify horticultural plants other than the landscape plants she had installed. Susan knew that upon graduation she might obtain a job in an area of horticulture that was different from landscaping. This job would surely require a working knowledge of the plants used in that job.

With this knowledge Susan began to think of ways to identify and classify the many different horticultural plants. She asked Mr. McQuary where she should start. Mr. McQuary told Susan that learning to identify plants was like solving a mystery. You examine *clues* and by collecting this information you begin to narrow down your choices. After all the possible clues are examined, you should have a clear picture of plant identification.

Mr. McQuary also suggested that as a first step Susan determine where the plant grows.

Where Plants Grow

Susan began to ponder the question of where plants grow. She concluded that plants grow indoors, outdoors, or in a greenhouse. Many plants are grown almost exclusively for use indoors as houseplants. The popularity of houseplants has increased the need for knowledge of their identities.

Figure 5-1. Plants grow under a variety of conditions and in a variety of sizes.

Many plants are grown indoors, primarily in greenhouses.

Susan knew that a great many plants grew outdoors. She was also aware that factors indicated where outdoors plants grew. For example, the hardiness of the plant places it in one of the two climatic hardiness zones (refer to Chapter 1.) Each climatic zone has an average minimum temperature that influences which plant can grow there. A knowledge of these zones and plant hardiness is very helpful in identifying horticultural plants. Primarily, the zones limit the number of plants grown in an area, which reduces the possible number to be identified.

Within each zone, plants are grown for specific purposes. For instance, certain plants are used for landscapes, such as the Taxus plants described in the chapter introduction. Other plants are used on fruit farms, in vegetable gardens, or on golf courses. The knowledge of where a plant is used also reduces the possible number of categories for identification.

Susan noted the various places where plants grow but she knew that this clue alone would not provide enough information for identifying horticultural plants. For example, if a person were purchasing a plant, the knowledge of where the plant would grow would not be sufficient in identifying the plant. And some plants grow both inside and outside. A coleus can be either an indoor or outdoor plant, depending upon the climate. Thus, she began to seek other plant characteristics that could be helpful in identifying plants.

Woody and Herbaceous Plants

Mr. McQuary told Susan that another clue involved determining whether the plant was *woody* or *herbaceous*. He explained that a woody plant has a coarse, hard stem that contains woody fibers. Woody plants normally grow for more than 1 year. Trees, shrubs, and vines are examples of woody plants.

Herbaceous plants do not have hard woodlike stems. Their stems are softer, and many of these plants do not grow for more than 1 year. Many flowers, most vegetables, and spring bulbs are examples of herbaceous plants.

Susan learned from Mr. McQuary that the woody-herbaceous distinction was quite commonly used in most horticultural jobs. He explained that many landscape plans would specify the specific use of a woody ornamental in a certain space and it was therefore very important to know woody ornamental plants.

Examining Plant Parts

The two clues thus far—where a plant grows and whether it is woody or herbaceous—do not help Susan narrow down all possible plants. Mr. McQuary told her that the next set of clues rested with each plant and its parts.

As she performed her duties with Mr. McQuary's firm, Susan began to examine plants to determine what characteristics could be used as clues to identify plants. Her first examining practice was to look at the plant as a whole.

The Whole Plant

When Susan looked at a plant, the first item was its size. Was the plant large, like a tree; medium, like a shrub; or small, like a houseplant? Susan discovered that the physical size of the plant could be very helpful in identifying plants.

The next item Susan noticed about plants was their unique shapes. She had learned to identify the American elm simply by observing its unique vaselike shape. Mr. McQuary related that the coconut palm and Norfolk Island pine also have such unique shapes that they are easily identified. In a similar manner, rhubarb, with it large fleshy leaves, has no confusing counterpart within the climatic zone in which it grows.

Susan knew that size and shape were helpful, but these factors alone would not enable her to distinguish among many similar plants. Size and shape provide clues, but if you only know that a plant has a certain size and/or shape, it will be difficult to distinguish just which specific plant is involved. For example, many arborvitae are quite similar in size and shape when young, yet in their mature size and shape are different. Thus, she became aware of the need to examine specific parts of the plant as a clue to the plant's identity.

Using Leaves to Identify Plants

Susan had noticed that Mr. McQuary often pointed to the type of leaves a plant had when he was identifying a specific plant. When she asked him about this, Mr. McQuary indicated that leaves are valuable in helping to identify most horticultural plants. He told Susan that there are four factors about leaves that are important as identifying characteristics: (1) the type of leaf, (2) the shape of the leaf, (3) the size of the leaf, and (4) the color of the leaf.

Leaf Groups. Susan discovered that leaves are often classified in two groups—broadleaf and needleleaf. Closely related to the leaf types is the word *deciduous,* which means to shed its leaves. In fact, the word deciduous means ''to fall off.'' Plants whose leaves do not fall off are referred to as evergreen. The evergreen gets its name from the fact that it stays green year round. Evergreens will not lose all their leaves at once.

Both deciduous trees and evergreens occur in both groups, although it is generally thought that broadleaf plants are deciduous. The notable exceptions are the American holly and the camellia, which are broadleaf but not deciduous. The baldcypress and larch are needle-leaved but deciduous.

Leaf Type.   Susan noticed that plants had different types of leaves. Mr. McQuary told her that the types of leaves are also important clues in the identity of plants. He said that the ability to recognize different leaf types is important for horticulturists.

A leaf may be attached to the stem as a single leaf blade, or it may be several leaf blades on a single petiole. When there is only one leaf blade on the petiole, the leaf type is called *simple.* Dogwood, oranges, and grapes are plants that have a simple leaf type (see Figure 5-2). When a leaf is made up of several leaf blades (called leaflets) attached to the petiole, the leaf type is called *compound.*

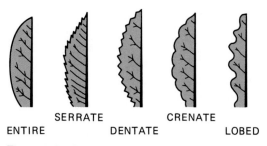

Figure 5-3. Some common leaf margins.

ENTIRE   SERRATE   DENTATE   CRENATE   LOBED

Compound leaves can be divided into two types—*pinnately compound* and *palmately compound.* The pinnately compound leaf type involves leaf blades that are arranged along both sides of a petiole. The palmately compound leaf type involves leaf blades that come from the tip of the petiole, just as your fingers branch out from the palm of your hand. Pecans and roses are plants with pinnately compound leaf type, while grape ivy, horse chestnut, and Washington palm are examples of palmately compound leaf types.

Leaf Margin.   While examining leaf type, Susan found that leaves had different shapes on the outer edges of the leaf. Mr. McQuary helped Susan to learn the different types of leaf margins.

Mr. McQuary told Susan that the magnolia has a smooth edge all the way around the leaf. This type of leaf margin is called *entire.* Birch tree leaves have a *serrated* margin. This type of margin is characterized by sawlike teeth that angle forward. Very similar to the serrated margin is the *dentate* margin, which has teeth that are at more of a right angle with the midrib of the leaf. The grape is an example of dentate leaf margin. A *crenate* leaf margin has rounded teeth with slight open spaces between them. Certain hollies have crenate leaf margins. Some leaf margins have indentations that go more than one-quarter of the way to the midrib of the leaf. This type of leaf margin is referred to as *lobed.* Sometimes a deeply lobed leaf will have huge

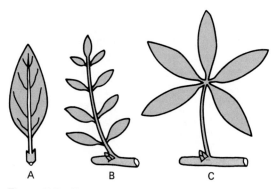

A   B   C

Figure 5-2. The three common leaf forms are: (a) simple, (b) pinnately compound and (c) palmately compound.

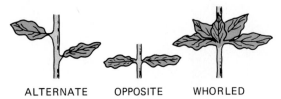

Figure 5-4. Three basic leaf arrangements are alternate, opposite, and whorled.

jagged edges that are indented almost all the way to the midrib. The Japanese maple and many oaks are examples of plants with lobed leaves.

Leaf Arrangement. Another identification characteristic of many plants is the way in which the leaves are arranged on the stem. There are three basic *leaf arrangements:* (1) alternate, (2) opposite, and (3) whorled. Figure 5-4 illustrates these three types of leaf arrangements.

Leaf Venation. Sometimes you can use the way the veins are arranged on the leaf as an aid in identifying horticultural plants. Many plants have *parallel* leaf venation. This means that the veins in the leaf are nearly equal in size and run parallel with (next to) one another from the base to the tip of the leaf (see Figure 5-5).

Pinnate leaf venation occurs when there is one main vein that forms the midrib, and small branched veins that branch off in a fairly even pattern from the one main vein. The American elm is an example of pinnate leaf venation.

When there are three or more major veins branching from one location at the base of the leaf, just like the fingers on your hand, the name *palmate* is used to describe the form of leaf venation. When a leaf has a palmate venation pattern, there is a network of little veins that branch off the larger vein, giving a netted appearance. All maples and grapes have this kind of venation pattern.

Leaf Size and Color. Mr. McQuary directed Susan to an area adjacent to the headquarters of the landscape firm. In this area he was growing shrubs and ornamental trees. Along one border of the nursery area were a couple of trees that were bluish in color. Mr. McQuary pointed out that the color of the leaf helped in identifying certain types of trees. This bluish tree was a blue Colorado spruce whose distinct blue color distinguished it from all other spruces.

The size of the leaf is sometimes a clue to the identity of plants. The longleaf pine, which grows in the Southeastern states, is readily distinguished from other pines simply by its long leaves. The large leaf of the American elm distinguishes it from the Chinese elm with its smaller leaf.

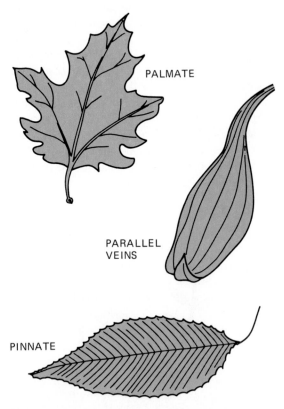

Figure 5-5. Leaf venation is another means of identification. There are three types of venation: (a) parallel, (b) pinnate, (c) palmate.

Figure 5-6. The waterlily is an example of a complete and perfect flower.

Susan was beginning to see that as many clues were examined she could begin to narrow down the choices in identifying a plant. Yet the clues discussed thus far do not guarantee that a plant can be adequately identified. Additional clues, such as the type of flower a plant produces, often provide the necessary help.

Using Flowers to Identify Plants

Mr. McQuary had emphasized that leaves play a very important part in identifying horticultural plants. He went on to say that another plant part, the flower, is also very helpful. Not all plants have recognizable flowers, but the size, shape, color, and time in which flowers are produced all aid in identifying plants.

However, a knowledge of flowers is helpful as an additional clue in identifying horticultural plants. There are two important flower characteristics that are helpful in identifying a plant: (1) reproductive characteristics, and (2) the way in which flowers are arranged on the plant.

Reproductive Characteristics. The flower has four basic parts: sepals, petals, stamens, and pistils. Susan recalled from her vocational horticulture classes that the stamens and the pistils were the male and female reproductive parts of a plant. Plants and individual flowers vary in the fact that some plants and flowers lack one or more of the four plant parts.

When the flower has all four parts, that is, sepals, petals, stamens, and pistils it is called a *complete flower*. If one of the parts is missing, the flower is called an *incomplete flower*. The lily is an example of a complete flower, while the pecan is incomplete in that stamens or pistils are absent on individual flowers.

Individual plants are categorized by the type of flowers they bear. For example, some plants bear perfect flowers. A flower is called *perfect* if it contains both pistils and stamens. A flower is called *imperfect* if it lacks either of these two parts. Susan was cautioned to distinguish between perfect and complete flowers on the basis of the fact that a complete flower has all four basic parts, whereas a perfect flower has at least stamens and pistils. The lily that has all four parts is both complete and perfect.

Mr. McQuary illustrated another reproductive characteristic of plants by examining the flower of a walnut. He pointed out that some walnut flowers do not contain stamens, while others do not contain pistils. However, he said that the flowers containing pistils and the flowers containing stamens occur on the same tree. Trees and other plants that have stamens and pistils in separate flowers but on the same plant are called *monoecious*. If, on the other hand, the stamen and pistils are in separate flowers and occur on separate plants, the plant is called *dioecious*. Plants that have perfect flowers are called *hermaphroditic*.

Flower Types. Just as individual flowers have an arrangement of parts, so a cluster of flowers has a pattern. The pattern of flowers on a plant is called an *inflorescence*.

The landscape firm planted many roses, car-

nations, and tulips. The type of flowers produced by these plants is called *solitary*. These plants produce only one blossom per stem. Susan learned that there are six other inflorescence patterns common to horticultural plants:

1. The *panicle* is a branch of inflorescence with flowers developing toward the tips of the branches. The spray chrysanthemum is an example of a plant with a panicle inflorescence.
2. *Spike* inflorescence occurs on gladiolus and is simple and elongated.
3. The snapdragon has a simple inflorescence with stalked flowers called *raceme*.
4. The *umbel* inflorescence has a flat-topped inflorescence that does not have a central axis. Queen Anne's lace is an example of umbel.
5. Candy tuft is a plant with a short, flat-topped flower cluster called a *corymb* inflorescence.
6. The *head* inflorescence is characterized by numerous short flowers crowded together. A daisy is an example.

Special Plant Identification Characteristics

Mr. McQuary reviewed with Susan the plant identification clues she had learned thus far: (1) where the plant grows, (2) the type of stem—woody or herbaceous, (3) examining the plant as a whole, (4) plant leaves, and (5) flowers. After this brief review he told Susan that there were other plant characteristics that could help identify plants under certain conditions. These special cases do not override the need to examine the five characteristics previously discussed, but help give additional clues.

Buds. When you attempt to identify deciduous plants in winter, it's difficult to distinguish between closely related plants. For example, dif-

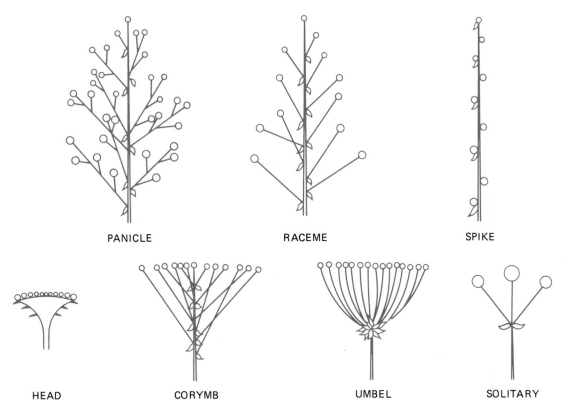

PANICLE RACEME SPIKE

HEAD CORYMB UMBEL SOLITARY

Figure 5-7. The common types of flower inflorescence.

ferences in buds and leaf scars can be used to distinguish between a red and a sugar maple.

Aroma. Mr. McQuary asked Susan whether she had every smelled honeysuckle, peppermint, or a sage plant and noticed the distinctive ôdor or aroma. These plants can often be identified by their aroma.

Fruit. The pineapple has such a distinctive shape that when the pineapple plant bears fruit it is easily identified. In a similar manner, the persistent orange or red clusters of fruit on firethorn (Pyracantha) permits one to identify this plant from distances of 50 or 100 feet.

Buds, aroma, and fruit are helpful clues to plant identity. However, these special charac-

teristics should be used as additional clues and not as primary identification characteristics.

Additional Plant Identification Information

Susan enjoyed her experience working with the landscape firm in the supervised occupational experience (SOE) program. Mr. McQuary, the firm's owner, had helped her to develop a system whereby she could learn to identify plants. As Susan gained additional experience in the firm she noticed that Mr. McQuary always used a technical name when he worked with the plants. Susan quizzed him as to why and he replied, "Common names vary from area to area, but the technical name remains the same everywhere." Thus, when you speak of *T. baccata rapandeus* you know it is the same plant regardless of its more common name, spreading English yew, which might vary from one geographic location to another. Another illustration of the use of technical terminology is when you hear two doctors talking about a patient's health. They use technical terminology rather than common language to avoid misunderstandings. Susan asked Mr. McQuary how the technical name was determined.

Technical Names of Plants

The technical names of plants come from the plant classification system used to categorize all plants. (Refer to Chapter 3 for a review of the plant classification system.) For the discussion of technical names you should recall that one *class* of plants is the angiosperms, which have two subclasses called monocots and dicots.

These two subclasses are then divided into *orders*. For example, one order of plants within the monocot subclass is Gramineae, which includes grasses and hedges. Orders are then divided into *families*. The family category is often used by plant scientists as a basic unit to study plants, because plants from one family often have similar characteristics. The family Graminaea includes our turfgrasses; bamboo; and our grains, such as wheat, corn, and oats.

The family is divided into *genus* and *species*. It is the genus and species name that gives a plant its technical name. For example, the genus for roses is *Rosa;* pines is *Pinus,* and maples is *Acer.* The second part of the technical name is the species. The white oak is *Quercus alba,* where *Quercus* is the genus referring to oaks and alba refers to a particular oak—the white oak.

Most horticulturists do not have to know the detailed process of how plants are classified. However, this is the origin of the technical name of plants, and is why technical names are used.

Plant Life Cycles

Mr. McQuary often referred to plants as being *annuals* or *perennials*. Susan asked him to explain what these descriptive terms meant. He told Susan that plants have three basic life cycles—annual, biennial, or perennial.

Annual. An annual plant completes its life cycle in one growing season. In other words, during one growing season the plant grows, flowers, produces seed, and dies. Annuals will not grow during the next season unless they are planted again (see Figure 5-8). Examples of annuals are cosmos, larkspur, spinach, marigolds, geraniums, and lettuce. Depending upon the geographic location, these plants would normally grow during a particular time of the year and would not develop during the next growing season without being replanted.

Biennial. Biennials are plants with a life cycle that takes two growing seasons to complete. During the first year a biennial may develop vegetative growth. During the second year it produces flowers and seeds and dies. Examples of biennials are celery, sweet William, and parsnip (see Figure 5-8).

Perennial. The third life cycle of plants is referred to as perennial. Perennial plants continue to grow from growing season to growing season. They may even take several years to reach maturity. While the life cycles of annuals and biennials end after they have produced seeds, the perennial will continue to grow and produce seeds for several years. Asparagus, rhubarb, raspberries, and fruit trees, as well as many ornamental trees and shrubs are examples of perennials (see Figure 5-8).

Figure 5-8. Marigolds are a popular example of an annual; cabbage is an example of a biennial; and rhubarb is an example of a perennial.

Identifying Horticultural Plants: A Review

During Susan's work experience with Mr. McQuary's landscape firm she had learned a great deal about how to identify horticultural plants. Much of her information had come directly from conversations with Mr. McQuary, but she had also learned much by actually installing plants in the landscape and maintaining those plants that had already been installed.

Susan became very familiar with the identification of woody and herbaceous plants. She knew the importance of leaves and flowers in identifying plants. She learned that several special characteristics, such as aroma, fruit, and shape, are helpful clues in identifying horticultural plants.

Susan learned how the technical name of plants was determined. More importantly, she learned why technical names were used.

Finally, Susan learned about the three life cycles—annual, biennial, and perennial—of plants.

As Susan worked with the landscape firm, she realized that all the basic information she was learning about plants was very important to the successful operation of the business.

THINKING IT THROUGH

1. Why is it important for a nursery worker to be able to identify ornamental shrubs?
2. List five basic features that aid in plant identification.
3. What is the difference between a woody plant and a herbaceous plant?
4. When you are looking at a plant as a whole, what basic characteristics are most noticeable?
5. What four characteristics of the leaf help make it useful in identifying plants?
6. Describe the difference between simple and compound leaf arrangements.
7. Define the following terms: (a) complete flower, (b) incomplete flower, (c) perfect flower, (d) imperfect flower, (e) monoecious, (f) dioecious, (g) hermaphroditic, and (h) inflorescence.
8. Describe the system by which plants get their technical names. Give an example.
9. What are the three life cycles of plants? Explain the basic differences among the three cycles.
10. Why do horticulturists use the technical name of plants instead of their common names?

THREE
Controlling Environmental Factors

Plants grow in many types of environments. A Swedish ivy is an indoor-environment plant that makes an excellent houseplant. The Japanese yew is an outdoor-environment plant that makes an attractive shrub for landscaping a building. A lily is grown in both environments, indoor and outdoor.

The person who grows plants must understand the control of environments for many types of plants. For example, the ivy must have certain light, temperature, and moisture requirements in order to grow properly. Cold temperature, not enough moisture, or too little sunlight will damage the ivy. The yew also grows best in good soil, when watered properly and provided with the proper plant nutrition, and when transplanted properly. The lily must be grown under the proper environmental conditions also, which includes the manipulation of cold temperatures to "force" the lily bulbs to grow at just the right time. In short, each type of plant has a set of environmental factors with which it grows best.

Horticulture workers must be able to control the environmental and related plant growth factors: soil, moisture, air, light, temperature, nutrients, growth regulators, pruning, and transplanting, to name a few.

A worker on a flower farm must know how and when to dig bulbous crops such as the lily for shipment to other areas of the country. A vegetable farmer must know the proper nutrition for each specific vegetable crop. A greenhouse worker must know how to rid soil of harmful soil microorganisms. A person growing poinsettias must be able to control the amount of light available to the growing crop. A tree surgeon must be able to prune a damaged or diseased tree properly. A landscape maintenance worker must be able to transplant seedlings by "pricking off."

If these workers are to produce quality crops of horticultural plants or to keep existing plants growing properly, they must understand the control of environmental factors and the effect that these factors have on plant growth.

Workers involved in selling plants must also understand the control of environmental factors. For example, a worker in a garden center might be asked by a potential customer to recommend a houseplant that requires minimal exposure to sunlight and infrequent watering. What houseplant would you recommend? Not only would you have to be able to identify the plant, as was discussed in Unit II, but you would have to understand which plant is affected by which set of environmental factors.

Unit III provides you with the basic skills for controlling environmental factors. Each chapter examines the many facets of the environment and related plant-growth control that influence the economic value of horticultural plants. Your mastery of these skills and knowledge are important and will assist you in obtaining and keeping a job in the horticulture industry.

6

Soil and Other Plant Growing Media

Soil is probably the single most important substance for horticulturists. Can you imagine the effect that a lack of soil would have on the horticulture industry? Where would trees, shrubs, flowers, and other plants grow? Sure, some plants might grow in a soilless environment, and in fact such is the case with hydroponics. And some plants can be grown in an artificial medium, as they are in some greenhouses. But not all plants could be grown in such environments.

There are as many different kinds of soils as there are plants. Many soil types receive publicity, such as the sand that covers beaches or fills the desert areas; the red clays that occur in many Southern states; or the deep, rich soils of the Midwestern prairies.

Horticulturists are very much concerned about the type of soil that is found in their areas. In fact, the soil that surrounds your community dictates the type of plants that grow best in your area. For example, cut flowers grow best in a coarse and porous soil. These types of soils have large particles and plenty of air spaces between the particles of soil. Almost all horticultural plants have a soil type in which they grow best.

CHAPTER GOALS

Your goals in this chapter are:

- To describe the physical characteristics of soil
- To relate the importance of soil to growing plants
- To identify the soil-water relationship
- To describe the use of soil mulches
- To identify the effect of soil organisms and the ways in which these organisms are controlled
- To describe what is meant by good soil tilth
- To identify methods of soilless cultures for growing plants

Soil is actually weathered rock. Some scientists believe that the earth was once a huge rock. Wind, rain, freezing and thawing, and other environmental factors began to weather away the rock, and over millions of years our layer of soil was formed.

Importance of Soil

Any item that is available in limited quantities soon becomes valuable. Coins, cars, and stereo records that were once produced in a limited quantity soon sell for many times their original value. Land (soil) is the same. There are only a fixed number of acres of land in the United States. And when you look at only the areas of land that can grow plants, the quantity becomes even less. More land cannot be created. Thus land becomes a very valuable resource.

Horticulturists need a share of this land for the plants they grow. Vegetable and fruit producers and nursery operators need good soils. Greenhouse owners need soil not only for the growing of plants, but also for the land that the greenhouse must occupy.

Soil is also important in determining where horticultural crops may grow best. For example, deciduous and evergreen fruit trees often grow best on sloped land. Many peach orchards are found in areas where the land is sloped or semi-hilly. However, vegetable crops need flat land areas where harvesting can be performed more efficiently and water distribution is more even.

Soil Function

Soils normally are composed of air, water, mineral matter, and organic matter. The composition of an average soil is as follows:

- 5 percent organic matter
- 25 percent air
- 25 percent water
- 45 percent mineral matter

Mineral matter is the weathered rock in the soil. Organic matter is the decomposed animal and plant materials.

Soils provide several crucial functions for growing plants. Plants obtain their water from the roots which absorb the water in the soil. Thus, a good soil must hold the sufficient amount of water for the plant's needs. In a similar manner the plant obtains many of its nutrients from the soil. A good soil therefore will provide those needed nutrients. A good soil also provides support to hold the plant upright while it is growing.

Physical Characteristics of Soil

A good soil that has the qualities mentioned in the above paragraph has certain characteristics.

Figure 6-1. Vineyards and orchards are developed in areas where the land is sloped or semihilly. Vegetables are grown on flat land.

In order for you to determine whether or not the soil is a good one you must become familiar with those characteristics. Horticultural plants grow best in good soils that provide adequate moisture, nutrition, and support. As a horticulturist you must be able to determine whether a soil is capable of providing these functions. If the soil is not capable you must be able to add certain elements to the soil that correct these problems. This management of soil is crucial to growing horticultural plants.

Soil Texture

Soil texture refers to the size of the various particles that make up the soil. These soil particles can range in size from gravel to the fine particles of clay. Sand, silt, and clay are the primary particles found in the soil (see Figure 6-2). Soils are usually made up of a mixture of these particles. It is the amount of the various particles that determines the texture (or feel) of the soil. You may have felt the sand on beaches.

The sand is very gritty because its particles are large. In fact, a sand particle is large enough to be seen with the naked eye. Since sand particles are so large, they do not hold water and nutrients very well. A soil composed primarily of sand will not provide enough water-holding ability or will not hold nutrients to grow good-quality

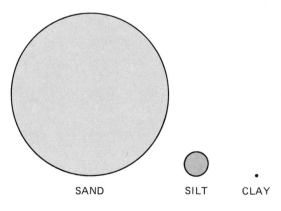

Figure 6-2. Soil particles vary greatly in size. Sand, silt, and clay are the primary particles.

horticulture crops. You will need to add silt or clay particles to a sand in order to improve its quality.

The size of silt particles is between that of sand and clay. Since silt is smaller, it holds water and nutrients better than sand. Silt particles are subject to erosion and during heavy rains tend to run together and form a crusty surface. Heavy or moderate erosion will influence the places where a plant can grow. Eroded soil does not provide the crucial functions needed to grow horticultural plants.

The smallest soil particles are clay. Clay particles are microscopic in size and tend to hold water and nutrients very well because they are so small and fit so closely together. In fact, a soil made only from clay particles may hold water and nutrients so tightly that even though the soil is wet the clay particles will not allow the plant to take up the essential water and nutrients. When a clay soil dries out, the soil shrinks and large cracks are left in the ground. Without other soil particles, a clay soil becomes cloddy and hard to manage.

It is important for a good soil to contain a combination of sand, silt, and clay particles. A good balance of these three particles will give a soil proper air and moisture movement, good support for the plant, and the capacity to hold nutrients.

To illustrate the problem of an extreme imbalance of one particular soil type, imagine attempting to grow a plant in a desert.

The desert soil is primarily sand. It does not hold moisture or nutrients and is so loose it provides poor support for the plant. However, if you were to take some of the desert sand and mix it with silt and clay, a good soil could be developed. Lots of water, nutrients, and additional plant support would have to be provided to adequately grow plants in the desert.

Determining Soil Texture. Since each soil particle has a different and distinctive feel, a horticulturist should be able to determine the texture of soil by feeling it. Soil texture provides a clue as to the soil's ability to hold water, and nutrients and provide support. Coarse-textured soil such

TABLE 6-1
NAMES OF VARIOUS TEXTURES*

Group	Soil Textural Classes
Fine-Textured	Clay
	Silty Clay
	Sandy Clay
Moderately Fine-Textured	Silty Clay Loam
	Clay Loam
	Sandy Clay Loam
Medium-Textured	Loam
	Silt Loam
	Silt
Coarse-Textured	Sandy Loam
	Loamy Sand
Very Coarse-Textured	Sand

* Courtesy of the Instructional Materials Laboratory, 10 Industrial Education Building, University of Missouri, Columbia, Missouri 65201.

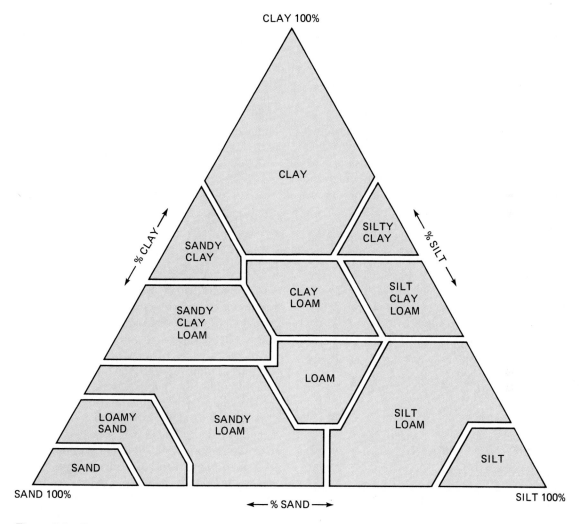

Figure 6-3. This textural triangle shows the various particle percentages and how textural classification is determined.

as sand is poor (in most cases) in the above three characteristics. Fine-textured soils with a heavy clay content provide good support, but are so tightly clumped together that they are difficult to work. Medium-textured soils (with sand, silt, and clay particles) are best, since they provide support and hold water and nutrients.

A simple way to determine a soil's texture is the *ribbon test*. The soil should be moist but not wet to determine the texture. A small amount

of soil is placed in the hand and worked between the thumb and index finger. A clay soil will hold together and form a ribbon about 1 in (inch) [2.5 cm (centimeters)] long. A silt soil will form a slight ribbon, then crumble. A coarse, sandy soil will not form a ribbon at all.

Most soils are made up of a combination of soil particles. The characteristics of the various soil classes and their names are listed in Table 6-1. A textural triangle as shown in Figure 6-3 shows the various particle percentages and what texture classification is given to the various soils, depending upon the percentage of particles they contain.

Soil Structure

The way individual soil particles are grouped together to form clusters of particles (aggregates) is called the *soil structure*. Soil scientists have named the shape of the structure in which the particles arrange themselves. Figure 6-4 shows several kinds of soil structures.

Soil structure is an important aspect of growing plants. Most plants grow best in soil struc-

tures that allow water and air to move about easily in the root area of the plant. The structure of the soil can be improved by adding organic matter such as composted leaves to the soil. Horticulturists also improve structure by mixing soil with inorganic substances such as vermiculite. These materials mix with the soil particles and actually help to change the way the various particles group together. Thus, a better soil structure is achieved.

Chemical Properties of Soil

Soil pH or Reaction

An important feature of soil is the soil pH. Soil has chemical properties that must be understood by the horticulturists to grow plants properly. Soil is a living substance and has a great impact on plant growth. Soil pH has to do with the level of acidity or alkalinity in the soil. The symbol "pH" is a chemical term that indicates the amount of hydrogen in the soil. The amount of acidity or alkalinity is referred to as the *soil reaction*. The soil reaction is indicated on a pH scale (see Figure 6-5). The scale is divided into 14 equal parts. The lower the number, the greater the acidity. Therefore, a pH scale reading of 0 would be maximum acidity, while a pH reading of 14 would be maximum alkalinity.

When the pH of the soil is 7, it is considered to be a neutral soil with equal amounts of elements that are acid and alkaline. Soils with a pH of 0 to 7 are said to be acidic or sour; soils with a pH of 7 to 14 are alkaline or sweet.

Why is it so important to know the soil pH? Different plants require nutrients in different amounts. The soil pH determines how much of a nutrient is available for a plant. For example, if the pH number is too low (or too acidic), elements such as aluminum and iron become increasingly available and many become toxic to certain plants. On the other hand, if the soil

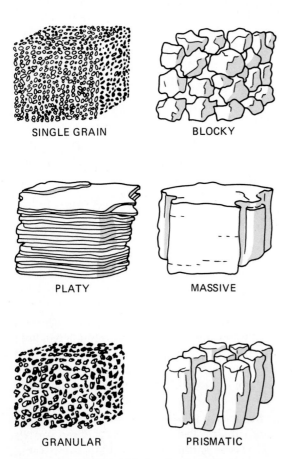

SINGLE GRAIN   BLOCKY

PLATY   MASSIVE

GRANULAR   PRISMATIC

Figure 6-4. Different kinds of soil structures.

PH VALUE

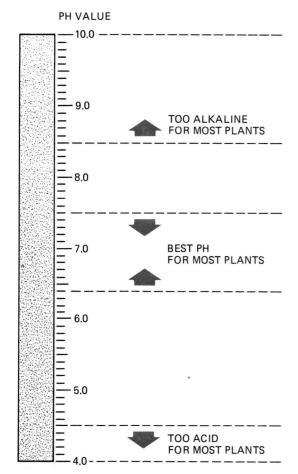

Figure 6-5. Soil pH reveals the level of acidity or alkalinity. This is known as the soil reaction.

pH is raised much above 7, nitrogen and phosphorus become unavailable to certain plants.

Some horticultural plants require a specific soil pH or the root hairs are unable to grow. If the root hairs are not present, the plant is without most of its nutrient absorbing surface.

Although there is no set pH that is absolutely perfect for all plants, there are a few generally known facts. Several plants grow and produce more in a slightly acid soil because more nutrients are available. Plant nutrients are most abundant at a pH of 5.5 to 8.5. When the pH

is 4.5, the soil is too acid; when the pH is 8.5 or more, the soil is too alkaline. Blueberries, dewberries, blackberries, birch trees, pin oaks, azaleas, camellias, and rhododendrons all require an acid pH. Cottonwood, camphor trees, catalpa, and eucalyptus are examples of trees that will tolerate a relatively high pH (alkaline).

Soil Fertility

A plant must have certain nutrients to grow properly. The following 16 elements have been found to be essential to plant growth: Carbon (C), hydrogen (H), oxygen (O), nitrogen (N), phosphorus (P), potassium (K), sulfur (S), calcium (Ca), magnesium (Mg), iron (Fe), boron (B), zinc (Zn), molybdenum (Mo), manganese (Mn), copper (Cu), and chlorine (Cl). In order to make the essential elements easy for you to remember the following saying using only the symbols of the elements has been worked out: C HOPKN'S CAFE Mg Mn Cu Mo Cl. What do all those letters say? *See Hopkin's Cafe managed by mine cousin Mo Cline.* If you add B and Zn to the saying, you can remember the symbols and have mastered the 16 essential elements (see Table 6-2).

Not all the elements are required in the same amounts by plants. Carbon, hydrogen, and oxygen are in the air and water, and their availability comes from the air and water supply available to the plants. Most plant tissue (94 to 99.5 percent) is made up of carbon, hydrogen, and oxygen. That leaves only 0.5 to 6 percent of the plant tissue for the other 13 elements.

Primary Nutrients. Nitrogen, phosphorus, and potassium are called *primary nutrients*. They are required by plants in relatively large amounts and are usually not found in sufficient amounts in soil. They must be *added* to the soil.

Nitrogen is probably the most important nutrient for plant growth. Nitrogen is needed for vegetative and reproductive growth. *Phosphorus* is important for plant functions such as cell

TABLE 6-2
THE SIXTEEN ESSENTIAL NUTRIENTS

PRIMARY NUTRIENTS		SECONDARY NUTRIENTS
Nitrogen		Calcium
Phosphorus		Magnesium
Potassium		Sulfur
TRACE ELEMENTS		**NUTRIENTS FROM WATER AND AIR**
Boron	Molybolenum	Carbon
Chlorine	Iron	Hydrogen
Copper	Zinc	Oxygen
Manganese		

division, development of seeds, flower formation, and creating healthy root systems. *Potassium* helps the plant remain healthy. It increases resistance to diseases and aids in both root formation and the manufacturing of food.

Secondary Nutrients. Calcium, magnesium, and sulfur are called *secondary nutrients*. The fact that these nutrients are called secondary does not mean that they are not essential to plant growth and development. Secondary nutrients are not required in amounts as great as the primary nutrients and are not often a crucial factor in plant growth.

Micronutrients. Micronutrients (boron, chlorine, copper, manganese, molybdenum, iron, and zinc) are also essential to plant growth but are only required in very small amounts. Micronutrients are needed in such small quantities that they are often given the name "trace elements" to emphasize that just a trace of these micronutrients will be found.

It is importrant that there be a balance between the nutrients, just as it is important that you received a balanced diet. Just as certain amounts of vitamins and minerals are required by the human body to help you grow properly, so plants also require a "balanced diet" of plant nutrients for proper growth.

Soil Testing

How does one know what nutrients are present in a soil? The best way to find out is by conducting a *soil test*. A soil test will show the total amount of the various nutrients in the soil. It is important to any horticulturist to know what nutrients are available in the soil so that time and money are not wasted by incorrectly applying fertilizer. Excessive amounts of nutrients can be as harmful to plants as not enough nutrient.

For a soil test to be an accurate measurement of the nutrients available, the sample must not be taken from just one spot. (see Figure 6-6). A nursery worker's soil may vary from spot to spot, with completely different soil and fertility within a few yards. Therefore it is essential that the sample be representative of the soil in the entire area. You should determine the areas you want tested and take several samples from the area. The several samples can then be mixed to make up samples for each area you want tested.

Your cooperative extension service office will assist you in getting a soil test made, so contact them so that they may instruct you on the proper procedure for the soil test.

Plant Analysis

Another method used to determine nutrient levels is plant analysis. *Plant analysis* involves

Figure 6-6. Soil samples must be taken from several spots to ensure that the sample is to be representative of the entire area.

studying the tissue of the plant to identify nutrient deficiencies.

A plant analysis can determine the nutrients that are limiting the growth and production of the living plant. Plant analysis can also be a useful tool for evaluating whether or not the fertilizer being used is doing the job.

Fertilizer

If a soil is found to be deficient, the next step is to add nutrients to the soil. Your soil test or plant analysis will tell you what nutrients are to be added. The process of adding nutrients to the soil is called *fertilizing.* The substance added to the soil is called fertilizer.

Fertilizers come from two sources—organic and inorganic. Organic fertilizers are produced by living plants and animals. Inorganic fertilizers come from minerals. If you purchase a fertilizer at your local garden center, you are most likely to buy an inorganic fertilizer.

Analysis

If you look at the bag or container the fertilizer comes in, you will see three numbers. Usually in a form such as 10-10-10 (see Figure 6-7). This is referred to as the fertilizer analysis. The first number refers to the percentage of nitrogen, the second to the percentage of phosphorus, and the third to the percentage of potassium. Thus, a 10-10-10 analysis would have 10 percent of each. Fertilizer can be purchased in many different analyses. You should read the label very carefully prior to purchasing a fertilizer to ensure that you are getting the correct percentage of nutrients.

Complete Fertilizer

A fertilizer is said to be complete if it has the needed percentage of the three primary nutrients. For example, complete fertilizers would be 10-10-10, 12-12-12, 5-10-10, 20-20-20. If a fertilizer is missing a primary nutrient, it is said to be incomplete. Ammonia nitrate (33-0-0) is an incomplete, as is triple phosphate (0-60-0). Complete fertilizers are used first, then incomplete fertilizers are used to provide a needed nutrient.

Slow-Release Fertilizer

Fertilizers are available in a form called *slow release*. This fertilizer works just as the name implies—through a slow release of the nutrients. Slow release-fertilizer is needed by plants where nutrients are needed over a longer period of time and where a quick-release fertilizer would be damaging to the plant.

Ratio

A horticulturist who fertilizers plants should understand the term *fertilizer ratio*. A ratio is a comparison of the percentages of the elements within a mixture. A 5-10-5 analysis fertilizer has a 1-2-1 ratio just as a 10-20-10 analysis. An 80-40-40 analysis would have a 2-1-1 ratio. If after taking a soil test a nursery worker chose to follow the recommendation of 80-40-40, how many pounds of a 20-10-10 analysis (2-1-1 ratio) would he or she need to apply? If you said 400 lb (pounds) you are right.

Fertilizer Application

Several methods may be used to apply fertilizer. The fertilizer must be placed close enough to the plant so that the roots can absorb the nutrients. Caution should be taken when applying the fertilizer to apply only the correct amount in the right location. Fertilizing too close or using too much fertilizer could cause severe damage to the plant. Following are several normally used fertilizer application methods.

Banding. Fertilizer is placed in bands beside, above, or below the seed. Placing the fertilizer 2 to 3 in (inches) [5 to 8 cm (centimeters)] from

Figure 6-7. Fertilizer analysis is shown on the label as three numbers.

the seed makes fertilizer available to the plant's root system just a few days after germination.

Broadcasting. Fertilizer is spread over the top of the soil and is normally worked into the soil. Nitrogen fertilizers are best adapted to the broadcasting method because nitrogen is able to penetrate the soil more easily than potassium or phosphorus.

Applying in Water. Nutrients such as nitrogen can be supplied in uniform amounts by adding the fertilizer to the water when watering or irrigating the plants

Foliage Spray. Necessary trace elements can be given to the plants directly through the leaves. This not a very efficient method of supplying primary nutrients, since nitrogen, phosphorus, and potassium are required in large amounts.

Aerial Applications. Dry fertilizer in the granular or pelleted form can be applied to rough, hilly land; dense forests; and large acreage by an airplane.

and may be damaged by freezing temperatures. In areas of warmer winter temperatures, a late summer application of fertilizer to Bermuda grass lawns helps to keep them green farther into the dormant period.

Container-grown plants are usually grown in a medium that is low in nutrients. These plants need to be fertilized often but with a *low-analysis* fertilizer. Unless you are using a slow-release fertilizer (providing a uniform supply of nutrients for an extended period of time), it is advisable to make monthly applications of the fertilizer.

Greenhouse flowers, bedding plants, pot plants, and container-grown woody ornamentals are grown in limited soil compared to field-grown nursery and field crops. While fertilizer recommendations for nursery and field crops are given in pounds per acre, recommendations for plants growing in a limited amount of soil are expressed in "parts per million" (ppm).

Always remember that fertilizer is not a substitute for sunlight and water. Fertilizer can only help if other conditions are satisfactory. Too much fertilizer can be as bad as not enough.

Fertilizer and Horticultural Plants

There is no general "fertilizer recipe" that can be followed for all horticultural plants. Most nursery crops tolerate a fairly wide range of soil pH, with a pH of 6 as optimum. You may choose to pay closer attention to the pH levels of plants such as camellias, rhododendrons, and azaleas, which are known to be acid loving. A pH of near 5.5 is necessary for these plants.

For nursery, vegetable, and orchard crops, fertilizing is likely to be necessary and you should be careful not to fertilize nursery and orchard crops late in the summer or early fall. The new growth that is encouraged by fertilizing late will not have time to harden before winter

Soil Water

Water serves as a solvent for many nutrients (including those in fertilizer), making them available to plants. Water itself is a nutrient and is essential to plant growth. An important function of soil is to provide plants with an adequate supply of water (moisture). Therefore, the relationship between soil and water is important.

Types of Soil Water

Water soaks into the soil through air spaces in the soil under the influence of gravity. Water that moves through the soil as a result of gravity is called *gravitational water*. Gravitational water is not usually used by plants, because it is not in the soil long enough to be absorbed by the roots.

Some water is held by soil particles. This water is called *capillary water* and is available to the plants.

The moisture content of a recently wetted soil after the gravitational water has had time to drain away is said to be at *field capacity*. When a soil is at field capacity, it is holding the greatest amount of water it can hold against gravity.

Hygroscopic water is water held so tightly by the soil particles that it is unavailable to plants. Hygroscopic water may be considered the moisture that is present in an air-dry soil.

Water and the Plant

A plant will wilt when the roots cannot take in water and deliver it to the rest of the plant as fast as the water is being transpired from the plant. When the plant recovers at night this is called *temporary wilting point*. The *permanent wilting point* is the moisture content where irreversible wilting occurs and the plant dies.

A large amount of water can only benefit a plant to certain limits. However, a plant can get too much water. It is not so much the water that is the problem, but when the soil becomes so saturated that air is no longer available to the plant roots, the plant begins to die. It is for this reason that a knowledge of soil characteristics is so important. Providing plants with a soil that drains well but yet holds enough water for growth functions is essential for success of growing plants.

Movement of Water in the Soil

How fast water moves through the soil depends to a great extent on the texture of the soil. The texture of clay is fine because of the small size of clay particles and because the spaces between the particles are small. Therefore, water moves through clay soils much more slowly than it does through sandy soils, which have larger pore spaces. Not only will the pore space determine how rapidly water will move through the soil, but it will also determine how much water a soil can hold.

As water moves down through the soil, it carries dissolved nutrients with it. This is called *leaching*. Leaching causes losses of nitrogen, potassium, and calcium and can be a serious problem in some soils. These nutrients must be frequently returned to the soil.

The amount of moisture in a soil can be measured by weighing the soil before and after it has been dried in an oven. The moisture content is commonly expressed as a percentage of the dry weight of the soil.

Modification of Soil

As you may recall, the function of a soil is to serve as a support for the plant, provide moisture and aeration, provide a source of nutrients for the plants, and have the correct pH for the plants being grown.

Many plant growers modify soil for the plants they grow. This modification is necessary because of the unavailability of good soil. Also, by modifying the soil the grower can change the soil texture, structure, and fertility for each plant or plants grown.

Modification Materials

Soil modification materials normally come in two forms: organic and inorganic. Examples of organic materials are peat, pine bark, hulls, and sawdust. Examples of inorganic materials are pelleted styrofoam vermiculite, calcined clay, and sand.

The decision as to which of the materials to use depends upon the needs of the soil. If drainage is needed, sand, pelleted styrofoam,

Figure 6-8. Organic soil modification materials are peat, pine bark, hulls, and sawdust. Inorganic materials are pelleted styrofoam, vermiculite, culcined clay, and sand.

peanut hulls, and pine bark are excellent soil modification materials. Good drainage is accomplished by increasing the number of large pores in the soil. Each of these materials will help increase the proportion of large pore spaces.

If water retention is needed, peat, vermiculite, and sawdust make excellent modification materials. Sandy-type soils usually need modification to improve their water-holding capacity. These materials give sandy soils a better balance of large and small pore spaces, which aids water retention.

Soil Mixes—Homemade

Almost all horticulturists have a mixture they feel is best. Very few of these are identical. The best mix is the one you feel most confident with and the one that best suits the plants.

Nursery Stock. Nursery stock requires a soil mixture that will give good support and has good water-holding capacity. Below are some examples of nursery stock mixtures:

1. ½ Sand
   ½ Peat
2. ⅓ Sand
   ⅓ Peat
   ⅓ Sawdust
3. ⅓ Bark
   ⅓ Peat
   ⅙ Pelleted styrofoam
   ⅙ Sand
4. ⅓ Sandy loam soil
   ⅓ Peat
   ⅓ Sawdust

Greenhouse Bedding Plants. Greenhouse bedding plants require a soil media that has good water-holding capacity. Support is not as critical as it is for nursery stock. Below are some recipes for mixtures for greenhouse bedding plants:

1. ⅓ Loamy soil
   ⅓ Peat
   ⅓ Pelleted styrofoam
2. ⅓ Top soil
   ⅓ Peat
   ⅓ Sand
3. ⅔ Top soil
   ⅓ Peat

Soil Mixtures—Commercially Made

Several commercially made growing media are now available. These mixes are used primarily for greenhouse crops. Some of the mixes do not have soil in them at all. Most of these mixes are already sterilized. The mixes are lightweight and provide good drainage, aeration, and good support for greenhouse plants.

These commercially made mixes usually contain a combination of peat and vermiculite. They come in a bag and are dry, thus water must be added to the materials as they are being used.

Several commercial mixtures are sold primarily for rooting seeds, transplants, and potted plants. The mixtures are popular because of their easy usage. And the grower is confident that each mix has the same recipe.

The importance of a good growing medium cannot be overemphasized. The medium used will partially be determined by how the plants are watered and fertilized. One point to keep in mind is that you should not use two different mixes on the same bench in the greenhouse. This makes watering very difficult.

Soil Mulches

What exactly is a mulch? Actually a mulch can be a number of materials that have the property of insulating the soil surface. What possible value could there be in putting a mulch on top of the soil?

One value of mulches such as the one the student is applying in Figure 6-9 is that they conserve moisture, because they reduce evaporation of moisture from the soil. Have you ever turned over a rock to find several insects crawling all around? Why were the insects under the rock? If you looked closer you probably noticed that the soil under the rock was more moist than the soil surrounding the rock. The rock served as a mulch for the soil and reduced the amount of moisture lost. The rock insulated the soil surface from the heat and drying effects of the sun.

When a mulch is applied to the soil surface, it also serves to reduce erosion. The mulch protects the soil from hard rain and the flow of water. It prevents the soil from washing away.

You may have seen the soil near construction sites simply turning to mud and washing into the street. Planting grass or some other type of cover in the bare soil can serve as a mulch because it insulates the soil from the effects of moving water. Mulches (such as straw) are often applied to newly planted lawns and seed beds, especially on slopes, to prevent the seeds from being washed away.

Mulches also help to control weed growth. When you turned over the rock, you probably noticed that there were no plants growing under it. In this instance the rock blocked the light from the seeds under it. In this manner, mulches such as wood chips discourage the growth of weeds.

Soil Organisms

A fertile soil is alive with all kinds of living beings. Insects and earthworms may be the most obvious living creatures (microorganisms) in the soil, but there are millions of microorganisms such as bacteria and fungi that also live in the soil. It is the fungi and bacteria present in soil

Figure 6-9. An employee carefully applies mulch to a bed of marigolds.

that are essential in the formation of organic matter. The decomposition of plant and animal material would not occur if it were not for the digestion processes of soil microorganisms.

Through the work of microorganisms the *humus* of soil is formed. Humus is a dark (almost black) material that is capable of absorbing and holding nutrients much like clay. Humus can hold more moisture than clay, but it does not cling as tightly to nutrients and water. Small amounts of humus can have great effects on soil structure and available nutrients.

While most soil bacteria and fungi are harmless, there are some that are not helpful to growing plants. Some soil microorganisms feed upon and give diseases to the plants that a horticulturist is attempting to grow. These harmful microorganisms are pathogenic (disease causing) and cause great losses of plant material.

Control of Soil Microorganisms

Because some microorganisms work in an antagonistic fashion, it is often necessary to control their growth. One method of controlling soil organisms is through *soil pasteurization* (sometimes known as soil sterilization). Some method of soil pasteurization is necessary not only in the control of disease organisms but in the control of insects and weed seeds as well.

Use of Heat. One method of soil pasteurization is through the use of heat. Steam is an effective use of heat for pasteurization of soils used in a greenhouse. The soil should be slightly moist, since either a very wet or dry soil heats slowly and many organisms may not be killed.

Steam pasteurization causes many changes in the soil. The soil often becomes granular as a result of soil aggregates' shrinking. It is important to water the soil thoroughly after pasteurizing so that all soil aggregates have a chance to swell back to their original size.

Another change that occurs is a change in the form of nitrogen. Pasteurization causes nitrogen to become a form that the plants are unable to use effectively and that may burn plant roots.

There are many changes in the amount of living organisms in the soil. Not all bacteria are killed even after 6 hours of steaming at 180° F [82° C], and it is doubtful that all organisms would ever be killed under the conditions when soil is steamed.

A temperature of 180° F [82° C] for 30 minutes is usually enough to kill most insect pests. If the soil is left alone it easily becomes infested with contaminants once again through dust particles in the air.

Steaming, along with careful watering and other good cultural practices, will help keep

Figure 6-10. Good soil tilth describes soil that is easily worked.

soil-borne diseases to a minimum. Soil can be used as soon as it is cool.

Another method of heat sterilization is the use of hot water. Boiling water can be used satisfactorily to sterilize pots and other equipment. However, soil sterilizing by this method is relatively ineffective because of the fact that the water loses heat once it is poured into the soil.

Baking the soil in an oven is another method of using heat to sterilize. This method is practical for only small amounts of soil.

Chemical Control. Formaldehyde is a chemical disinfectant that can be used to sterilize soil outdoors when steam is not available. The soil is loosened up well and is then saturated with the formaldehyde solution. After the solution is applied, the ground should be covered with plastic for 24 hours and then allowed to dry

and air out. It may take as long as 2 weeks for the formaldehyde to escape from the ground. As long as you smell formaldehyde, it is not safe to plant.

Tear gas has also been used to sterilize the soil when steam is not available. Tear gas is applied to the soil by injecting it with a special applicator 4 to 6 in [10 to 15 cm] deep. Tear gas will kill living plants and is therefore hard to apply in a greenhouse unless there are no plants in it. The vapors of tear gas are not pleasant and a gas mask is necessary.

Methyl bromide is a soil fumigant effectively used for killing many pests in unrotted roots. A soil fumigant kills pests through its fumes or vapors. It is injected into the soil in much the same way as tear gas is. Methyl bromide is odorless and toxic to people. It is essential to use a gas mask and exercise extreme caution when sterilizing with this chemical. In some states you need a special permit and license to use methyl bromide on a commercial level.

Some chemicals such as Truban with Benlate, or Banrot by itself are very effective in controlling fungus problems.

Vapam, Vorlex, Mylone, Terraclor (PXNB), and Dexon are other chemicals used to control soil-borne diseases.

Seed Treatment. Another method of eliminating disease organisms is through seed treatment. Seed treatment involves coating the seed with a fungicide such as mercuric chloride, calcium hypochlorite, or cuperous oxide. There are a number of commercial compounds available for seed treatment. You should carefully check local or state recommendations for use of seed treatments, especially mercury compounds.

The idea behind seed treatment is to either eliminate organisms from the seed or to protect the seedling if it is planted in soil that is infested with harmful organisms.

Some vegetable seeds have been treated with hot water (122° F [50° C] for 15–30 minutes) to remove seed-borne diseases. Whatever treatment is used on seeds, be sure to use it carefully so that you do not injure the seed.

Soil Preparation

The physical condition or tilth of soil is significant in its relationship to plant growth. Soil *tilth* is a term used to describe a soil that is easily worked (see Figure 6-10). Good soil tilth is essential to healthy root growth; good soil structure and texture; water-holding capacity, and aeration of the soil. When soils become crusty or puddled, normally the situation results from poor soil tilth.

The best way to alter the tilth of the soil and restore the soil to good condition is through the addition of organic matter. Organic matter is formed through the decomposition of plant and animal materials. Organic matter affects not only the physical condition of the soil but the fertility as well. Organic matter is a storehouse for nitrogen and other nutritients. The physical condition of the soil is improved because organic matter increases the water-holding capac-ity of the soil. This is very important to sandy soils. Organic matter also increases the aeration of clay soils, which is important.

The best method for increasing long-term organic matter is through providing constant ground cover. One of the best covers is grass sod. Organic matter is added to the soil through decomposition of old roots. Orchard growers keep their soil in good tilth by using sod. The sod helps to protect the soil from the compaction that can result when heavy equipment moves over the soil.

Sometimes a crop will be grown and then plowed under while still growing. The crop is being used as a *green manure* crop. The rapid breakdown of the green manure crop adds organic matter to the soil. Sweet clover, which has a high nitrogen content, makes an excellent green-manure crop.

Soilless Cultures

Can plants be grown without soil or a mixture of materials? Yes, many plants are now being grown in what is called a soilless culture.

Figure 6-11. Using a water culture, plants can be grown without soil.

There are two methods by which plants are grown without soil. These are called water culture and gravel culture.

In the water culture method called *hydroponics,* a wire mesh is used to support the plant with the roots suspended in water. Nutrients are added to the water. Air is added to the water to replace that used by the roots. Water culture systems must have an aerator, which constantly aerates the water and provides oxygen to the plant.

In the gravel-culture system, the roots are placed in gravel or very coarse sand. Nutrients are supplied to the plant by being pumped into the gravel or sand. Water is added in the same manner.

Although it has not been proven that hydroponics produces healthier, sturdier, and larger plants, there are some advantages to these cultures. Plants can be grown in environments where no soil is available, such as on ships or in areas where the soil is frozen year round. Another advantage is that the water and gravel cultures can be standardized, whereas soil is not as capable of such uniform control. By using hydroponics a grower can maintain a uniform growing culture. Immediate changes in the nutrients available to the plants can be made.

Hydroponics also gives the scientists easy cultures in which to study plant growth.

Soil and Other Media: A Review

Soil is one of our most precious resources. The horticulturists are very concerned about good soil, because of the delicate crops they grow.

As a prospective horticulture employee, you too must learn about soil and other plant-growing media. Soil texture, structure, tilth, and fertilization are soil properties that you may deal with everyday. Also, you must be aware of the relationship between good soil and the growing of quality horticultural plants.

Unless you are working in a place where hydroponics is used, you will most likely be working with the soil in which plants are being grown. As a greenhouse employee, one of your jobs could be the pasteurization of the soil to control microorganisms. As a nursery worker, you might be concerned with the preparation of a field where nursery stock is being grown. Many other examples could be given, but the point is the same—as a horticulturist, you must know about soils and other growing media in order to grow plants properly.

THINKING IT THROUGH

1. What is the importance of soil to growing plants?
2. Describe what is meant by the terms "soil texture" and "soil structure."
3. What is soil pH? Give some examples of horticultural plants that grow best in acid and alkaline pH.
4. What are the 16 essential nutrients?
5. What are the three primary plant nutrients? Describe the function each primary nutrient performs.
6. What is the purpose of soil testing?
7. What is meant by a fertilizer analysis of 20-10-15?
8. What is the relationship between soil and water?
9. Name three organic and three inorganic soil modification materials.
10. Why are soil mulches used in horticulture?
11. Describe three methods that could be used to control harmful soil microorganisms.
12. What is meant by good soil tilth?
13. Describe two methods of growing plants without soil.

7

Moisture, Light, Temperature, and Air

A vocational horticulture class was growing a crop of poinsettias in the greenhouse for a class project. The plants were propagated in the early fall, and they were planning to have the plants ready for sale just prior to the Christmas holiday season. However, when the plants were to have developed their popular red foliage nothing happened! The plants remained totally green.

The teacher and the class tried to solve the problem in a logical manner by reviewing the environmental conditions under which the plants were grown. First they examined the plant for clues to nutrient deficiencies and disease/pest problems. These were ruled out. Second, they reviewed their watering practices and did not find any problems. Third, they checked to see whether the greenhouse temperature could be the problem. They concluded that the temperature did not cause the problem. Fourth, the class discussed the possibility that the oxygen and/or carbon dioxide supply could have caused the problem. This too was ruled out. Finally, they examined the light requirements. The teacher reminded them that the poinsettia was a short-day, long-night plant—meaning the plants require 12–14 hours of continuous darkness to flower. If more light was allowed to reach the plants, vegetative (green growth), not reproductive growth (flowering), would occur. The class at once agreed that excessive light was the problem.

The class listed possible sources of the excessive light. These included the greenhouse lights' being accidentally turned on at night perhaps by the janitorial staff, and the security lights around the school were lit at night. Also the greenhouse was located next to the football field, thus during home games the lights illuminated the greenhouse. Actually, it was any or all of these light sources that caused the problem with the poinsettias.

This case history points out that environmental factors do have exacting effects on growing horticultural plants. Moisture, air, and temperature also can greatly influence the growth of quality horticultural plants. One important point to remember is that *all* horticultural plants need these four factors—moisture (water), light, temperature, and air in proper amounts.

CHAPTER GOALS

Your goals in this chapter are:

- To describe the importance of water on plant growth
- To define the effect that light quantity, light quality, and light duration have on plants
- To identify the effect of temperature on plant growth and the vegetative/reproductive balance
- To describe the unfavorable effects of temperature on plants
- To describe the relationship that oxygen and carbon dioxide have on plant growth

Water

Anybody can water plants! Have you heard that, or maybe said it yourself? However, nothing could be farther from the truth. The most common cause of plant growth problems is ineffective watering practices. Too much water as well as too little water can cause problems in growing plants.

Functions Of Water

Since proper watering is so important you might ask "just what does water do for the plant?" We often take it for granted that plants must have water, but we may not understand why. The answer to the question is that water serves five basic functions: (1) water acts as a solvent, (2) water provides a transportation medium for moving plant materials within the plant, (3) water helps maintain plant cell turgor, (4) water helps to cool the plant, and (5) water is necessary for basic plant growth processes (see Figure 7-1).

Solvent. Did you know that plants are about 80 percent water? This large volume of water acts as a medium to dissolve, absorb, and combine with other essential elements within the plant. The soft drink you had recently was mostly water. The water acts as a solvent to "hold" the ingredients of the soft drink. In fact, water is often called the universal solvent.

Transportation. In addition to acting as a solvent, water moves throughout the plant by way

FUNCTIONS OF WATER

SOLVENT
TRANSPORTATION
CELL TURGOR
COOLING
GROWTH

Figure 7-1. The five functions of water in growing plants.

of the xylem and phloem tissues. This movement transports plant materials to and from various parts of the plant.

Turgor. Water also helps plants to maintain their turgor. Turgor in a plant is similar to blowing up a balloon. As air is blown into the balloon it becomes taut and firm. In a similar manner as the plant absorbs water, individual cells (like miniature balloons) are blown up and become taut and firm.

One important aspect of cell turgor relates to the guard cells, located primarily on the leaves around the stomatal openings. These openings allow water vapor and oxygen to escape and carbon dioxide to enter. If the plant is water deficient, the guard cells are not taut and firm. They then open and close improperly, or not at all. If enough cells lose their turgor, the plant will wilt. Turgor, therefore, is very critical in helping the plant to remain in a suitable state so that it can grow properly.

Cooling. Water also acts as a cooling agent for plants. During times of excessive heat and/or sunlight, the water, in the form of water vapor, escapes from the stomatal openings. This tends to "cool" the plant. Adequate water in the plant's total system is necessary. Also, during high-evaporation times, the guard cells may close, thus conserving water within the plant.

Plant Growth. As discussed in Chapter 4, water is necessary for photosynthesis. In fact, without water, photosynthesis would come to a complete standstill. Also, water is important in the respiration process. Thus without water, the two most important plant growth processes would not occur. Plant growth would be stopped.

Absorption of Water

Plants obtain most of their water from the soil. The plant root system absorbs water. Very little water goes directly into the plant through the leaves and stems.

Roots. The amount of water the roots can absorb depends on two factors: (1) the depth and density of the roots, and (2) the ability of the roots to absorb the water.

If you were to dig a hole in your yard, you would discover that the deeper you dug, the more moist the soil would become. If you dug deep enough, the hole would become filled with water. Of course, the depth you had to dig to reach soil saturated with water would depend on the area of the country in which you live. The depth of the *water table* will vary in different areas of the country.

The water table is the position in the soil where the soil is saturated with water as shown in Figure 7-2. In some areas, the water table

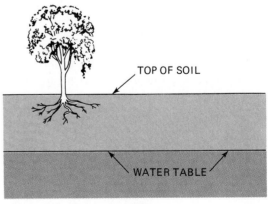

Figure 7-2. The water table.

can be reached at depths of 3 to 6 ft (feet) [0.9 to 1.83 m (meters)] whereas in some others, you would have to dig hundreds of feet deep.

The depth of the water table has a direct effect on which plants grow best in that area. This is because the root systems of plants differ in their depth and volume. Some plants are called "deep rooted" because their roots grow very deep. For example, alfalfa, a common farm crop, is a deep-rooted plant. Its roots may grow more than 6 ft deep.

Some plants have a great number of small roots or a massive root structure. This great volume of roots allows for greater absorption of water. This is especially important for plants that you are transplanting. If they have a large number of roots, the chances of survival are much better, since they are able to absorb enough water to support the plant tops.

The actual absorption of water in most plants occurs through the root hairs. Each root hair is surrounded by water held in the soil. As the plant grows and needs water, a pressure is built up inside the plant, which in turn builds up a pressure to draw water from the roots. This pressure becomes greater inside the root than outside in the soil area. This pressure actually causes the water to be "pulled" into the root. Once inside, the water begins its journey up the plant for use by stems, leaves, and flowers.

Soil Factors. The absorption of water is also influenced by the soil conditions. First, the amount of water in the soil is crucial. If the soil is at *field capacity* (refer to Chapter 6, page 82), the plants can more easily absorb water. If only hygroscopic water (water held so tightly by the soil particles that it cannot be adsorbed by the plant roots) is available, then no water can be absorbed and the plant begins to wilt.

The texture and the type of soil have a great influence on moisture availability. Coarse-textured soils that contain primarily sand particles do not hold water as well as medium-textured soils. If you have ever stood on a sandy beach and watched the waves splash water on the shore and noted how quickly the water disappeared into the sand, you can realize how quickly water absorbs into a coarse textured soil. Conversely, fine-textured soils hold water for a long period of time. In fact, some very fine-textured soils are said to be wet all the time.

If plants are grown in a heavy soil (large portion of clay particles), they can withstand drought much better that plants that are grown in a lighter soil type.

The amount of organic matter also affects water availability. Soils high in organic matter have a greater capacity to hold water, thus making more water available for the plant. This is one reason why we add organic matter to soils.

The distance water moves through the soil influences how much is actually available to the plant. Normally, water moves very short distances in the soil—no more than 2 ft [0.61 m]. Thus, roots only can absorb available water in their immediate area.

Soil temperature also influences the amount of water available. In cold soils, the water does not move as freely as in warm soils.

Water and Vegetative/Reproductive Growth

What would happen to a crop of greenhouse tomatoes if they were heavily watered from the time they were planted until the plants reached maturity? The answer is quite simple. The plants would mainly develop vegetative growth—leaves and stems—but would not develop adequate reproductive growth—flowers and fruit. Why?

Assume that all other factors for plant growth are present in an optimum supply—proper temperature, air, nutrients, and light—and that water is the only environmental factor that you can vary.

Under these conditions, if large quantities of water were provided, the vegetative growth would dominate reproductive growth. The plant would be larger, be more succulent or softer, have more leaves and stems, and would bear less fruit. The fruit production would occur at a later date than under normal conditions. Just the opposite would occur if the amount of water provided for the plant were limited. In this instance, smaller plants with an earlier fruit-bearing date would produce fruit.

By controlling these factors, the plant grower can use the water supply to adjust plant growth. The experienced grower might regulate moisture to produce fruit such as tomatoes in the green-house, at a specific time. How would this happen?

If the grower reduced the water supply, the photosynthesis rate would slow down. This means less vegetative growth would occur, and the plant would begin to produce reproductive growth. Hence, by reducing the water supply, the grower can shift the vegetative/reproductive balance.

Another example of moisture control is in the production of snapdragons. Research has shown that stronger stems develop in snapdragons if the soil is allowed to dry slightly between waterings.

Figure 7-3. Careful watering of plants is critical to the production of high-quality plants.

Techniques of Hand Watering

The knowledge of water and its influence on plants is only part of the water story. An important principle to remember is that the quality of the horticultural crops is equal to the quality of the job of watering done, assuming that all other environmental factors are at optimum levels. These are some basic watering techniques to remember when hand-watering plants.

First, the biggest problem with watering is *overwatering*. Plants drown when they are overwatered. The roots must have air and too much water creates a shortage of air in the soil.

Second, check the soil before watering. If the soil feels moist most likely the plants do not need watering. You should water only when the soil feels dry. You water only when the plants need watering. There will also be a less likely chance for disease and insects if plants are watered properly.

Third, when you water do it thoroughly. A point to remember is to apply enough water so that it drains through the soil and to the bottom of the pot. If necessary water the plant again after allowing time for drainage.

Fourth, use a water breaker on the hose. This helps to reduce soil compaction and breakage of stems, leaves, and flowers.

Light

You learned in a previous chapter that light is required for plant growth. The primary source of light is the sun. Plants soak up sunlight just as you do on a sunny day at the lake or on the beach. This light comes from the sun in waves, which to us appear to be white. However, sunlight is composed of many different colors or rays. The rays of the most importance to the horticulturist are the red, blue, and green rays. Red and blue rays are absorbed by the plant and used in plant-growing processes (photosynthesis). Green rays are reflected by the plant and help give the plant its green appearance. As we examine the effect that light has on plant growth, three characteristics must be reviewed: light quantity, light quality, and light duration.

Light Quantity

The quantity of light from the sun is often referred to as the sun's intensity or concentration. First, light quantity varies according to the season of the year. Because of the position of the sun during the different seasons of the year, the quantity of sunlight varies. For example, in North America the maximum quantity of sunlight received occurs in the summer months when the sun's position allows the greatest quantity of sunlight to shine directly on the plants. In other words, the sun is at its "highest" position in the sky then.

The quantity of sunlight also varies throughout the day as the sun "moves" through the sky. For example, there is not as much sunlight during the early morning and late afternoon hours, whereas there is more sunlight during midday.

Researchers have demonstrated that the quantity of sunlight directly influences plant growth. As you recall, photosynthesis requires sunlight to occur. The more sunlight the plant receives, the better capability it has to produce more plant food through photosynthesis. As the sunlight quantity is decreased, the photosynthesis

Figure 7-4. Light quantity has a dramatic effect on a plant's growth rate.

Figure 7-5. Artificial light such as being used in this experimental situation, can be used to provide long-day conditions.

process of the plant is decreased proportionately. You could illustrate this by growing some plants and covering some of them with layers of cheesecloth. You would note that the plants that were partially shaded from the sun will have a less-rapid growth rate than those plants that had full exposure to sunlight.

Light quantity is measured in footcandles. The more intense the quantity, the higher the footcandles. In its simplest form a footcandle is the amount of light given off by a candle at a distance of 1 ft.

Light quantity can be controlled by covering the plants with cheesecloth, which will shield the plants from a good portion of the sun's direct rays.

Light Quality

Have you ever been outdoors on a cloudy day either working or enjoying some type of recreational activity, and noticed that after a period of time in the sun that you were sunburnt? You might wonder how you could be sunburnt on a day when the sun was not bright. As noted earlier, the rays from the sun are the source of energy and these rays still reach the earth even on a cloudy day. These rays also reach green plants and are absorbed by the plant for use in the plant growth processes. The clouds act as a partial filter that does not stop all the rays from the sun. Thus, the rays that would tan and burn our skin come right on through the clouds and are absorbed directly by our skin, or, in the case of a green plant, by the plant's leaves and stems.

The red and blue rays from the sun are the ones that are most beneficial to plant growth. Hence, these rays are the *quality* aspect of light.

Light Duration

Not only does the quantity and quality of sunlight that a plant receives affect its growth, but the *duration* of the sunlight also affects its growth. The amount of time that a plant is exposed to sunlight is called the *photoperiod*. Normally plants fall into three categories when

TABLE 7-1
A PARTIAL LISTING OF PLANTS
CLASSIFIED ACCORDING TO DAY LENGTH

Type of Plant	Short-day	Long-day	Day-neutral
Fruits	Strawberry	. . . . .	. . . . .
Vegetables	Potato	Spinach, radish, lettuce	Tomato, pepper
Ornamentals	Mums, poinsettia	Gardenia, aster	Carnation

classified by photoperiod. They are short-day, long-day, and day-neutral plants (see Table 7-1). Actually this means that plants grow and produce best with a short number of hours of sunlight per day, or a long number of hours of sunlight per day, or that some plants are not affected by the length of day.

Actually, it is the length of the dark or night period that influences plant growth, but for practical reasons, we use the short-day, long-day terminology.

Chrysanthemums are called a short-day plant. While growing in long light days, they produce foliage or vegetative growth. But as soon as short days are provided, they begin to produce flowers or reproductive growth. Mums grown in natural conditions outside the greenhouse will flower only in the fall after shortening fall days provide the optimum day length. In a similar manner, flowering of mums can be delayed by artificially providing long days through electrical lighting.

An experienced greenhouse grower will manipulate long and short days of mums to obtain an optimum balance between vegetative and reproductive growth. In this manner the grower ensures that the plants flower at just the right time.

But how does a grower control the length of the day? You probably can guess that a covering can be placed over the plants to shut out the light. The grower shortens the day by pulling a dense black cloth over the plants, and can also provide artificial light to lengthen the day. Thus, by the use of coverings and artificial lights, the grower can provide optimum day lengths.

The skill of controlling day length is very crucial for the greenhouse operator. A miscalculation could completely damage the flowering of a crop of plants. As you can imagine, Easter lilies that flower after Easter are impossible to sell.

Temperature

Horticultural plants should be grown as nearly as possible at their optimum temperature. In a general sense, temperature is proportional to rate of cell division (growth). This means that as temperature varies above and below the optimum for a particular plant, the rate of cell division increases and/or decreases. This changing rate of growth means more or less development of leaves and stems.

Temperature control is a crucial skill for a horticulturist. A grower who produces crops in the greenhouse must try to maintain the optimum temperature in the area where plants are being grown. For example, the gardenia needs an optimum daytime temperature of 68 to 77° F [20 to 25° C]. The buds and flowers will wilt if exposed to temperatures a few degrees above this optimum range.

To help the greenhouse grower control the temperature in the greenhouse, thermostats are used. These devices are set at a particular temperature, that is, 68° F [20° C], and when the temperature in the greenhouse drops below this setting, the thermostat automatically starts the greenhouse heating system. In warmer weather, the thermostat automatically starts the cooling system or opens the ventilation, which lowers the greenhouse air temperature.

Temperature And Vegetative Reproductive Growth

Temperature will affect the reproductiveness of the plant, depending on whether it is basically a warm- or cool-season crop. A higher-than-optimum temperature will cause a cool-season crop such as spinach to flower. In contrast, a lower-than-optimum temperature in tomatoes will cause them to grow more vegetation and set less fruit. When dealing with temperature, you usually need to know the optimum and normal growing temperature for the crop you are growing.

There are specific instances when temperature is used to *force* flowers or bulbs. The chrysanthemum produces proper flower buds in controlled temperature. The mums will flower longer if kept in an environment where the daytime temperature is 59° F [15° C]. Of course, this is a very cool temperature, especially if the mum is being kept inside a house.

Daffodils are also forced in greenhouses, using temperature control. Usually in early October the daffodil bulbs are planted in boxes or planters. These containers are then stored in cool barns or sheds, or in coolers at 35 to 40° F [2 to 4° C]. The cool temperatures allow the bulb to mature and allows for growth to occur. The bulbs are then transferred into the greenhouse in midwinter, when growth begins. A greenhouse temperature of 55 to 60° F [13° C to 15° C] allows for optimum growth. The flowers are then ready for cutting in 3 to 4 weeks.

Thus, an experienced greenhouse operator must use optimum temperature to control the vegetative/reproductive growth cycles of greenhouse crops. Improper temperature could result in a *lost* crop, which mean lost income and low profit.

Unfavorable Effects Of Temperature

The unfavorable effect of temperature on horticulture crops occurs in two instances: temperatures above the optimum range, and temperatures below the optimum range.

High Temperatures. Temperatures higher than the optimum will normally cause a loss in yield of the crop. In higher temperatures, the respiration rate of the plant is greater than the photosynthesis rate. This means that the products of photosynthesis are being used faster than they are being produced. In other words, the plant food normally being produced for plant growth is not available. Thus yield is reduced. Remember that for a plant to grow and develop, photosynthesis must be greater than respiration.

Temperatures above the optimum can affect many horticultural crops. For example, carnations produce weak stems if grown in higher-than-optimum temperatures (above 50° F [10° C] at night, 55° F [13° C] on cloudy days, and 60° F[15° C] on bright days). Azaleas produce flowers best in greenhouses at a temperature of 60° F [15° C]. The poinsettia, however, needs a temperature of 70° F [21° C] to flower.

If you were working in a greenhouse and producing these flowers, you would have to carefully control the temperature.

Low Temperatures. Low temperatures also produce low yields. Since optimum temperature is needed for optimum photosynthesis rate, at low temperatures the photosynthesis process is slowed. This results in slow growth, which means reduced yields.

Low temperatures restrict the production of many horticultural crops to certain geographic areas of the United States. A classic example is the location of citrus crops in Florida, Texas, and California; cool-season berry crops such as the blueberry in Northern areas; and dates to certain areas in California. Also, the variation in temperatures, both summer and winter, also dictates where a crop can be grown outdoors.

Winter Temperature. Winter temperatures normally have two opposite effects on plants. Cold temperatures can be harmful and fatal to many plants. On the other hand, some plants require a certain number of days of a low temperature in order to grow properly. This is especially important in crops adapted to temperate climates, where the plants go dormant for a certain period of time.

Harmful Effects. Plants are normally classified as hardy—able to withstand cold temperatures—and nonhardy—unable to withstand cold temperatures.

Winter injuries to nonhardy plants can result from damage to immature tissues. The new growth areas (immature tissues) of the plant are more prone to freezing temperatures than are mature tissues. That is why we do not recommend pruning ornamental trees and shrubs in the fall. This encourages late fall immature growth, which usually suffers winter injury. Also, the mature tissues of the plant have a specific freezing point at which the water in the cells actually turns to ice, thus rupturing the cells and killing them. If enough are killed, the plant can die. Peaches can usually not tolerate temperatures below −15° F (−26° C). Entire peach orchards have been lost in extremely cold weather in the central part of the United States.

Winter injury can also result from winter drought. Just as in other seasons of the year, plants need adequate water supply during the winter. When the soil is frozen, the movement of water up into the plant is severely restricted. On a windy wintery day, a broadleaf evergreen can become water deficient in a few minutes and the leaves can become brown. Have you ever seen someone watering plants in winter? Probably not, but there are some practices, such as windbreaks, which improve the water-holding capacity of the soil; and late-fall watering, which can help avoid winter drought.

Unseasonable variations in temperature can also be harmful to plants. Orchards in mid-South areas often experience this type of damage (particularly the flower buds) when later-winter warm spells are followed by freezing temperatures. The warm temperatures cause the buds to begin to break, thus making them more vulnerable to the following colder temperatures.

Beneficial Effects. Many plants require specific periods of cold temperature in order for the plant to grow and produce properly. Peaches are a prime example, because they require 700 to 1,000 hours below 45° F [7° C] before they break their rest period and begin growth. Lilies need 6 weeks of temperatures at 33° F [1° C] before being planted. This period of rest time is referred to as the dormant period. Plants that must have a specific dormant period will not produce properly if that period is omitted or shortened.

Temperature and Disease

Did you know that powdery mildew can grow to epidemic stages in a slight water vapor film on the leaves of roses? But it can be prevented if the grower maintains an optimum and constant temperature control.

Although temperature may directly affect the growth of diseases, it more often causes an indirect effect from its influence on air and soil moisture. For example, with powdery mildew the water vapor is deposited on the rose leaves in humid conditions, but by maintaining proper

Figure 7-6. The thermostat is used to maintain optimum temperature ranges in the greenhouse.

temperature control, the humidity can be reduced and less water vapor will form on the rose leaves. Then the growing environment for powdery mildew is reduced.

Like plants, diseases also have optimum temperatures at which they grow best. In fact, most plant diseases can be labeled as either cool-season or warm-season diseases.

Plant growers, both indoors and out, recognize the relationship between temperature control and plant disease. A cool, damp growing season in the flower-growing area of California usually means a possible disease problem for the grower. This results from the cool-season disease organisms that thrive in such weather and are more of a problem than in a warmer, dryer growing season. An indoor plant grower has similar temperature-disease-related problems, but can usually manipulate the environment more closely if he or she so chooses.

Air

Plants must have air for growth, primarily for the oxygen and carbon dioxide. Both these elements are necessary for photosynthesis, the unique plant growth process.

Oxygen

Both the top (aerial) and bottom (root) portions of the plant need oxygen. The top portion obtains oxygen directly from the air through the stomates. Also, oxygen must be present in the soil in order for water absorption to occur. Research has shown that if the soil pore space is saturated with water, mineral and water absorption by the roots is slowed and may stop altogether. The water occupies all the soil pore space and the oxygen supply is soon depleted. In effect plants can *drown* by overwatering.

Carbon Dioxide

In an earlier chapter you learned that carbon dioxide (CO_2) was available to plants from the air (.03 percent CO_2). There are instances where CO_2 levels are too low to adequately allow for proper photosynthesis. The most common occurrence of this is when greenhouses are tightly closed on bright sunny days and the CO_2 in the atmosphere is depleted.

When this occurs, CO_2 must be added to the air through outside ventilation or by artificial means. During cold weather, the ventilation cannot be opened, so CO_2 must be added.

Dry-ice converters are the simplest systems for adding CO_2 to the greenhouse. The CO_2 is then circulated by fans to ensure an even distribution.

Adding carbon dioxide above the .03 percent level by artificial means has proven to increase the growth and productivity of some plants and is a common practice in some tomato, carnation, rose, and other indoor-crop cultures.

Figure 7-7. Pollution has a harmful effect on the growth of plants. This potato plant has been severely damaged by air pollution.

Pollution

Environmental pollution is also affecting plant growth. Areas that have a high incidence of air pollution have unique problems that affect the growth and production of certain plants.

Air pollution influences plants by filtering the sunshine and contaminating plants with harmful air contaminants. Some plants are harmed by minute qualities of air pollutants, and so cannot be grown in highly industrialized areas. The grower must always guard against leaks in heating systems and the improper use of paints, pesticides, and other chemicals. Water pollution causes economic problems for many horticulturists, since the horticulturists must bear the expense of obtaining pure water. Fluorine, a common additive to many municipal watering systems, will cause injury to a number of foliage plants. The grower must also use soil that is uncontaminated by herbicides.

Moisture, Light, Temperature, and Air: A Review

The four environmental factors—moisture, light, temperature, and air—all affect plant growth. A horticulturist works with each of these factors in producing quality horticultural crops. Each crop has optimum ranges within which it grows best. A separate optimum range exists for each of these four environmental factors. Consequently, as a worker in the horticulture industry you will consistently consider and use these factors to grow quality plants.

As a greenhouse worker you will be monitoring moisture, light, temperature, and air on a hourly basis to be certain that the plants being grown have the optimum environment.

THINKING IT THROUGH

1. What are the five basic functions of water in plant growth?
2. What effects do soil and plant factors have on water absorption?
3. Describe the relationship between water and the vegetative/reproductive balance of growing plants.
4. What effect does light quantity, quality, and duration have on plant growth?
5. Describe the relationship between temperature and the vegetative/reproductive balance in plants.
6. What effect does temperature have on:
   (a) disease organisms?
   (b) plants growing outdoors in winter?
7. How can pollution affect plant growth?

8

Controlling Plant Growth

Have you ever driven by or seen a picture of an orchard or a vineyard and noticed the trees and vines? Most likely two things impressed you: all the plants were approximately the same size and shape, and the plants were equally spaced and growing in neat rows. This arrangement of the plants did not occur by accident. The horticulturists growing the trees or vines control the growth of these plants to ensure uniform size, shape, and spacing.

Principles of controlling plant growth are very crucial for all horticulturists. Plant growth is controlled by physical means using a technique called pruning. The orchard grower prunes the trees, the vineyard operator prunes the grape vines, the homeowner prunes the hedge, and the nursery operator prunes young trees and shrubs.

In addition to physical control, plant growth is controlled by biological and chemical control.

Controlling plant growth requires a great deal of skill. For example, a person who is pruning fruit trees must prune at the proper time of the year, must know what branches to prune, and must know how to prune limbs so as not to damage the tree. A poorly pruned tree seldom can be corrected, and in many cases can be damaged to the extent that it is not physically attractive or economically useful.

Controlling plant growth is often described as an art form. In fact, some shrubs are pruned to specific shapes to be used for specific uses. Consider the animal-shaped shrubs used to landscape Disneyland/Disneyworld grounds for aesthetic purposes.

CHAPTER GOALS

Your goals in this chapter are:

- To list the reasons for pruning
- To demonstrate the correct procedure for pruning a fruit tree
- To describe the use of cold temperatures to control the growth of bulbous crops
- To describe the use of chemicals (plant hormones) in regulating the growth of certain horticultural crops

Physical Control of Plant Growth

Horticulturists spend a great deal of their time physically controlling the growth of horticultural plants. Shape, size, attractiveness, and sometimes color are factors that make plants valuable. Horticultural plants in their natural state do not always grow to the right shape or size without help from the grower.

Physical control of plant growth requires that the grower physically control growth of specific parts of the plant in order for a certain type of plant characteristic to occur. The basic method for physically controlling plant growth is called pruning. Pruning is the control of plant growth by removal of parts of the plant so that a specific type of growth will result.

Reasons for Pruning

One cardinal rule when controlling plant growth by pruning is to remember that you *never* make a cut or remove a part of a plant without a *reason*. Plants are not simply pruned for the sake of pruning. Far too many horticulturists prune their plants in the spring simply because "you always prune plants each spring." *You never prune unless there is a reason.*

There are many reasons why plants are pruned. Basically these reasons can be grouped into four broad categories: (1) economic, (2) aesthetic, (3) special effects, and (4) public safety.

Economic. Many horticultural plants grow best when they are properly pruned. And when a plant is growing at its best, its economic value is also greater. For example, fruit trees are pruned to establish a sturdy plant that is capable of bearing quality fruit. An unpruned tree will produce too many limbs, which can inhibit the production of a quantity of quality fruit. Removing excess limbs actually enhances quality fruit production.

When pruning for economic reasons, the grower will be attempting to control the growth for several purposes. Pruning is often done to modify the *apical dominance* of a particular plant. In most plants the growth of the plant is controlled by the apical bud. This bud has dominance over the lateral buds, which grow beneath it on the plant. If you will observe the growth of a plant, you will note that it grows at the apex (top or end) of the stems. This growth results from the apical bud. By removing the apical bud, growers stimulate the growth of the lateral buds. The lateral buds produce stems

Figure 8-1. Fruit trees must be pruned to avoid the growth of too many limbs.

absorb enough water and nutrients for proper growth. In these cases either the top of the plant is pruned or the roots are pruned to achieve proper balance.

Fruit trees are pruned to achieve a specific structure. Many fruit trees produce such a heavy load of fruit that unless strong limbs exist, the fruit load can break the limb. Another problem with fruit trees is that too many limbs can actually reduce fruit production. Thus some limbs are removed (usually starting when the tree is quite young), which allows for the growth of other limbs so that they grow larger and stronger.

Also, when fruit trees are being grown the quality of the fruit can be improved by removing excess fruit. It has been shown that if you remove some of the fruit from a tree, the remaining fruit grows larger and thus becomes more economically valuable.

Finally, pruning is necessary when a plant has a damaged or diseased limb. The removal of these inflicted parts increases the possibility that the entire plant will live and remain healthy.

and leaves, which gives the plant a balanced growth. The apical dominance principle varies with different species.

Another economic reason for pruning a plant is to *rejuvenate* its growth. This is effective with older plants. Rejuvenation refers to making the plant young and vigorous again by removing many of the older plant stems and allowing new growth to occur in their place. An example of a plant often pruned to rejuvenate its growth is the forsythia. The older, more mature stems are removed usually over a period of years, and young, more vigorous stems replace them. Thus the plant keeps its youthful appearance and growth.

Another reason why plants are pruned is to keep the top of the plant in balance with the root system, or vice versa. If the top becomes too large for the roots, then the plant's growth can be drastically reduced since the roots cannot

Aesthetic. Many horticultural plants are grown for their aesthetic attractiveness. The grower physically prunes these plants in order to retain their beauty. For example, a hedge such as the Regel Privet is pruned to keep its shape, size, and attractiveness. An unpruned or improperly pruned hedge is distracting and aesthetically unattractive.

Many ornamental plants, such as the forsythia or the Common Bay, are pruned to keep their shape and size in line with the landscape. Shrubs used to landscape a house must be pruned so that they add to the beauty of the house and do not distract from it. The pruning of these shrubs, such as the upright Japanese yew or the spreading yew, involves the removal of excessive growth.

The utilization of space is also crucial to a well-planned landscape. Plants are often pruned to help utilize the space properly. Uncontrolled

Figure 8-2. Bonsai trees are intentionally dwarfed by careful pruning and nutritional control.

growth can take over valuable living or recreational area. Thus, pruning of these plants can ensure that space utilization is maximized.

Special Effects.   Many plants are pruned to achieve special effects. Two of the most famous examples are the topiary work that is used to landscape Disneyland/Disneyworld and the Bonsai work.

Topiary is the work done to sculpt shrubs in the shape of animals. If you have been to either of the Disney parks or seen pictures of them, you have noticed that the shrubs were pruned to look like animals. This is using a form of pruning to achieve a special effect with plants.

Many homeowners prune shrubs or hedges around their house to achieve a certain effect. Sidewalks are framed with a low, square row of shrubs. Family initials are spelled out with shrubs or plants.

Bonsai is the growth of miniature potted trees. These trees are dwarfed by pruning and nutritional control. They may be many years old and yet are potted in a small container (see Figure 8-2).

Public Safety.   Often trees, hedges, and large shrubs are pruned for public safety. Power companies are required by law in many states to remove limbs that touch or interfere with power lines and poles. Also, high trees and shrubs are pruned around intersections where they might affect the visibility of pedestrians and on-coming cars.

Techniques of Pruning

Now that you have the reasons for pruning in mind, you must become skilled in the proper techniques. There is no single technique for pruning. The technique depends upon the plant and the reason for pruning. Proper technique for pruning also includes the usage of the proper pruning tools and processes.

Training.   A pruning technique very familiar to fruit growers is called *training*. Fruit trees need strong limbs to hold heavy loads of fruit. To develop these strong limbs, you must begin pruning the trees while they are young. This allows the main unpruned limbs to grow strong. It is also easier to remove unwanted limbs while they are quite small.

When removing these unwanted limbs, it is important to know exactly which limbs to prune. Figure 8-3 provides illustrations of two limbs

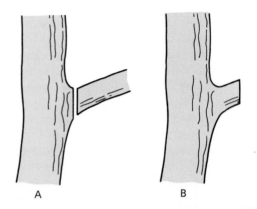

Figure 8-3. Which limb has been pruned correctly?

that were pruned. Which one was pruned correctly? If you said the one on the left, you are correct. Why? The cut should be made as close to the tree trunk as possible. The pruned limb on the right is too far away from the trunk, and leaves a stub which can decay and later cause the main tree trunk to deteriorate. The stubs are also a haven for diseases.

When training a fruit tree, there are three ways fruit trees are pruned. Figure 8-4 illustrates each method. To prune a fruit tree properly requires the removal of a considerable portion of the limbs. Often the person who is removing the limbs is reluctant to prune enough. Strong limbs will result if proper pruning is followed.

The central leader as shown in Figure 8-4 has one main central axis. The branches grow off this axis. The open-center method of pruning cuts this central axis and allows for limbs to grow in a vaselike fashion. The modified version is a combination of the central leader and the open center.

Once the tree has been properly trained, maintenance pruning occurs to keep the tree's shape intact. A knowledge of the fruiting habit of each crop is important for proper pruning to occur. For example, some apple trees produce fruit on special short stems called spurs. Thus, heavy pruning that removes these spurs would

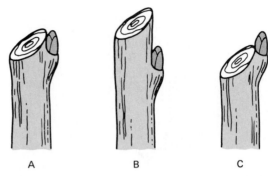

Figure 8-5. Thinning cuts should be made ½ inch above the active bud. A is a good cut. B is cut too far away. C is cut too close.

cut fruit formation. This would happen if an inexperienced worker pruned off these spurs.

Training is also used in pruning grapevines. To do this, the major limbs (called canes) are allowed to become very vigorous while being supported by guide wires. Other canes or potential canes have been removed by pruning. Research has shown that this type of pruning greatly improves the production of grapes.

Thinning. It is necessary to keep the size and shape of ornamental plants consistent with the landscape. A *thinning* process is used to help many ornamental plants keep a proper shape. An attractive landscape requires plants that are in proportion to the structure around which they are planted. Allowing shrubs such as junipers to become overgrown can detract from the appearance of the landscape.

When pruning plants using a thinning process, you should avoid giving the plant a "haircut." By thinning limbs at varying lengths, you reduce the crowding of inner stems.

Thinning cuts, like many pruning cuts, are made ¼ in (inch) [0.6 cm (centimeters)] above active buds (see Figure 8-5). This technique promotes new growth. As you prune a shrub after the new growth develops, you should trim back about one-third of the new growth. This produces a uniform growth pattern.

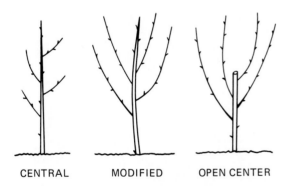

CENTRAL   MODIFIED   OPEN CENTER

Figure 8-4. The three methods used in pruning fruit trees are: (a) central, (b) modified, and (c) open center.

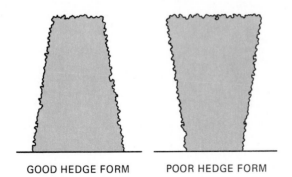

GOOD HEDGE FORM   POOR HEDGE FORM

Figure 8-6. A well shaped hedge will be wider at the base than at the top.

Pruning evergreens also requires the use of thinning. Care should be taken to keep the natural shape of the evergreen. One important point to remember when thinning evergreens is that the nongreen portion does not put out new branches; thus, cutting back these older portions of the evergreen can greatly alter the plant's appearance.

Heading Back. A hedge is often described as a gardener's nightmare. A well-pruned and shaped hedge requires a lot of work with the pruning shears. A good hedge is dense from top to bottom and has a uniform growth.

To properly prune a hedge, a *heading back* technique is used. As the hedge grows, it is pruned back to uniform heights. New growth of 8 in (20 cm) may be uniformly pruned 4 to 6 in (10 to 15 cm). This technique allows for uniform growth. It also allows the gardener to develop a dense hedge from the base to the top.

A common problem with hedges is improper shaping. Ideally a hedge should be wider at the base than at the top (see Figure 8-6). Pruned in this manner, the top of the hedge does not shade the bottom and uniform growth can occur, keeping the hedge full.

Disbudding. Disbudding is a specialized form of pruning. An example of the use of disbudding is with the large-flowered chrysanthemum. Side

Figure 8-7. A greenhouse employee disbuds a growing chrysanthemum.

or lateral buds on this plant are removed so that only the one main flower bud is allowed to develop. This results in one large flower and the plant is more valuable economically. In Figure 8-7, a greenhouse operator is disbudding a growing chrysanthemum.

Pinching. In addition to disbudding, many plants are pinched to force the growth of additional stems and flowers. The carnation is an example of a plant where the stem is pinched back. About 5 weeks after planting the carnation, stems start to elongate. These stems should be pinched at the first internode of this elongated growth. The pinch is simply breaking the stem to the side. This specialized pruning process is quite common with many herbaceous plants grown in greenhouses where the grower wants a continual supply of flowers from the same plant.

Vegetable plants grown in greenhouses are pruned to maximize the growth and production. Tomato and cucumber vines are thinned and pinched to allow for maximum production in a limited land area. Usually, the side shoots are removed and the plants are trained on a series of wires or twine.

Pruning Roots. A landscaping crew was challenged with the job of moving a 15-ft-tall shrub. The removal of large shrubs and trees is not easy and often takes 2 years. It takes that long because moving the shrub could cause stress from root loss. To minimize the stress involved, a technique called root pruning is necessary. The crew that is to move this shrub to a new location must prune the roots one year, then move the plant the following year, hence the 2-year time involved.

You might ask yourself what effect does root pruning have on the plant? How does it reduce stress? And how do you prune roots?

Pruning roots has many of the same effects on the plants as does stem pruning. A plant that has been root pruned will have a reduced vegetative growth, somewhat increased reproductive growth, and a general reduction in total plant growth. This reduced growth results be-

Figure 8-8. Power-driven equipment is used to prune roots of large trees and shrubs.

cause when the roots are cut back, the amount of water and nutrients absorbed is understandably less, since there are fewer roots for absorption. The root system itself grows to produce new roots in the area close to the main unpruned roots of the plant. This vigorous new growth of the root system helps the plant to survive the transplanting move better.

The techniques of root pruning depend on the size of the plants to be pruned. A small plant can be pruned by using a spade and making a deep cut around the plant. This technique simply cuts the roots around the plant and stimulates new root production in the area around the plant.

If the plant were large, a power-driven piece of heavy equipment especially designed for cutting roots would be used. This piece of equipment has a blade-type structure that penetrates the ground and severs the roots around the plant.

Environmental Control of Plant Growth

There are a number of ways other than pruning whereby the horticulturist can control the plant growth. A student who takes a job in a greenhouse that specializes in the production of flowers and potted plants will be confronted with the use of environmental manipulation to control plant growth.

Use of Temperature

One form of environmental manipulation is the use of cold temperatures by greenhouse operators for specific plants to control the plant's growth. Cold temperature helps keep the plant dormant until the grower is ready to force it to grow. The student working in the greenhouse operation might see this technique applied in the production of Easter lilies.

Lilies are grown by a method of environmental manipulation called *cold temperature forcing.* In this method lilies are produced from bulbs that are field grown, primarily on the West Coast, dug in September, and shipped to local storage centers throughout the country. They are then cooled at 40° F [4° C] for 6 weeks while they are in the cases. About 17 weeks before Easter, they are potted and grown in greenhouses using a 60° F [16° C] night temperature. Root growth will begin in a few days, and stems will break through in about a week.

An alternate method is called *controlled temperature forcing.* The bulbs are potted, then stored warm 65° F [18° C] for 3 weeks. Then they are moved to a cooler and held there for 6 weeks at 40° F [4° C]. They are then placed in the greenhouse. This method usually produces the best quality flower crop.

Similar cold treatment is also used for other bulbous crops such as daffodils and irises.

Another use of cold temperatures to control plant growth is called *hardening.* With this process, a plant is slowly subjected to cooler temperatures. The effect of the cold temperatures is to toughen the plant by changing the utilization of plant carbohydrates. As plants are exposed to cooler temperatures, the plant changes its physiological processes and the plant actually becomes more tolerant of cooler environments.

Hardening is used to prepare plants for a period of cold weather such as the outside weather during winter months. In areas where seasonal weather changes occur, hardening occurs automatically as the temperature becomes cooler and the season changes from summer to fall to winter. Hardening is used by the grower to prepare plants that are grown indoors for planting outside where the temperature is cooler.

Plants that are to be placed outdoors are slowly exposed to cooler temperatures so that they are more tolerant of the cool environment when placed outdoors. You can actually harden your own plants by placing them in their pots outdoors for short periods of time and slowly increasing the amount of time they are outdoors until they are ready for planting.

Different plants react differently to the effects of hardening. A plant such as cabbage transplants that have been hardening might show injury at a temperature at freezing, 28 to 32° F [− 2 to 0° C], whereas a group of similar plants that have been hardening might not show injury at temperatures of 20 to 25° F [− 6 to − 4° C]. The few degrees (32 to 25) might just be enough to save a crop, which would save many dollars of lost revenue. Thus, hardening plants can be an excellent investment of time and money.

Chemical Control of Plant Growth

Physical and environmental control of plant growth is not always the most economical or practical method to regulate the growth of plants. And there are many growth processes that cannot be regulated by these two methods.

Chemicals are used by plant scientists to control the growth of plants. Farmers use chemicals to control the weeds in their crops. Homeowners use chemicals to kill dandelions on their lawns. Horticulturists also use chemicals to control the growth habits of certain horticultural plants.

Growth Regulators

The promotion or stimulation of plant growth is of prime importance to the horticulturist. Many crops and plants are more valuable if grown to their maximum, while others are more valuable if they are inhibited in their growth.

One of the first growth regulators discovered was *auxin*. This substance is produced naturally by the plant and is actually a plant hormone. The most commonly known natural auxin is called indoleacetic acid (IAA). After discovery

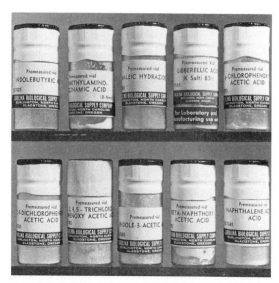

Figure 8-9. These growth regulators can be used to stimulate or inhibit plant growth.

that auxin was produced naturally by the plants, scientists were able to produce a synthetic auxin. Synthetic auxins are used by horticulturists in a variety of growth regulation functions.

Root promotion is one of the best-understood uses of auxins. Rooting compounds are made from synthetic auxins. Research shows that when cuttings are treated with a rooting compound, the formation of new roots is stimulated. Geranium cuttings are treated with a rooting compound before placement in a medium.

Cell division and enlargement are greatly influenced by growth regulators. Gibberellin, another plant hormone, when applied to celery plants will stimulate cell division and enlargement. This results in a longer celery stalk. And, as you know, the longer, larger celery stalk costs more at the grocery store, which influences the price paid to the celery grower.

Although auxins and gibberellins are used to promote growth, many horticultural plants are actually worth more if their growth has been inhibited. While this may be hard to believe, the use of chemicals to retard the growth of many plants is a common practice among horticulturists. One common example is the use of a growth-inhibiting hormone called *B-Nine* to treat chrysanthemum so that a shorter, more compact plant is produced. There is a tendency for mums to grow tall. Thus, a .25 percent solution of B-Nine sprayed on certain varieties about 2 weeks after they are pinched will result in a shorter, more compact plant, which is more valuable economically. B-Nine is also used in the production of azaleas to produce a more compact plant and at the same time stimulate the production of additional flowering buds. This results in an azalea plant that is more valuable.

Plant hormones exist for about any control purpose needed by the horticulturists. Dormins are used to promote dormancy in buds and seeds. Kinns are combined with auxins to stimulate root development. Ethylenes are used to achieve ripening of bananas and the uniform flowering of pineapples.

These many plant hormones are very crucial to the horticulture industry. The horticulturists must know how to use the chemicals correctly and in many cases safely.

Controlling Plant Growth: A Review

If you were employed in a nursery, one of your jobs would be to control the growth of the many different plants being grown. Young trees and shrubs would be pruned to ensure that they grow properly. On the other hand, if you were working in a greenhouse you might be using temperature- or growth-regulation chemicals to alter the growth of plants. Temperature would be used to *force* lily growth at just the right time, while a chemical such as B-Nine would be used to treat chrysanthemums so that a shorter, more compact plant was produced. Lilies grown at just the right time and a shorter, compact mum are more valuable, and hence are important to your employer.

Plant growth control methods require skills and knowledge as outlined in this chapter. Your basic understanding of these skills is crucial to potential employers. And with experience you will become competent in the various ways to manipulate a plant's growth to achieve the greatest economic value.

THINKING IT THROUGH

1. List four objectives of pruning horticultural plants.
2. Describe how you would prune a fruit tree to have an open center.
3. Define the following terms in relation to pruning:
   (a) thinning
   (b) haircut
   (c) heading back
   (d) disbudding
   (e) pinching
4. Under what conditions would you root-prune a plant?
5. How are cold temperatures used to environmentally manipulate the growth of bulbous crops?
6. What is hardening?
7. What is a plant hormone?
8. What plant hormones are used to promote and regulate the growth of plants?

9

Propagating Horticulture Plants

Every time you walk through the woods or fields, you find yourself surrounded by plants. Have you ever wondered where they came from? As you look around, you can see a world covered with trees, shrubs, grasses, flowers, etc., as far as the eye can see. Have you wondered how a flower became a flower or a tree became a tree? How does one plant reproduce another exactly like it? Why do we see other plants with all kinds of flower colors, leaf types, or sizes?

The bursting forth of new plants and plant life in the spring is living proof that existing plants somehow produce new plants to keep the world populated with their species. Even desolate areas such as the desert break into a glorious springtime show. Each area of the world has species of plants that somehow are able to grow, reproduce themselves, and then die. Sometimes this cycle is completed in a few weeks, often it takes years. The process whereby you start with one plant and somehow multiply it to end up with two or more plants is called propagation or reproduction.

Horticulturists are involved with plant propagation. Almost every type of horticulture business is involved directly or indirectly in propagating or increasing the number of horticultural plants.

The greenhouse operator propagates plants by planting seeds, which is called sexual propagation; or by taking cuttings from existing plants, which is called asexual propagation. The nursery operator, turfgrass specialist, vegetable grower, and pomologist all depend upon propagation. In fact, you might say that all of horticulture starts with propagation.

As a prospective horticulture employee, you must develop basic skills in plant propagation. These skills will be critical to your job. For example, a plant grower must know how and when to germinate the annual and vegetable seeds. A nursery worker must know how and when to take cuttings from narrowleaf evergreens and to graft fruit trees. A rose nursery worker must know when and how to bud the number of roses that she or he will need to sell in the future.

CHAPTER GOALS

Your goals in this chapter are:

- To differentiate between sexual and asexual plant propagation
- To describe the process by which a plant grows from a seed
- To demonstrate the types of asexual propagation of plants: cuttings, layerage, budding and grafting, and specialized structures
- To describe which method of asexual propagation is used for specific plants and why

Did you know that the Washington navel orange does not have seeds? That fact in and of itself does not mean a great deal until you consider just how a seedless orange reproduces itself. This variety of orange can reproduce itself even though there are no seeds, by means of asexual or vegetative reproduction.

Plants differ from most other forms of life because of their ability to reproduce from both sexual and asexual processes. Sexual propagation refers to the reproduction of the plant from a seed. Almost all plants produce seed. When you plant a garden or seed your lawn, you are reproducing a plant by sexual propagation.

Asexual propagation refers to the reproduction of a plant in a vegetative manner. A portion of the existing plant is removed (such as a stem) and placed in an environment where it can develop roots and grow into a plant identical to the plant from which the portion was taken.

Since most plants produce seed, you might wonder why asexual propagation would be used to reproduce a plant. There are advantages to using each of these types of propagation:

1. Sexual propagation works best when a large volume of plants must be produced at a minimal cost. For example, seeding a garden or a lawn requires far more plants than could economically or practically be reproduced asexually.
2. Seeds are an excellent way to store plants over a period of time.
3. Seeds provide an easy way to store the plant from one year to the next.

One disadvantage of sexual propagation is the long time it takes for many plants to grow from a seed to a mature plant.

Asexual propagation is used, as in the case of the navel orange, to grow plants that do not produce seeds. Asexual propagation is often easier and faster than producing a plant from seed. And when using asexual propagation, seed dormancy problems (when a high percentage—under 90 percent—of the seeds do not germinate) are reduced.

Sexual Propagation

Sexual propagation starts with flower formation. As with animal reproduction, there must be male and female parts. However, in most plants the male and female parts are contained on the same plant and usually in the same flower.

Sexual reproduction involves the union of the pollen cell and the egg cells and the resulting formation of a seed. This seed is really a miniature plant. The pollen cell contains all the parts of a mature plant, and has the capability of producing a new mature plant upon germination. It will also contain a supply of food that is sufficient for germinating and establishing the new plant. Producing new plants from seeds requires skill and knowledge. Seeds are also an efficient means of producing large numbers of plants, and they are easily stored until needed.

The successful propagation of plants by seeds requires that you be familiar with the specific requirements of the plants you plan to grow.

For example garden peas are planted in gardens about 1 in [2.54 cm] deep in rows 2–4 ft [0.610–1.220 m] apart. The planting date for peas varies with the area. For example, winter peas grow best in California, Arizona, and along the coast in the Carolinas. Spring and summer varieties grow best in cooler Northern areas such as Wisconsin.

Cabbage plants should be placed 12 in (30.48 cm) apart. Cauliflower grows best in a cool, moist climate.

You must have knowledge of planting dates, depth of planting (some seeds need light), moisture levels, germination temperatures, spacing, and special soil mixes needed.

Seeds

The seed is a miniature plant enclosed in a protective coat (see Figure 9-1). The miniature plant consists of stems, roots, and leaves, with a supply of food to keep the seedling growing until it can produce its own food supply.

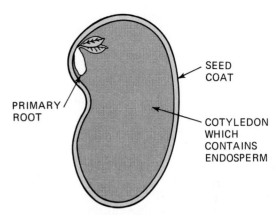

Figure 9-1. The parts of a seed.

All seeds are made up of three parts: (1) the seed coat, (2) the embryo, and (3) the endosperm. The seed coat protects the seed from mechanical damage and prevents the inner parts from drying out. The seed coat is composed of one or more layers, which may be thin and easily broken or thick and hard, depending on the seed. The embryo is the immature plant found inside the seed. The endosperm is the food storage tissue inside the seed. It is composed mostly of starch, with some oil and simple proteins. All these parts together can, under proper germinating conditions, produce a viable plant.

Germination. A seed would not be of much use to us if it always stayed in the same structure and did not produce a plant. The process by which a seed changes into a developing seedling is called *germination*.

For germination to occur, certain environmental conditions must be present. These conditions include specific levels of moisture, temperature, air, and light. It is crucial that the seeds being germinated receive the proper level for each of these environmental conditions.

For all seeds, moisture is the single most important factor involved in the germination process. Many seeds lie dormant for years at normal room conditions. We have all purchased

TABLE 9-1
FACTORS INFLUENCING GERMINATION

Seed
Soil media
Moisture
Temperature
Oxygen

packages of dormant flower or vegetable seeds. When they are removed from this environment (the package) and planted in the proper environmental conditions, they germinate in a few days. Often this can happen even if the only change in the environment is the addition of moisture. The amount of moisture present is very important as the seed continues to germinate. If a newly developing seedling is given too much or too little moisture, it will die.

Temperature is the second most important environmental factor involved in seed germination. The temperature requirements needed for germination often are as critical as the needed moisture requirements. Often where the soil and air temperatures are cooler than optimum, the seed uses all its stored food before it can germinate and grow. The optimum temperature for germination can vary from 65 to 80° F [18 to 27° C]. For best results, you should use the optimum temperature for the specific kind of seed you are germinating.

A temperature of 70° F [21° C] is needed during germination. Once the seedling emerges, the temperature should be reduced to the optimum growing temperature for that particular crop. For example, in Chapter 4 you learned that tomatoes grow best in nighttime temperatures of 60 to 70° F [16 to 21° C].

Aeration is an important environmental condition for successful seed germination. There must be ample air or oxygen around the seed so that it can "breathe." Aeration is controlled by not overwatering the soil. If soil is waterlogged, the water will occupy the air space,

there will be no oxygen available, and the seed will die and eventually rot.

Light is not required for the germination of most seeds. However, research in recent years has indicated that light may be beneficial in various stages of germination or for some specific seeds, such as lettuce.

In some cases the reason for a seed's not germinating is in the seed itself. Seeds have their own internal mechanisms that prevent germination, a situation that is referred to as seed *dormancy*. Some seeds have seed coats that are too hard to permit the absorption of water. With a seed like this, you will need to crack the seed coat mechanically to allow water to be absorbed. This condition is usually more prevalent in woody trees and shrubs. One type of seed, that of the jack pine, will not germinate without high levels of heat. This is a built-in mechanism of the plant, so that if a forest were burned down, the seeds of the pine would germinate and repopulate the area. Other seeds need such things as acid treatments, moist chilling, and mechanical damage to the seed coat to obtain germination. Honey locust trees are usually given a 30-minute bath of concentrated sulfuric acid, while seeds of *Taxus cuspidata* (Japanese yew) require three months of storage in a moist medium. It is very important to remember that seeds vary in their germination requirements. The best source of information about the germination of a particular seed is the seed package. Read and follow its directions carefully. It will tell you the temperature, moisture, light, and germinating times necessary. The success of your efforts will depend on how well you do your seed propagation. In seed propagation:

1. The seeds should produce the plant you want to grow.
2. The seeds must be capable of germinating. Do not use damaged or diseased seeds.
3. You must use viable seed and must supply proper amounts of moisture, temperature, light, and air to the seed and the resulting

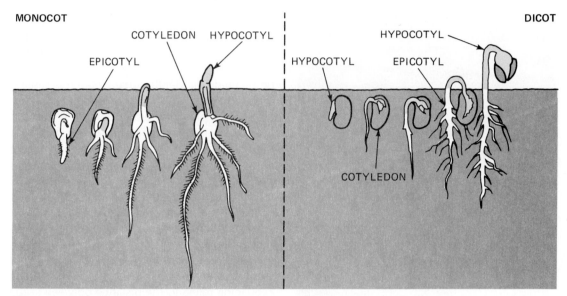

Figure 9-2. The developing seedlings in monocots and dicots.

seedlings. Remember, you have not successfully propagated the seed until the seedlings have been established in their permanent location.

The Seedling. At the time when all these environmental conditions are at their optimum for the seed, it will leave its dormant state and start growing.

Germination starts when the seed soaks up water. With the absorption of water, the seed will soften and enlarge. This enlargement of the seed will put pressure on the seed coat and cause it to split open. Through this split in the seed coat will emerge the primary root, which will later form the root system of the plant. Next will emerge the stem, which, in a short period of time, will develop leaves that produce the food needed for the new actively growing plant (see Figure 9-2).

Techniques of Seeding. Do you know the proper techniques for seeding? You can either plant your seeds by field seeding directly (plant-ing the seed in a field location), or by planting the seed indoors in containers called flats.

Field Seeding. Direct seeding in the field requires that you be able to perform certain functions and to know specific items in relation to seeding. Good seeding, which will result in a good crop of seedlings, is enhanced by a good *seed bed*. A good seed bed has the following characteristics:

1. It is free of debris.
2. It has adequate moisture for germination.
3. It is loose enough to provide for aeration, yet is not full of clods.
4. It is level.
5. It has a soil mix that allows for the seeds to contact the soil particles.

Before you begin to plant the seed, you should examine the seed bed to check for these factors and take steps to see that deficient items are corrected.

After the seed bed is prepared, you must select a good-quality seed to plant. Good seeds

Figure 9-3. Planting the seeds or reseeding a lawn must be done carefully and accurately to ensure a high percentage of plants.

important factors to consider. The depth at which you plant depends upon the size, type, and moisture availability. Many people determine depth by a simple rule: seeds should only be planted at a depth equal to three times the seed's diameter. Also, many seeds are so thin that simply placing them on the top of the soil and allowing the watering process to sink them into the soil is adequate. If when planted the seeds are too shallow, the chances of the seed's not getting enough moisture to germinate is greater, because the upper soil surface dries out sooner. If the seeds are planted too deep, the seedling's food supply will be used up before it reaches the light. You should examine the seed package and follow label directions in planting seeds.

Spacing refers to the distance between seeds when they are planted. Depending upon the use planned for the seedlings, spacing will vary. For instance, if you plan to transplant the seedlings immediately, they can be planted thickly and then thinned while transplanting. If, however, the plants are to be field grown, providing a space between each seed is crucial. For example, in planting squash in a garden, it is often recommended that 2 to 3 ft [60 to 90 cm] be left between each hill of squash seeds. Although this seems like a lot of space while planting, the mature plants need this room to grow properly.

Seeding can be accomplished either by hand or by machine. If you are planting a small area like a family garden, or sowing seeds in a flower garden around your home, seeding by hand is very effective and efficient. However, in larger jobs seeding machines prove efficient and effective for seeding. These machines can be set to allow for proper spacing, depth, and seeding rate (the number of seeds planted).

After the seeds are planted, you should add a mulch. A mulch is any loose material placed on top of the soil. The primary advantages of using the mulch are to conserve moisture, control weeds, and prevent soil erosion. In field seeding, the common types of natural mulch are

are those that have a high germination percentage, are not damaged, are free of insects, and are healthy and of proper size. The package in which the seeds are purchased will list the germination percentage. If the germination percentage is 90 percent, this means that 90 out of every 100 seeds planted should germinate.

Planting the seed is the next skill that is crucial. Depth of planting and spacing are straw, peat moss, bark, leaves, wood chips, cotton seed hulls, and corn cobs. Artificial mulches, such as sheets of black plastic, are also used.

Indoor Seeding. Seeding indoors normally refers to the seeding into flats. If you purchase young flowers or garden transplants at your garden center, most likely these plants were seeded and grown in a flat in a greenhouse. The flat is a wood, plastic, or metal container. These flats have drainage holes to allow for moisture flow. The size of flats is normally from 11 × 20 × 2½ in to 14 × 22 × 3½ in. [28 × 50 × 6.35 cm to 36 × 56 × 8.8 cm].

When you are seeding in flats, the first item you should prepare is the propagating medium. The medium used should be free of disease and insects, which is accomplished by pasteurizing the materials. The material used should be lightweight and porous. This will allow for aeration and moisture-holding capacity. The material should also supply ample nutrients for good growth. A review of the chapter on soil media will provide you with an excellent assortment of material recipes that can be used as a medium for seeding flats.

When seeding plants in flats, there are several precautions you should take:

1. Label all flats properly so that you know what plants were seeded.
2. Label the dates planted and seeding rates.
3. Space the seeds in the flat in rows to facilitate transplanting later.
4. Cover the seeds in a way that is similar to outdoor seeding, three times the diameter of the seed.
5. After seeding, water gently with a fine mist. Do not spray with a hose since the force of the water will wash the seeds out of their rows and/or sink them too deep in the soil to germinate.
6. Cover the flat with a mulch such as burlap, or a plastic bag.

After the seeds germinate, move the flat to a cooler place in the greenhouse (10° F, or −12° C cooler) and place where adequate light can reach the emerging seedlings.

Watering seedlings properly is very important. On bright, sunny days the seedlings may need watering more than once a day. Do not allow the flat to dry out. Follow a regular schedule for watering the seedlings, and check the moisture of the flat at regular intervals. A good rule is to water enough so that soil will be dry before dark.

The temperature for seedlings varies with the individual crop. Plants that require cool temperatures for proper growth should grow in an environment that does not get below 50° F to 55° F [10 to 13° C] at night, and 60° F to 70° F [15 to 21° C] during the day. Warmer temperature plants should have temperatures from 50 to 60° F [10 to 15° C] at night and 70° F [21° C] during the day.

Transplanting Seedlings. After the seedlings have emerged and have reached a height that will allow each seedling to be handled easily (usually 1 to 3 in, or 2.5 to 7.5 cm), the seedling is ready to be transplanted. Transplanting refers to moving the individual seedling from the flat to its growing container or pot.

The types of pots vary, and depend in many cases upon the personal preference of the person growing the plants. Plastic, peat, and clay are the common types. Cost of the pots, appearance, and strength are important factors to consider when selecting pots.

After the plants have been transplanted, the individual plant should be shaded for a few days and held at a slightly cooler temperature. Water them through a nozzle that produces a fine mist, to keep from washing the seedlings out. Fertilize once a week with 2 tablespoons (Tbsp) [30 ml (milliliters)] of 20-20-20 per 2 gallons (gal) [7.6 l (liters)] of water. After the seedlings are transplanted, grow for 3 to 4 weeks at 60° F [15° C]. After the seedlings root move to 40° F [4° C] for 2 to 4 weeks. This allows the seedling to become stronger and stockier. You will recall from the chapter on controlling plant growth that this is the hardening process.

Sexual propagation as described in the preceding sections is used to grow many different varieties of plants. However, many other plants are propagated by asexual methods.

Asexual Propagation

Propagating plants asexually involves use of only a vegetative part or parts of the plant (the root, stem, and/or leaf). This is possible because the vegetative parts of many plants have the capacity for *regeneration* not only of a similar part, but of the missing plant parts. For example, a single leaf of a begonia has the capacity to produce a new plant complete with leaves, stems, and roots. In contrast, a single leaf of a tomato will root, but will not grow any other leaves. In this case, you need the leaf and a growing bud on the cutting. The existence of many valuable fruit and ornamental varieties of plants depends upon our ability to reproduce them asexually. If this method of reproduction were not possible, many of our plants would be lost because they would not come "true" from seed production. The only way to preserve the Red Delicious apple or Chrysler Imperial rose is through asexual production.

The next time you eat a banana, notice that it does not contain developed seeds like those of the apple. Over many years, the fruit has

TABLE 9-2
FACTORS AFFECTING PROPAGATION

| Propagating media |
| Temperature |
| Moisture |
| Growth harmones |
| Equipment |

been developed so that eating it will be more enjoyable. With the undeveloped seed, the banana cannot be reproduced sexually, so the only method of reproducing a banana is asexually. This is also the case with certain figs, grapes, and navel oranges.

Factors Affecting Asexual Propagation

To further your chances of success in asexual propagation, you need to know which factors will affect your success. These factors include the medium, the temperature, the moisture, growth hormones, and equipment.

To ensure good propagation, the most important factor after determining the proper handling method is the medium or soil in which the plant is being propagated. The medium must provide sufficient amounts of moisture and oxygen and be completely disease free. Various mixes are recommended and used, as was discussed in Chapter 6. They may contain one or more of the following materials: soil, sand, peat, vermiculite (expanded mica), and pelleted styrofoam. The mixture of one-half each peat and pelleted styrofoam is highly recommended due to its good water-holding capacity and drainage, and it is free from disease. Sand is often used for easy-to-pot cuttings.

All rooting media should be pasteurized to kill all disease organisms. In commercial operations, the media are heated to a temperature of 180 to 220° F [82 to 105° C] for 20 to 30 minutes. Always use clean flats, pots, labels, etc., and do not recontaminate them.

To increase the ability to grow better with both seeds and cuttings, bottom heat is often used. Recommended temperatures range from 70 to 75° F [21 to 23° C]. The heat helps by stimulating cell division in the rooting area of the stem.

Moisture is very important for the development and growth of the plant. Too much water will cause the seedling or cutting, to rot and die. Too little water will inhibit germination and

Figure 9-4. Mist systems are used in greenhouses to provide the necessary moisture for cuttings and other delicate plants.

prevent the seedling's development. A common cause of death for cuttings is the lack of moisture in both the atmosphere surrounding the cuttings and the rooting medium. Mist is often used to increase the humidity and also to reduce the temperature around the leaf in order to decrease water loss. Many operations now are covering cuttings with plastic opaque sheets to provide a moist rooting atmosphere.

Often for hard-to-root cuttings, plant hormones are added to the cut surface to increase rooting. Some commonly used rooting hormones are auxins and gibberellins (refer to Chapter 8). The basal end (bottom end) of a cutting is dipped into the hormone and then planted in the soil. Note that too much hormone can inhibit root formation.

The last factor to consider is the equipment. You do not need elaborate facilities or equipment to be successful at asexual propagation, but your equipment should be clean and disease free. Equipment such as misting systems can be used to increase your success in asexual propagation. Other methods, such as plastic coverings, may be used to decrease propagating failures. Determine which techniques are best suited for each plant you wish to propagate.

Types of Asexual Propagation

There are several methods of asexual propagation: cuttings, layerage, budding, grafting, and specialized structures.

Cuttings. There are three basic parts to all plants, the leaves, stems, and roots. All three of these parts can be used in asexual propagation, but you have to know which part you must use to propagate the specific plant involved. The three types of cuttings are referred to as leaf cuttings, stem cuttings, and root cuttings.

The stem cutting is the most commonly used kind in the horticulture industry today. A stem cutting is made by cutting a segment or piece of a growing limb or shoot from the parent plant. It usually contains at least one node and internode and a growing point (bud). The cutting is then placed in a rooting medium where it will develop new roots and leaves.

To have your stem cuttings root successfully, you must be aware of four important keys:

1. The time of year that the cutting is made. As we have just shown, different types of cuttings need to be taken at different times of the year. Cuttings taken at the wrong time of

the year often will not root. For example, lilac cuttings must be taken at a specific stage of softness.

2. The age of the cutting. Cuttings should always be taken from the newly grown portion of a plant. This increases the likelihood of rooting.

3. The position on the plant from which the cutting is taken is very important. All cuttings must be taken just below a node. It is well known that roots tend to form at a node where food materials and internal plant hormones tend to accumulate.

4. The plant should be healthy, with enough nutrients and food to sustain it until roots have formed. Always select healthy branches for cuttings. Do not select inner, slender branches.

If these simple factors are followed, stem cuttings are an easy, fast way of producing a large number of identical new plants.

As you look at the plants around you, you can see that different plants have different stems. The stem found on a juniper is different from that of a magnolia or coleus. With each different kind of stem, you usually need to use a specific type of stem material. Normally, the types of stem materials are classified as herbaceous, softwood, semi-hardwood, and hardwood.

Herbaceous Cuttings. The herbaceous cuttings (often called slips) are made from plants whose stems do not turn woody, such as the coleus, chrysanthemum, geranium, and carnation. This type of cutting may be taken at any time of year, so long as the herbaceous plant is rapidly growing and has a relatively high water content. With experience you can determine water content of the plant by examining the soil and by pinching a leaf petiole to see whether moisture seeps out. To make this type of cutting, you must make the cut just below a node (the thickened area on the stem where a leaf develops). Often these cuttings can just be snapped off by hand. The leaves should be removed on

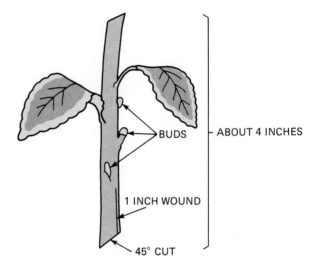

Figure 9-5. A herbaceous cutting taken from a coleus plant.

the lower portion of the cutting. The cutting should then be placed in the soil deep enough to support it, at which time you should make sure that the node is in the soil. Usually no rooting hormone is needed to stimulate roots. Roots tend to develop at the node in about 1 to 4 weeks.

Softwood Cuttings. Softwood cuttings are made from new growth in the spring or early summer, when that part of the plant is growing rapidly. The cutting is taken in the same manner as herbaceous cuttings are. Gather the cuttings during the early part of the day when the water content in the stem is high. It is important to keep the cuttings moist to ensure rooting, so gather them in a plastic bag. A rooting hormone that can be purchased is used to stimulate root production. Softwood cuttings are usually rooted in a moist area, with rooting occurring in 4 to 6 weeks. Plants propagated by softwood cuttings include pyracantha, magnolia, and spirea.

Semi-Hardwood Cuttings. Semi-hardwood cuttings are usually made from narrowleaf and

SOFTWOOD CUTTING

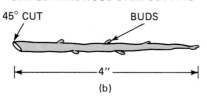

5"

SLANT
BASAL
CUT

REMOVE
BOTTOM
LEAVES

1/2" IN
DIAMETER

(a)

SIMPLE HARDWOOD STEM CUTTING

45° CUT   BUDS

|← 4" →|

(b)

Figure 9-6. (Top) Softwood cuttings are taken in the spring or early summer. (Bottom) Hardwood cuttings are usually taken in the winter months.

ergreens such as *Taxus* and juniper, azaleas, and holly.

Hardwood Cuttings. There are two types of hardwood cuttings, those taken from deciduous plants (plants that have lost their leaves) such as the apple and maple; and the narrowleaf evergreens, such as yews or junipers. The hardwood cuttings are made in the winter months when the plant is in the dormant stage. This is the main difference between semi-hardwood and hardwood cuttings. They are also dipped in a rooting hormone, and placed in a cool, moist area. They usually take several months to root.

Leaf Cuttings. Leaf cuttings can be taken on certain specific plants that have the ability to reproduce both new stems and new roots from a single leaf or a leaf and its petiole. Not all plants have this ability. There are four types of leaf cuttings: leaf blade, leaf-petiole, leaf section, and leaf bud.

Leaf blade cuttings can be used on such plants as sedum or jade plant. These plants have little or no apparent leaf-petiole. The leaf is broken off or cut with a sharp knife from the parent plant. The cut-leaf edge is inserted into the rooting medium to hold the leaf erect. A new growing point, and hence new leaves and roots, will form at the base of the leaf. Rarely will the original leaf become part of the new plant. After the cutting develops its own leaves, roots and stems, the old leaf will die.

The leaf-petiole cutting is used when the leaf and attached petiole are removed from the mother plant and planted. Be certain to set the petiole end deep enough in the rooting medium so that the cutting will stand erect. The new plant will develop at the base of the petiole. Peperomia and African violets are propagated by this method.

Leaves of the sanseveria (snake plant) are propagated by the leaf section method. To prepare a leaf section, cut the leaf into pieces

broadleaf evergreens. The cuttings are taken from the latest growth on the plant after it has finished the rapid summer growth. The cuttings are made about 3 to 6 in [7.5 to 15 cm] long. The leaves on top of the cutting are left, but the lower ones are removed. The cuttings are dipped in a rooting hormone and placed in a moist area. They may root in 4 to 6 weeks or take several months to root. Plants propagated by semi-hardwood cuttings include euonymus, ev-

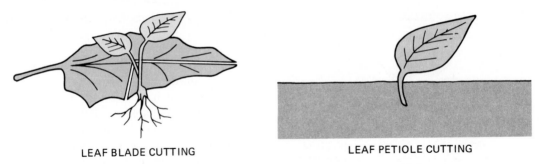

LEAF BLADE CUTTING   LEAF PETIOLE CUTTING

Figure 9-7. Examples of a leaf blade cutting and a leaf petiole cutting.

about 2 to 3 in [5 to 7.5 cm] long. Insert the bottom (basal) end into the soil. Always keep the basal end down, because the cutting will not root properly otherwise. The new plant will develop near the base.

The last leaf-cutting method is the leaf bud cutting, which consists of a leaf plus the bud on a section of the stem. The bud is found at the base of the leaf in the leaf axil. The small bud will give rise to a new stem. Roots will form at this area. The attached bud is necessary. Otherwise, without the bud, the cutting will root but no shoot growth will occur. Plants such as grape ivy, English ivy, and philodendron are propagated by this method.

Root Cuttings. The final type of cutting is called root cutting, which is taken by using a 2- to-4-in [5-to-10-cm] section of the root and placing it in a moist rooting medium. This method works with only a few types of plants, such as red raspberry, horseradish, and Oriental poppies. The cuttings should be taken in late winter or early spring, when the roots are well supplied with stored foods but before new growth starts. Late spring should be avoided, since the plant is actively making new shoots and the food supply is low in the roots.

Layerage. Layerage is a method of propagation whereby the plant develops roots on its stem while it is still attached to the mother plant. The

rooted plant is then detached or cut from the parent plant after it has developed roots, and it becomes a new plant. Because the new plant produces roots while it is still attached to its mother plant, it is one of the surest methods of propagation. The new plant is able to obtain the needed nutrients from the mother plant while producing roots. For this reason, it is commonly used to propagate plants that are difficult to propagate by any other method. The amount of time required will depend upon the type of plant being used. Almost all plants can be layered. Foliage plants such as the rubber plant will root in 4 to 6 weeks, whereas woody plants such as the forsythia may take several months.

On a small scale, layering must be used to good advantage. The layers do not require the close attention to watering, humidity, and temperature that other methods of propagating do.

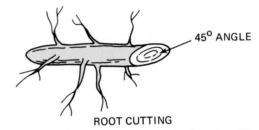

45° ANGLE

ROOT CUTTING

Figure 9-8. The root-cutting method works with only a few types of plants, such as Oriental poppies or garden phlox.

As long as you meet the growing conditions of the mother plant, the layers will form. It is not always practical on a large scale because of the space needed for the large number of mother plants. Also, the parent plants produce a small number of new plants (usually one new plant per layer). The three main types of layering are tip, trench, and air layering.

Tip Layering. Tip layering is the simplest of all methods of layering. The tip of the branch is placed in the soil, usually pointed downward, and covered to a depth of 2 to 3 in [5 to 7.5 cm]. The branch may be anchored with a wire or a wooden peg, and the soil is packed lightly to hold the branch securely in place. The shoot tip will begin to grow downward into the soil, but will curve upward to cause a sharp bend in the stem. At this sharp bend the roots will develop. This method of reproduction often occurs in nature. For example, if a branch of a raspberry or forsythia breaks and touches the ground, it will grow and produce roots, thus producing a new plant called a layer. Another variation is to allow the branch tip to be uncovered, which means that it has already started to grow upward. The branches may also be girded or notched to induce rooting. Tip layering is usually practiced on plants with long, pendulous branches.

Trench Layering. The next method of layerage is referred to as trench layering. With this method of propagation, the entire long branch or stem is bent over and covered with soil. When you are placing the branch in the trench, it should be covered with a shallow layer of soil, not to exceed 2 to 5 in [5 to 12.5 cm]. The branch will also need to be anchored to hold it securely in place. The new plants will grow from buds at the nodes. With the entire branch covered and plants growing from every node, this method has the advantage of producing a larger number of plants. Trench layering is usually used in the propagation of fruit and nut trees, which root with difficulty by cuttings.

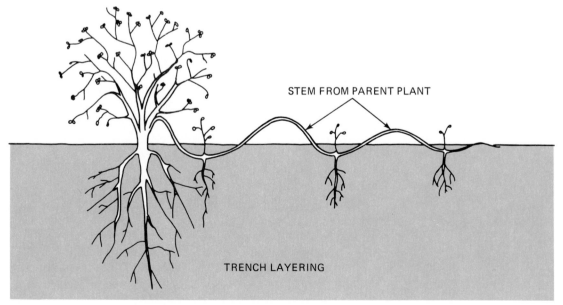

STEM FROM PARENT PLANT

TRENCH LAYERING

Figure 9-9. Trench layering produces a large number of plants.

Air Layering. Air layering is a method commonly used on foliage plants such as the rubber tree, shefflera, and croton. It is the oldest method now known, and it was called "marcottage" by the ancient Chinese. There are two methods for air layering. One is the girdle cut and the other is the cleft cut. In the girdle-cut method, two parallel cuts are made around the stem, about 1 in [2.5 cm] apart. The cuts are made deep enough to remove the bark layer. You will note the shiny green cambium right under the bark. After the girdle cut has been made, the cut area is dusted with a rooting hormone and then covered with moist sphagnum peatmoss, which is tied in place and covered with a piece of plastic. Secure both ends. The plastic retains a constant level of moisture in the area where the new roots are to be formed, and you can see the new roots forming. When the roots have reached the outside of the moist peatmoss, the plastic is removed and the new plant cut off just below the area where the new roots are formed. Plant the new plant and protect it from extreme heat and sun until the plant has developed a more fully developed root system.

The cleft cut is similar to the girdle cut, except that only one cut is made all the way through the stem. In the cleft cut, a portion of the bark may or may not be taken off on one side of the stem. This cut is made just below a node. Usually the roots will only form on the side where the cut was made. A toothpick may be placed in the notch to hold it apart. With both methods, it is very important to ensure that the sphagnum peatmoss stays moist. Drying out will prevent root formation.

Grafting. Many plants are propagated by a method called grafting. *Grafting* is defined as inserting buds, shoots from one plant, into the stems or roots of another plant. When the graft involves the implanting of a bud, it is commonly referred to as *budding*. If a twig or shoot is used, it is commonly referred to as a graft.

Graftage is commonly used in horticulture to combine plant varieties that have an undesirable characteristic with plant varieties that have a desirable characteristic. For example, certain varieties of peach trees are resistant to soil *nematodes* (a small soil organism that attacks the roots of certain plants, causing damage to the plant). Grafting the top of a peach tree that produces the type of fruit desired to the root stock of a tree variety that is nematode resistant will result in a tree that produces quality fruit in soil areas that are infected with nematodes.

When more than one part of the plant is involved (as opposed to budding), a *scion* is added to a *stock*. The scion is the piece cut from one plant and used to graft to another plant. The stock is the plant that provides the roots and may be either the roots or an entire plant.

Budding. When bud grafting is used, only one bud and a small piece of bark from one plant

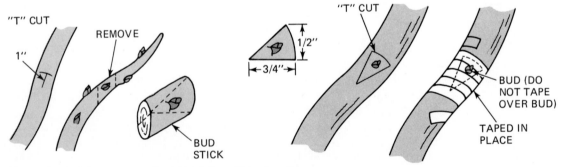

Figure 9-10. The steps involved in taking a T-bud graft.

are grafted onto the stem of another plant. Before budding is attempted, you should ensure that the plants are compatible. Determining compatibility is difficult. Generally, plants within the same species are compatible for grafts. When you graft between species, unless you are experienced and knowledgeable, you should consult professional help. In addition to involving compatible plants, most budding also involves young plants, primarily those whose branches are not more than 1 in [2.5 cm] in diameter. Budding is primarily done during the spring and summer when the plant is young.

The techniques of budding depend on the type of bud graft: (1) T-budding, and (2) patch budding.

The steps in making a T-bud graft are to select a limb on the stock tree that is 2 to 3 in [5 to 7.5 cm] in diameter. A very sharp knife is used to make a T-cut on the limb (see Figure 9-10). Using the knife again, remove a 1-in [2.5-cm] piece of bark (scion) from the budstick. Cut the piece of bark as shown in Figure 9-10. Carefully lift the edges of the T-cut on the limb and insert the piece of bark. Be sure that the inserted piece of bark fits snugly so that good cambium contact occurs. Using the grafting tape, fasten the bark into the T-cut, but do not cover the bud. Try to place the bud on the limb opposite the prevailing winds to prevent wind damage.

Patch budding is conducted in a manner similar to T-budding, except that a patch is taken from one limb and placed with another patch of similar size. It is taped in a manner similar to T-budding.

Other Grafts. The graft union is the most important aspect of grafting. This union is the intermingling of the tissue produced from the stock and scion. This interlocking of tissues is very complex, yet there are certain things that must be understood. For a graft to be successful, good contact between the cambium layers of the scion and the stock are essential. The cambium is the layer of cells (tissue) on woody plants that falls between the bark and wood in the stems. When the scion is placed in the stock and the cambiums are intermingled, the union that results gives rise to new cambium and results in new tissues' formation. This cements the union of the scion and the stock, resulting in a successful graft. Not all plants are able to be grafted. Dicots, which have distinct tissue patterns, are able to graft successfully; however, most monocots where the tissue patterns are dispersed are not prone to successful grafting.

Depending upon the plants involved, there are several types of grafts. The most common types of grafts are: (1) tongue-and-whip graft, (2) approach graft, (3) cleft graft, (4) bark graft, (5) notch graft, and (6) wedge graft. The first two types of grafts are used when the diameter of the stock and the scion are similar. The latter four are used when the diameter of the stock is greater than the scion.

The *tongue-and-whip graft* is used to graft a scion to a stock that is approximately the same size. Figure 9-11 illustrates an example of the use of the tongue-and-whip graft.

The *approach graft* is used to join stems of plants to their own root systems. This process as illustrated in Figure 9-12 is primarily used on conifers and evergreens.

The *cleft graft* is one of the more important forms of grafting. To complete a cleft graft, you need a sharp knife, grafting wax, the stock plant, and the scion. The knife is used to split the stock plant as shown in Figure 9-13. Concurrently, the scions are beveled as is also shown in Figure 9-13. The cleft cut in the stock is opened and the scion is placed in the stock. Care is taken to ensure that the cambium tissues touch. The scions are then inserted as shown and covered with a grafting wax (compound). In many cases, two scions are placed in the cleft, and if both grow, one is subsequently removed.

Bark grafting is accomplished by splitting the bark of the stock. The scion is placed in the slit and held there by nailing. Wax is then placed over the cut surface.

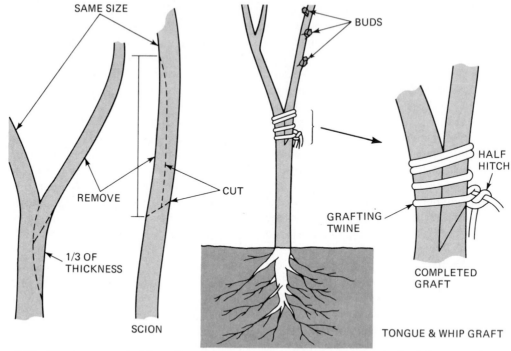

Figure 9-11. The tongue-and-whip graft method is used to graft a scion to a stock of the same size.

Notch grafting is related to the bark graft; however, instead of the bark's being slit, a notch is made. The scion is then placed in the notch and held in place by nailing, then covered with wax.

Wedge grafting consists of removing a wedge of tissue from the middle of the stock. Then the scion is cut in a V-taper and inserted into the wedge of the stock. When the scion is inserted, care should be taken to ensure that the cambium of the stock and the scion touch. A grafting wax is then used to seal the graft.

Grafting waxes are commercially produced. A good wax is free of toxic substances and will not melt or crack at ordinary temperatures. The wax serves to protect the wood surface from

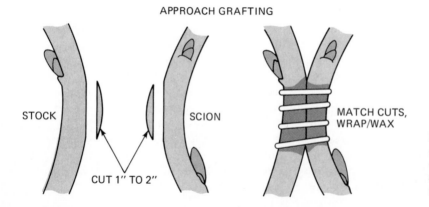

Figure 9-12. To join stems of plants to the root sytem of another plant, the approach-graft method is used.

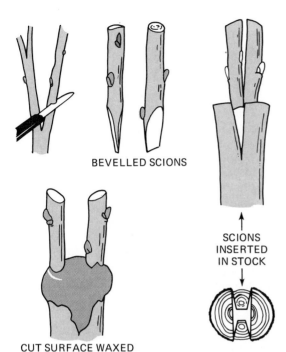

BEVELLED SCIONS

↑
SCIONS
INSERTED
IN STOCK

CUT SURFACE WAXED

Figure 9-13. Cleft grafting is one of the important techniques of grafting.

decay, to retard transpiration of the tissues, and to permit a normal interchange of oxygen and carbon dioxide.

Specialized Structures. Many plants have specialized structures such as rhizomes, runners, bulbs, and corms. The primary function of these plant parts is to store food, but they are also very useful for vegetative reproduction. These parts are usually divided to become new plants.

When a rhizome is manually cut into sections, it is called a *division*. In the case of the runner, bulb, and corm, when a portion is naturally divided, it is called *separation*. We will first discuss division of the rhizome.

The main portion of this plant part grows horizontally at or just below the soil level. The rhizome resembles a root, but it is actually a specialized stem because it has nodes, internodes, buds, and sometimes leaves. Examples of the rhizome are the iris, lily-of-the-valley, and many turf grasses. Propagation of the rhi-

zome occurs by manually cutting it into sections, making sure that each section has at least one eye or node. The new roots and shoots will develop from the nodes. New plants are established in a relatively short period of time with this method.

CORM SEPARATION

CUT FOLIAGE 1 – 2"
ABOVE NEW CORM

NEW CORMS

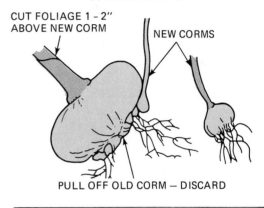

PULL OFF OLD CORM — DISCARD

SEPARATION OF BULBLETS FROM BULB

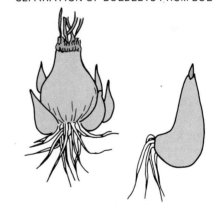

TUBER DIVISION

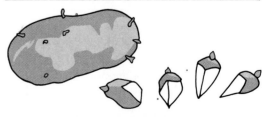

Figure 9-14. Some specialized structures in plants enable propagation by separation.

The runner is also a specialized stem. It also runs horizontally, but usually above the soil level. It forms a new plant at each node or second node, depending on the specific plant. In contrast to the rhizome being propagated by division, the runner is propagated by separating it from the mother plant after it has first formed roots. The new plantlets will stay attached unless they are cut off. They can usually be removed and rooted separately, as are cuttings. Examples of this method of propagation are the strawberry and strawberry begonia.

Bulbs and corms look similar to each other, but their structure is different. The bulb is a specialized bud with fleshy leaves surrounding it, whereas the corm is a solid condensed stem. They are both propagated by separation. A *bulb* is a single large bud with a small stem at its lower end. There are many fleshy scalelike leaves growing from the upper surface of the small stem. Roots emerge from the bottom of the stem. From the large bud, usually several new buds appear, each producing a new small bulb called a *bulblet*. In the commercial industries, there are specific methods for injuring this bud so that it will produce a larger number of new buds, hence, new bulbs. Plants with bulbs are the onion, narcissus, lily, and tulip.

In contrast, the *corm* is mostly fleshy stem tissue containing nodes and internodes. Roots form from the lower surface of the stem. The new cormels form at the nodes. Gladiolus and crocus are both examples of corms.

Propagating Horticultural Plants: A Review

All plants are propagated either by sexual or asexual means. As a horticulturist, you too will most likely deal with the propagation of plants.

Sexual propagation involves the production of plants from seeds. You must know how to seed properly, which involves preparation of the seed bed, depth and spacing of planting, watering, fertilization, and transplanting. This seeding process also requires the careful monitoring of environmental factors such as temperature and light.

Asexual propagation requires knowledge in producing a new plant by using a portion of an existing plant. Cuttings, layerage, budding, grafting, and specialized structures are the different methods of asexual propagation. Asexual propagation requires that you know the how and when and where in producing the new plant. Cuttings must be taken at the correct time, treated with the proper growth hormone, and placed in the correct rooting medium. Similar crucial techniques are also used for the many other types of asexual propagation.

Propagation is an important activity for horticulturists. Without proper propagation, quality plants cannot be produced and without proper plants, the modern horticulture industry is economically hurt. Proper propagation results in quality plants and a healthy horticulture industry.

THINKING IT THROUGH

1. Define sexual and asexual propagation.
2. Describe the process by which seeds germinate and produce plants.
3. What factors affect germination?
4. List and describe the various kinds of asexual propagation.
5. What are the three basic types of cuttings?
6. What is layerage?
7. What is grafting?
8. How would you determine the correct method of asexual propagation for a particular plant?
9. Describe how the T-bud and cleft grafts are made.
10. What "specialized plant structures" are used in propagating certain plants?

10

Horticulture Plant Operations

A manager of a landscape firm was called by a homeowner and asked to come and examine some pyracantha shrubs. The homeowner had installed the shrubs earlier in the spring, and they were brown and showing signs of dying. The manager's inspection of the pyracantha and subsequent quizzing of the homeowner ruled out several causes of the problem. Since planting, the shrubs had been provided with excellent care, yet were dying.

After examining the shrubs and talking further with the homeowner, the manager discovered the problem. The shrubs had been improperly planted, and all the good care after the poor job of planting could not save the shrubs.

The owner had removed the pyracantha from their containers and removed all the soil from the roots. The shrubs had been planted very shallow (the ground had been very hard, which made hole digging a chore). The owner hadn't watered the shrubs thoroughly while planting. In short, several crucial plant-handling procedures were violated.

If the homeowner had known the proper technique of planting (transplanting) shrubs, several dollars could have been saved.

On the other hand, had the landscape firm installed the shrubs, they would have had to replace them at their own expense. This would result in lost revenue. Thus it is crucial that the manager of a landscape firm and the firm's employees know the proper method of handling plants.

There are many other cases in the modern horticulture industry where the handling of plants is crucial. Your knowledge of these plant-handling operations is important as you seek employment in horticulture.

CHAPTER GOALS

Your goals in this chapter are:

- To demonstrate the correct procedure for transplanting a tree or shrub outdoors
- To describe the procedure for balling and burlapping a tree or a shrub
- To demonstrate the procedure for removing the container and planting container-grown plants
- To describe the procedure for guying and staking a tree
- To demonstrate the procedure for "pricking off" transplants
- To demonstrate the procedure for repotting a plant

Handling Outdoor Plants

You have just purchased a Colorado blue spruce, which you plan to plant in your yard. The tree, similar to the one pictured in Figure 10-1, is in a peat container. How do you plant this tree properly to avoid the problems encountered by the homeowner mentioned in the chapter introduction?

The survival rate of trees such as the blue spruce is directly related to your care and skill in setting the plant out properly.

Planting Trees and Shrubs

The blue spruce that you are to plant will require you to employ most of the common plant-handling techniques for outdoor plants. Even though in this example the plant is in a peat pot (a containter-grown stock), the basic principles of planting all types of trees and shrubs grown in other containers apply. Differences in planting will be noted later in the chapter.

Figure 10-1. The care and skill used in planting the Colorado blue spruce will determine its health and survival.

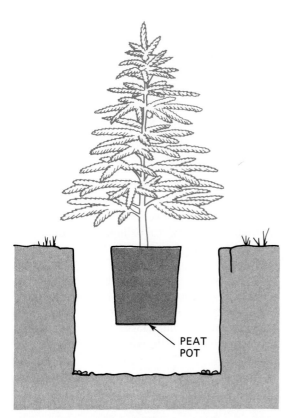

Figure 10-2. Make sure the hole dug is twice the volume of the container.

Site Preparation.   As you choose the place for your tree you should examine the site carefully. Is the site so close to buildings or other structures that its future growth will be inhibited or it will interfere with these structures? For instance, is the tree being planted close (within 1 ft, or 30 cm) to a sidewalk? If so, as the tree gets larger it will obstruct the sidewalk and may eventually have to be moved. The site should be such that the plant will blend in with the existing landscape. For instance, will the blue spruce add to the landscape or will it be out of place with the existing plants and structures?

After a suitable site has been selected, the major task is to dig the hole. As a general rule the hole you dig should be twice the volume of the container that is to be planted. Also, the hole should be both deeper and wider than the container (see Figure 10-2).

When digging the hole, you should make sure that the sides of the hole are not smooth. Uneven side edges make it easier for roots from the plant to penetrate the soil.

As you dig the hole, save the dirt that is removed so that it can be replaced around the tree as the tree is replanted. The soil can be placed on a plastic sheet or canvas as it is removed.

Planting.   Once the hole is dug to the proper depth and width, you are now ready to place the plant in the hole. It is at this point that a major plant-handling error is often made—*planting the tree crooked.*

Place the plant in the hole to check for proper depth and width. The top of the container should be level with the ground. Some of the loose soil removed from the hole should be placed in the bottom of the hole (see Figure 10-3).

Once you are satisfied that the hole is correct, you are now ready to place your blue spruce in the hole. Since your tree is already in a peat pot, it is not necessary to remove the pot, because the peat will dissolve in the ground. Metal and other containers will need removing.

The peat pot is often penetrated in several places to allow the roots of the tree easier access and contact with the soil.

Now place your tree in the hole and check to see that it is straight. Have someone check it with you. The tree can be straightened more easily now than at any other time.

Finishing The Job.   The final step in the process is to fill the hole with the loose soil. Place the soil around the pot, tamping it firmly as you go. When you have filled the hole approximately one-half full, water the plant thoroughly, allowing for the water to sink into the hole and the soil. Then finish refilling the hole, using the leftover soil.

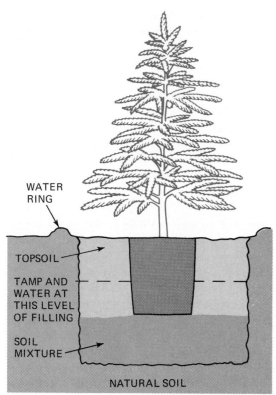

Figure 10-3. Loose soil removed from the hole should be placed at the bottom of the hole. Make sure the top of the container is level with the ground.

A mulch is placed on the soil to help conserve moisture by keeping the soil from drying out too fast. Some of the leftover soil can be used to form a water ring to aid in future waterings (see Figure 10-3).

This procedure applies to planting a tree in a homesite and also around a rectangular area, a commercial building, or an industrial site. If you work for a landscape firm or a grounds maintenance crew, you will probably spend much of your time planting and transplanting trees and shrubs.

Balling and Burlapping

The spruce tree you have just planted was in a peat pot. If you looked when you purchased your plant, you observed that there are other ways to prepare a tree or shrub for sale besides using a peat pot. Many such plants are sold in burlap balls. As pictured in Figure 10-4, these plants are actually wrapped in burlap bagging and tied. Soil and roots as found in the ground where the plant was growing are actually sold with the plant. The process of preparing plants for sale in a burlap bag is called *balling and burlapping*.

The skills associated with balling and burlapping are crucial for horticulture workers in nurseries and in landscaping firms. A great deal of time is devoted to balling and burlapping trees and shrubs.

Digging Plants. When you dig a tree or shrub for balling and burlapping, you must make sure that it is done properly in order to reduce stress on the plant. The initial task is to determine how large to make the ball, which tells you how far

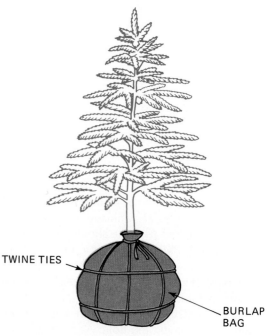

Figure 10-4. Some plants are wrapped in burlap.

to dig around the tree. Generally, the circle around the tree should be about one-quarter of the height of the tree. For example, a 4 ft [1.2 m] tall tree or shrub should be dug with a 1-ft [30-cm] wide circle (see Figure 10-5).

Next, dig a trench around the circle about 1 ft [30 cm] deep. The bigger the plant to be balled and burlapped, the deeper the trench to be dug. Generally you want to ensure that enough of the root system remains intact to support the tree in its new location. As the trench is dug around the plant, you use your spade to penetrate the area under the plant. A sheet of burlap is placed in the trench and slid under the root system. You should take care not to break the soil ball. The burlap is secured around the ball as noted in Figure 10-4. Twine and nails are used to secure the burlap. Several passes with the twine are needed to securely fasten it.

You are now ready to remove the ball (the tree) from the hole. *Do not lift the plant by the trunk!* The spades used to dig the hole can be used to lift the tree out of the hole to the point where you can lift by the burlap ball and move the tree.

Container-Grown Plants

Many plant growers are growing trees and shrubs in containers that can easily be sold. The boxwood is a small shrub that is easier to handle if it is grown in a container. This removes the need for balling and burlapping. Several different kinds of containers are used. Peat pots, metal pots, and baskets are the most common ones.

Peat pots are very popular, as a result of their lower cost and the fact that they decay in the soil when planted. The latter advantage ends the difficulty of removing the container during planting, and greatly reduces the risk of breaking the soil ball when transplanting.

Metal containers are strong, but must be removed before transplanting and generally cost more per unit than peat pots. Baskets are used to a lesser degree than the peat or metal pots,

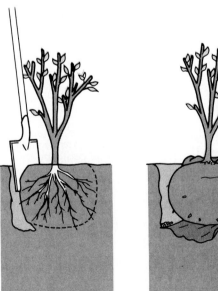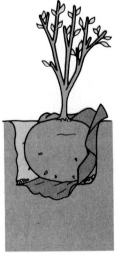

Figure 10-5. The steps involved in digging a plant.

primarily because of their lack of availability in many areas.

Remaining Containers. If you purchase a tree or shrub in a container that must be removed before planting, care should be taken to remove the container without disturbing the root system. Likewise, if you are planting a tree or shrub on a commercial site, the container would have to be removed with care.

Smaller plants (containers 1 ft, or 30 cm, in diameter and smaller) can be removed by gently tapping the container on a stationary object such as a sidewalk and removing the plant. Larger plants (containers over 1 ft, or 30 cm, in diameter) require more effort to remove.

To remove the larger containers, cut the bottom out of the container. Then place the container in the hole that was dug in the same manner noted earlier. Using cutters, slit down the side of the container. The container can then be removed from the root system.

If a metal container is being removed you should take care not to cut your hands on the jagged edges of the container.

Handling Bare-Rooted Plants

In addition to handling plants in containers and burlap balls, many horticulture businesses sell plants such as ornamental vines and shrubs in packages with bare roots. Syringa (lilac), hydrangea, and spiraea are some common plants that are sold in bare-root form. In other words, the plants are packaged without soil on the root system. If you plant these, you should follow a different procedure for proper planting, as was described earlier.

Many of these plants are packaged with directions for planting printed on the package. However, the package may become torn or you might plant a bare-rooted plant that does not have directions available—hence the need for general directions in planting bare-rooted plants.

Your first step in planting bare-rooted plants such as the spiraea is to examine carefully the root system. Damaged roots should be removed. However, remove these roots gently so as not to damage healthy roots.

If possible you should soak the root system overnight in water. Soaking for more than 12 hours, however, could cause some harm to the plant. Do not allow the roots to dry out completely.

Dig the hole in the same manner as in planting container-grown plants. However, since no pot or burlap ball exists, you must dig the hole to allow the roots to spread out. Generally this means digging the hole about 6 in [15 cm] deeper and 1 ft [30 cm] wider than the spread of the roots (see Figure 10-6).

Mix the soil that was removed from the hole and replace 6 in [15 cm] of the soil in the hole. Then place the plant in the center of the hole

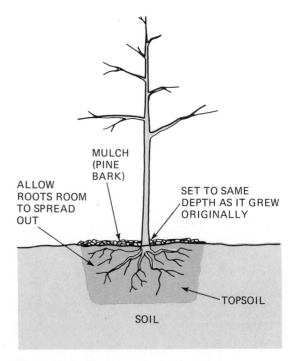

Figure 10-6. Planting a bare-rooted plant.

and firmly tamp the removed soil in around the roots while the hole is about one-half full. Then water the soil in the hole thoroughly. After the water has soaked in, continue to replace the soil. Remember to check constantly to be sure that the plant is straight. When the hole is filled within 2 to 3 in [5 to 7.5 cm] of the ground level, a water and fertilizer solution as used when planting container-grown stock should be applied.

Finally, a mulch is added to the remainder of the hole as shown in Figure 10-6. As the job is completed, remember to clean around the area by removing all trash and unused soil.

Guying and Staking

Even with the best handling technique possible, there are times when the transplanted tree or shrub grows in a slanted or crooked manner. Often landscape firms are hired to straighten trees that have become crooked or damaged because previous plant-handling techniques were poor. As a worker in this type of firm you would have to be able to guy and stake these trees. The process of mechanically straightening the plant is referred to as *guying* and *staking*.

Guying and staking are normally used for the following reasons: (1) to straighten a tree or shrub, (2) to protect a plant from animals, or (3) to repair a damaged tree or shrub.

Guying and staking should be done as early in a plant's life as possible, generally as soon as the problem is noticed. During the plant's early growing stages it is much easier to correct the problem. Figure 10-7 provides examples of guying and staking used to correct or prevent damage to trees and shrubs.

Figure 10-7. Guying and staking are used to correct or prevent damage.

cussed. Plants propagated indoors by sexual means are produced by seeds normally planted in flats. Asexual propagation, especially cuttings, is accomplished by means of rooting in a propagation bed. In both cases, once the seedling or rooted plant develops, it must be moved from the flat or bed into a more permanent place.

Handling Indoor Plants

Many plants are propagated indoors. In Chapter 9, sexual and asexual propagation were dis-

Transplanting Seedlings

Greenhouse workers spend a lot of time planting seeds. The seeds sown in flats develop into a mass of seedlings. The seedlings are ready to be

HOLD BY THE LEAVES OF THE
TRANSPLANT NOT BY THE STEM

Figure 10-8. Hold a transplant by the leaves, not by the stem.

transplanted once the first true leaves develop. Figure 10-8 illustrates the development of the first true leaves. Transplants are easily harmed and susceptible to shock. Care should be taken in handling the transplant. As a worker in the greenhouse you will be handling many transplants. The proper technique is very important in order to have a large percentage of the transplants survive the stress of transplanting.

The process of removing the individual seedling from the mass produced in the flat is called *pricking off*. As individual seedlings are pricked off, they should be handled by the leaves and not by the stem (see Figure 10-8).

Green peppers are a common vegetable crop that is grown in flats then pricked off as seedlings and sold as transplants. The process of transplanting green peppers begins as you prick off the seedlings and place the peppers into their new container immediately. The longer the roots of the transplants are exposed to the environment the greater the chance for damage and the more likely it is that the young seedling will not survive the transplant.

When you are planting the transplant in its new container, care should be taken to plant the seedling to the proper depth. The transplant should not be planted too deep. As a general rule, the transplant shoud be planted to a depth of about 1 in [2.5 cm] below the first true leaves, and at about the same level as the seedling was growing in its original flat. The green pepper has easily identifiable first true leaves and is easily transplanted.

A pencil can be used to make a small hole in which to place the transplant. The soil around the transplant should be firmed as the seedling is transplanted. The transplant should be watered.

While you are transplanting it is advisable to discard weak seedlings. Also, as you transplant the seedling, larger, more vigorous seedlings should be grouped. After all, when you go to a garden center to purchase some green pepper transplants, you want the most vigorous plants. This will help as you sort plants for sale, and as you price the plants. Specific details for care of the transplant were presented in Chapter 9.

Potting and Repotting

Many transplants and cuttings are transferred from the flat or propagation bed directly into the pot. The plant is then grown in this pot and sold. Other transplants are planted in temporary pots and sold in this manner so that the purchaser will transplant the plant in a permanent place later. For example, a cabbage plant grown indoors is pricked off from the flat and placed into a small peat pot, where it is sold to a customer. The customer will plant it in a garden. Geranium cuttings might go from the propagation bed directly into the pot where they are sold, not necessarily to be transplanted again.

The above cases point to a need to know how to pot and repot plants. As an employee of a garden center, you will be assigned the task of potting plants. The skill you develop in adequately potting plants is very important to your employer. Hence, you need to know the correct techniques for potting plants.

As transplants grow they outgrow the pot and

Figure 10-9. The ideal time to repot a plant is just before the plant's active season begins.

need repotting. For example, you may wish to repot your Swedish ivy in a more attactive pot.

Potting and repotting can occur at any time, but it is normally best to repot a plant just before its active season begins. The active season refers to the flowering or other growth period. At this time the roots are better able to handle the stress. Also the plant is not actively producing flowers or foliage, which could add to the stress of transplanting.

As you begin the repotting process you should remember the following: (1) the roots should be handled with care and allowed to spread out in a natural manner; (2) do not mash or damage the foliage; and (3) a potting soil such as those recommended in Chapter 6 should be used.

You should choose a place to work where you have ample room. The pot should be partially inverted and tapped gently. The root ball and the plant should slide out. If the plant has been watered properly it will slide out with little difficulty. If the root sytem is too wet, the plant will stick to the pot. If it is too dry, it will fall out immediately, which could damage the plant. Thus, check the water supply to the plant before planning your repotting process.

Many times the plant will be *pot bound*. This occurs when the plant has grown to the extent that the plant's root system has grown to the shape of the pot. If the roots of the plants you are repotting are pot bound, you should secure a slightly larger pot and carefully separate some of the root system so that it is as large as the pot shape. Then, using a good potting soil, repot the plant in the larger pot.

After repotting, prune some of the foliage back. The reduction of the foliage and the freshly repotted roots will result in a vigorous growth period and a healthy plant.

After you have potted or repotted the plant it should be watered thoroughly. More frequent waterings should be given the plant than normal until the plant has settled in properly, which will occur within a couple of weeks.

Plant Handling Operations: A Review

If you followed the directions as outlined in the chapter, your blue spruce should be properly planted. Good care and attention should help your tree grow to a normal and healthy size. Plant handling is a very important skill for you to master. Many of the jobs that you will take in the horticulture industry will require you to handle plants.

Landscape firms constantly plant and transplant shrubs, trees, vines, and other ornamentals. Nurseries grow container-grown stock, balled and burlapped trees and shrubs, and many bare-root vines. Greenhouse operators spend a great deal of personnel time in pricking off seedlings. A retail florist might repot plants in attractive and decorative pots to add value to the plants.

If you took a job in any of the types of firms mentioned above, you too would be challenged to learn the proper techniques for handling plants.

THINKING IT THROUGH

1. Outline the step-by-step procedure that you would follow in planting a tree such as a 6-ft [180-cm] tall Colorado blue spruce.
2. Describe the procedure that must be followed to ball and burlap a plant properly.
3. What is the correct method for removing the container from a metal-container-grown plant that is to be transplanted?
4. What is meant by guying and staking a tree?
5. How do you prick off a seedling?
6. How would you repot a plant that has outgrown its present pot and has become pot bound?

FOUR
Horticultural Plant Health

Have you ever had the flu or even a cold? If you are like most young people, you have probably had a cold or the flu sometime in your life. Do you remember how you felt? Your nose may have been runny, you might have had a cough, your sinuses may have been blocked, and you might have had a headache. Do you know what causes such illnesses? Small microscopic organisms called *bacteria* or *viruses* are responsible for most colds, the flu, and a number of other diseases.

Did you know that plants can catch diseases just like humans? Bacteria and viruses are often responsible for plant diseases. Though a sick plant does not develop a runny nose and a cough, it does show signs of being sick. The plant may change color and the leaves may fall off. The plant may become limp.

Diseases are not the only factors that may cause a plant to be unhealthy. Insects and pests may also cause unhealthiness. Often a plant will become sick if it is not being cared for properly. A plant needs a certain amount of water, temperature, air, light, and nutrients to grow properly. If one of these environmental ingredients is out of balance (too much or too little), the plant may become sick. Have you ever eaten too much food and developed a stomachache? You were sick but you did not have a disease. Your intake of food was out of balance, which resulted in your being sick. Plants often have a similar problem with too many or too few nutrients, or the incorrect amount of water, air, or light. The environment has to be in balance for proper plant growth.

The entire horticulture industry needs healthy plants. It is important for horticulturists to provide the ideal growing environment for plants and protect them from diseases and pests. Fruit and vegetable producers are expected to produce high-quality produce that is free from disease. Golfers like to play golf on lush green fairways. Nursery owners would have trouble selling unhealthy shrubs and trees. Florists would not be able to produce quality floral arrangements with sick flowers. Plant health is important in all phases of horticulture.

In this unit, your goal is to learn how to help plants become healthy. You will discover what a healthy plant should look like and the procedure to follow in determining what is wrong with an unhealthy plant. The type of environment needed by plants will also be described. You should be able to identify common horticultural pests and diseases, and to control them, when you finish this unit.

11

Identifying and Controlling Plant Disorders

Have you ever wanted to be a detective? As a horticulturist you will probably have the opportunity to be a detective. Instead of solving crimes you will be solving plant health problems. A detective carefully examines the situation, searches for clues, asks questions, and then solves the mystery.

Pat works at a retail garden and flower store. One summer day an angry customer carried a sickly looking potted philodendron into the store. "I want my money back," the customer demanded. "I bought this plant here a month ago. Look at it now. The leaves are turning brown and the plant has stopped growing. This plant obviously has a disease."

Do you agree the plant obviously has a disease? Could there be other reasons for the plant's problem? Pat picked up the plant and carefully examined the leaves and stem and felt the soil. Pat then asked the customer how the philodendron was cared for at home.

The customer explained, ''I placed this plant on a table in front of a window on the south side of my house. This way it would get plenty of sunlight. I also watered it every day, so it certainly had enough water.''

''I think I know what the problems are,'' Pat stated. Pat then turned the plant over, removed the pot, and examined the roots and soil. Pat's suspicions were confirmed.

Have you figured out what the problem was? It was not a disease and it was not an insect problem. The problem was caused by improper plant care. The customer was not providing the right growing environment for a philodendron. Too much water and excessive sunlight was causing the problem. Many people blame pests and diseases for plant problems when the real problem is in the care of the plant.

Pat suggested that the plant be moved to an east or north window, or behind a curtain in a south window, so that the light would be less intense. New growth would then be normal, but previously damaged leaves would never recover. Pat also said that the soil surface should be felt daily, and only on the day that it feels dry to the touch should the plant be thoroughly watered (until the point where some runs out the drain hole). Watering should not be repeated until the soil is again dry to the touch. Pat asked the customer to call the shop in 2 weeks if the plant had not improved.

As a horticulturist you will be expected to identify plant problems and suggest remedies. If you work in the retail (selling) phase of horticulture you will be asked by customers to identify plant problems and suggest solutions for the problem. This is part of the job. If you are engaged in the growing of plants you must be constantly alert to problems that would cause plants to grow poorly.

In this chapter you will explore what happens to plants when the five environmental factors required for plant growth (moisture, temperature, light, air, and nutrients) are out of balance, and you will learn how to identify injuries.

CHAPTER GOALS

Your goals in this chapter are:

- To describe the characteristics of a healthy plant
- To develop a system for diagnosing plant disorders
- To identify symptoms of plant disorders created by imbalances of water, temperature, air, light, and nutrients
- To identify physical damage to plants
- To prevent plant disorders that are caused by environmental factors and physical damage
- To develop a plant management plan

The Healthy Plant

How do you think Pat determined what was wrong with the philodendron? It was no lucky guess. Pat had a procedure for determining what was wrong with the plant. A good plant detective will have a standard plan for identifying plant disorders, just as a doctor has a procedure for diagnosing human disorders. First, Pat needed to know what a healthy plant should look like.

Characteristics of a Healthy Plant

Could you describe the characteristics of a healthy plant? What size should it be? Should it have an odor? Describing the characteristics of a healthy plant is not as easy as it appears at first. Even though each plant species is different, there are some general characteristics of healthy plants. It might help if we compared the characteristics of a healthy human with the characteristics of a healthy plant.

Figure 11-1. You can be a plant detective.

Erectness. A human who is feeling well normally stands erect. If the human does not feel well, he or she may slump down. An unhealthy plant will also slump. The plant looks wilted. With a few exceptions, such as the weeping willow tree, a plant that is droopy or wilted is unhealthy. In the case of plants that grow along the ground, such as watermelon vines and some ivies, the leaves will be erect when the plant is healthy. They will not be lying flat on the ground. Therefore one general characteristic of most healthy plants is that they or their leaves will appear to be erect.

Color. The color of a human may change when he or she is unhealthy. Often the skin will become pale. In the case of certain diseases, such as measles, splotches will appear on the skin.

The color of plants will also change when they become unhealthy. A common symptom of many plant disorders is that the leaves turn yellow or brown. Leaves may also turn other colors, depending upon the problem. Unusual white splotches on the leaves usually indicate a disease or pest problem. A healthy plant will have the coloration that is normal for that plant. In many cases this color will be green. However, there are a number of plants, such the striped inch plant, the caladium bicolor, and variegated wandering Jews, whose leaves are green with a mixture of other colors such as white. Therefore, before you can tell whether or not a plant is healthy by looking at its coloration, you have to know what the plant is supposed to look like.

Leaves. Have you ever known a person to suddenly start losing hair or have skin that starts to flake off? Often in early summer, people will stay outside too long and get too much sun. The result is a sunburn and flaking skin. Falling hair or flaking skin is often an indicator of a disease or environmental problem, such as too much sunlight, in humans.

Plants do not have hair to lose, but they do lose leaves. Other than in the fall of the year, when deciduous trees and shrubs shed their leaves, plants should not be losing large amounts of leaves. If a plant is constantly dropping leaves, this indicates that there is some type of disorder. A healthy plant does not constantly lose leaves.

Size. Did you ever see a picture of a person who was suffering from malnutrition? A person who is 10 or 12 years old and suffering from malnutrition may be the size of a 4-year-old. A person who is extremely small for his or her age may have a health problem. The same is true in plants. An unhealthy plant may be stunted or growing poorly. The leaves may be smaller than normal. One exception to this is the kind of plant that is grown bonsai style. *Bonsai* is the Oriental art of regulating plant growth through pruning to deliberately cause a plant to remain

Figure 11-2. Unhealthy plants may have droopy or missing leaves, may be of abnormal size or color, and may exhibit signs of physical damage.

small. A healthy plant should be the same size as other plants of the same species and age.

Odor. At times, a person who is unhealthy may have an odor. Certain diseases in humans produce a distinctive smell. Diseases and other disorders in plants will also produce an odor. The decomposition and rotting of plant cells has a distinctive smell. When plants are watered too much, the soil around the roots may smell sour or musty.

It should be noted that a number of plants have an unusual fragrance that is characteristic of that plant. If you tear a leaf on a geranium you will notice a distinct odor. That is a characteristic of the plant and is not an indication of a plant disorder. A healthy plant will have an odor that is usual for that plant.

Wounds. If you have ever cut yourself you probably noticed it immediately. There was an opening in the skin and it may have bled. This type of problem is easy to see because in humans the symptoms are visible.

Plants can be wounded also. Insects and pests may bite holes in leaves. They may also pierce the stem and suck liquid out of the plant. Sap can ooze from the holes, just as blood can ooze from humans, but it will not be red. Insects and pests crawling around on a plant or visible wounds on a plant indicate that a plant is unhealthy or soon may be. A plant could also suffer physical damage from careless humans or other forces such as the wind.

The Appearance of a Healthy Plant

A healthy plant will be of a size and color typical of other plants of the same species. It will be erect, or the leaves will be erect. It will have a full set of leaves, except for deciduous plants in the dormant season. It will not have insects or pests, and will not exhibit signs of physical damage.

A Plan for Diagnosing Plant Disorders

Now that you know what a healthy plant looks like, you are ready to develop a plan for determining plant disorders. What would you do first? What did Pat do first? The procedure followed by Pat was:

I. Examine the plant.
   A. Look at the leaves.
      1. Look on the top of the leaves.
      2. Look at the underside of the leaves.
      3. Compare the leaves at the top of the plant with the leaves at the bottom of the plant.
   B. Look at the stem.
      1. Examine the main stem.
      2. Examine axillary stems.
   C. Look at the roots and soil.
      1. Examine roots and soil.
      2. Smell the roots and soil.
   D. Shake the plant.

II. Determine how the plant was cared for.
   A. Moisture
      1. How often was the plant watered?
      2. How much water does the plant receive?
      3. How humid is the environment surrounding the plant?
   B. Light
      1. How much light does the plant receive?
      2. What type of light does it receive?

Figure 11-3. You should develop a systematic plan for diagnosing plant disorders.

C. Temperature
1. What is the temperature in the area where the plant is?
2. Where is the plant located in relation to sources of heat or cold?
D. Air
1. Is the plant located where the flow of air is excessive or uneven?
2. Is the plant getting enough air?
E. Nutrients
1. How often is the plant fertilized?
2. What type of fertilizer is used?
F. Chemicals
1. Have pesticides or other chemicals been used near the plant?
2. How close are factories or oceans to the plant?

If this plan is followed, there is a good chance that you will be able to determine what is wrong with a plant. In the remainder of this chapter, you will learn why you ask the questions that are listed in the plan and what you are looking for when you examine the plant. A chart diagnosing plant disorders is found in Figure 11-7.

Examine the Plant

The first step in determining what is wrong with a plant is to give it a careful examination. Look closely at the leaves and the stem. Look at both the tops and undersides of the leaves. Examine the leaves on the top of the plant and near the bottom. You are looking for several things. The first thing to look for is insect damage. Are there holes or tears in the leaf caused by chewing insects? If you do not see insect damage, look closely for insects. Many insects are very small and hard to see. To find insects you should shake the plant over a piece of white paper and then look for insects on the paper. A hand magnifying lens will help you find the insects. These small insects usually suck sap from the plant and will not leave visible symptoms such

as eaten leaves. Instead the leaves will often be a sickly yellow or brown. (For details identifying pests and their damage, see Chapter 12.) If you cannot find insects or signs of their damage, you are ready to look for other problems.

The next factor that should be considered is disease. As you examine the leaves and stem look for unusual color variations. Are the leaves and stems shiny and green, or do they have odd-looking streaks and spots that are yellow, brown, ringed in colors, or dead? If they do, a disease may be present. A white dusty appearance could be powdery mildew (a fungus disease). Sometimes insects and diseases attack open flowers or flower buds. (Diseases are discussed in Chapter 13.)

How is the Plant Cared For?

Before you can decide whether disease is the problem you must consider how the plant has been cared for. A number of diseases have symptoms that are nearly identical to the symptoms of plants that have been improperly cared for. Improper plant care can even favor the growth of certain diseases. In a great majority of cases, the problem with unhealthy plants is improper care. This occurs when one of the environmental factors needed for plant growth gets out of balance.

How do you determine whether the plant has been cared for properly? You could keep a log or a diary with information about the temperature, watering rate, fertilizer or pest control applications, and cultural practices used for the plants. When the plants begin to show signs of a problem, you can look in the diary for clues. This will help you to determine whether an environmental factor may be responsible for the problem.

A greenhouse operator noticed several *Sedum morganianum* (donkey tail) plants that started turning brown on the edges, as if they had been burnt. The grower did not know why this had happened until the log was consulted. Two days earlier, the greenhouse had been fumigated with the chemical parathion. Parathion will cause succulent plants such as donkey tails to "burn" if it is applied in too strong a concentration. By keeping a diary the greenhouse operator was able to determine what had gone wrong.

A person who has a few plants at home should also consider keeping a diary. For example, the plants in Lynn's home did not seem to grow very well. Lynn could not figure out what the problem was until a diary was kept. Lynn then discovered that the way in which the plants were being watered was the problem. During one week the plants were watered four times, and in another week the plants were not watered, and in a third week the plants were only watered once. As a result of the diary, Lynn started watering the plants on a regular schedule and the plants started growing better.

If you work in a retail plant shop helping customers with their plant problems, the procedures will be different. You will have to ask the customer questions. What type of questions should you ask? Do you remember what Pat asked? Some questions that should be asked are:

* How do you water the plant? Frequently? How much water do you use?
* Did you fertilize the plant? When? With what?
* Where is the plant located in relation to windows and artificial light? Does it get direct sunlight?
* What are the daytime and nighttime temperatures in the room in which the plant is kept?
* Was the plant suddenly placed near an open window or in a draft?
* Have you sprayed the plant with anything? If so, what and when?

The answers to these questions will help you to determine whether the plant has a disease, an insect problem, or a problem from an environmental condition.

Environmental Factors that Cause Plant Disorder

Do you remember the environmental factors that are needed for plant growth, which were discussed in Chapters 6 and 7? They are moisture, light, temperature, air, and nutrients. Plants need certain amounts of each. If there is an excess or lack of any of these, the plant will respond with poor growth. As a plant detective you will need to identify the symptoms created by an excess or lack of each environmental factor. As you read about each factor remember the episode of Pat and the customer.

Moisture

Plants need appropriate moisture to grow properly. Too much or too little water will cause plants to become sick. Too-frequent watering does not allow enough air to reach the roots and the plants actually suffocate just as people do when they drown—people drown because of the lack of air! On the other hand, if the soil dries completely, this kills the delicate root hairs (the main water-absorbing organ) and the plant is unable to absorb water when it finally has been applied—too late!

Generally, plants that grow rapidly or produce heavy crops of flowers, fruits, or vegetables need considerable amounts of water. Also, plants with many leaves generally need more water than plants with few leaves. You may wish to refer to the chart on pages 160 and 161 to check the water requirements for various plants.

A common mistake made by many people is to overwater plants. More plants are killed by overwatering than by underwatering.

Overwatering. There are several symptoms to look for in determining whether a plant has been overwatered. Often the leaves will turn yellow, may be wilted, and may even drop from the plant. Since this could also be a symptom of underwatering, you should feel the soil. If the plant was watered just before you examined it, while the usual care was to let it dry too much before watering, you would have trouble telling whether the improper care was overwatering or underwatering. If you suspect overwatering, turn the plant upside down, place one hand at the point where the stem enters the soil, and tap the pot lightly against a bench. This will loosen the ball of soil. Remove the pot. If the ball of soil is wet and the roots are soggy or mushy, then overwatering is probably the problem. Smell the soil. Does it have a mushy, moldy or acrid aroma? If the soil is not presently wet, this will indicate overwatering. If overwatering is the problem the plant may need to be repotted, using new soil. If the plant is still fairly healthy, then proper watering will be the solution.

Underwatering. Have you ever noticed a tomato plant that was droopy on a dry hot day? It probably needed water. Plants that are not receiving enough water wilt or droop. The soil of such plants will normally be warm and dry to the touch. Plants that are wilted need water quickly, or they will be severely damaged. After a period of time without water, the leaves will turn yellow or brown and growth will stop. Even if the situation is corrected it may be 2 weeks before the plant resumes normal growth. Severe wilting is a shock comparable to trauma in humans.

When to Water. Is there a way to determine the correct time to water a plant without waiting for the leaves to wilt and the plant to droop? Yes! The most common procedure for determining when to water potted plants is to feel the surface of the soil. If the soil is damp the plant should not be watered. If the soil is dry then the plant should be watered. Plants should be checked daily to see whether they need water. During hot days it may be wise to check potted plants in the morning and in the afternoon to determine whether water is needed.

Another method that is being used in horticulture to determine whether plants need water is a water meter. The meter normally has a small battery that produces an electric current. The electrical current flows through the soil. Since water conducts electricity, the more moisture in the soil the greater the flow of electricity. The electrical flow is measured and displayed on a dial that shows the amount of water in the soil.

Watering the Plant. The way in which plants are watered has an effect on plant disorders. Spraying water on the leaves of plants generally should be avoided because it may cause spotting or discoloration of the leaves. Drops of water on rose, camellia, and orchid petals may produce brown spots, especially in hot weather.

Two methods of watering potted plants are recommended. One method is to water the plant from below. In order to use this method the potted plant must have a tray or saucer under it. Water is poured into the tray. The water is then drawn up into the pot through the holes in the pot by capillary action. *Capillary action* occurs when water rises up, generally in a tube or stem. An example of capillary action would be putting the tip of a paper towel into a puddle of water. The entire lower portion of the paper towel will soon be wet, because the water travels upward as a result of capillary action. After the soil becomes moist, the excess water in the tray should then be poured off. This method requires that the soil must never become very dry, or the capillary action will be interrupted and the water will not move upward through the soil. The remedy is to water the top of the soil in order to reestablish capillarity. This watering method works particularly well with African violets, but may be used with a wide variety of plants.

Commercially, foliage plants are grown in a large-scale adaptation of the capillary system, which makes use of a moisture-holding mat (like a rug) on which the pots are set.

a. WATERING FROM ABOVE   b. WATERING FROM BELOW

Figure 11-4. Two methods of watering plants.

The most common method of watering plants is from above. Water is added to the potted plant by using a hose or watering can. The water is poured onto the top of the soil and not onto the plant. The Chapin or "Spaghetti" system is also used in a number of greenhouses. This involves many small plastic tubes, which are hooked to a water line. The end of each tube is placed on top of the soil in a pot. The grower turns the water on when plants need water. This system is also used for watering container nursery stock. A flat-spray system that puts water only onto the soil surface for cut flowers grown in beds or benches is used in some greenhouses.

Relative Humidity

Along with the need for water in the soil, a number of plants need moisture in the air. Moisture in the air is called relative humidity. Many of the plants grown indoors originated in tropical jungles where the air was moist. These plants include the velvet plant, anthurium, caladium, prayer plant or rabbit tracks, kohleria, orchids, and some ferns. When these plants are placed in dry houses or flower shops away from the greenhouse, problems may develop. This is especially true in winter when artificial heat is used. Artificial heat tends to dry out the air.

Leaves on some plants will turn yellow or brownish and flower buds may drop from flowering plants if the humidity is too low. The leaves may become dry and brittle and die.

Several procedures can be used to raise the humidity. In floral shops, humidity trays are commonly used to increase humidity. A humidity tray is a large, flat container made of a waterproof material such as plastic or metal. The tray is filled with small pebbles, and water is then added to the tray. The potted plants are placed on top of the pebbles. This prevents the roots from sitting in water. The evaporating moisture increases the relative humidity in the air in the immediate vicinity of the plants. Many amateur orchid growers use this technique. Small hand-held plastic or metal misters are sold in many plant stores. They are used to spray a fine mist on plants. Spraying the tops of plants to increase relative humidity is successful for only very brief periods, and may result in disease problem.

Figure 11-5. Plants not receiving sufficient light will grow tall and spindly, and have pale, greenish-yellow leaves. (Left) In the laboratory, two bean plants are compared. (Right) In nature, the spindly tree on the left suffers in the dark shadow of the strong tree on right.

Light

Do you remember what was wrong with the philodendron mentioned earlier in this chapter? One problem was too much water. The other problem was excessive light. Plants need light to grow. Too much light, though, may damage some kinds of plants. On the other hand, if the light intensity is too low, photosynthesis occurs at a low rate and plants will not grow.

Insufficient Light. Generally, there are three symptoms that indicate when a plant is not getting enough light. One is tall, spindly plants. You may have noticed this condition in a home where a number of plants are grown in a small area. Another symptom of insufficient light is pale greenish-yellow leaves and stem. The plant is not receiving enough light to adequately carry on photosynthesis (producing enough food to grow properly). This results in a pale green color. The third symptom of inadequate light is the dropping of leaves from the lower parts of

the plant. The top leaves will intercept whatever light is available. The bottom leaves will not receive enough light and will drop off.

The solution for the problem of lack of light is to move the plant to an area where it will receive more light, or use artificial lighting.

Excessive Light. Too much light can also create problems with plants. The symptoms that indicate when the plant is receiving too much light are also symptoms for other problems. Because of this, questions must be asked. By asking questions you, as a plant detective, can determine which factor is causing the problem. Some of the symptoms of too much light for greenhouse and houseplants are:

1. Short, stocky growth
2. Wilting easily
3. Exposed leaf surfaces that turn pale or even scorch and have dead areas, especially near the centers of the leaf blades

By asking where the plant is located and remembering these symptoms, you should be able to determine whether the plant is getting too much light. This is how Pat determined that the philodendron was getting too much light. Pat knew that the philodendron was a medium-light plant. When the customer mentioned that the plant had been placed at a window on the south side of the house, Pat had a pretty good idea that too much light was one of the problems. This diagnosis was confirmed when Pat saw that the leaves were turning pale and had scorched areas.

A plant that is receiving too much light should be moved to a location where less light is available. Houseplants that are placed outside for the summer should be under a tree or on a shaded porch to protect them from direct sunlight. In the case of commercial plants grown outside, a shade house or lathhouse can be used to protect the plant from too much light.

Temperature

Are you comfortable in a room that is hot or cold? Probably not. Generally, plants are the same as humans. They like moderate temperatures the same as you do. If you are comfortable in a room, the chances are good that your plants will be comfortable, too. Houseplants that are exposed to excessive heat may grow slowly and dry quickly. In severe cases, the leaves may turn yellow or brown. Here again, questions should be asked to determine how the plant is being cared for. If it is placed near a heat duct, on top of the TV, or near a stove the temperature may be too hot. The plant should be moved to a cooler location. The temperature should also be constant, since plants do not respond well to abrupt temperature changes.

High Temperatures. Plants in the greenhouse could become too hot in the summer and show the same symptoms of excess heat as houseplants. Commercial greenhouses use systems to reduce the heat. Often, during the summer, greenhouse coverings are also coated with shading compound to reduce the heat.

On golf courses, on excessively hot days, the greens may have to be "syringed" (sprinkled overhead) to reduce the temperature of the leaves and prevent scorching, which would result in an unsightly brown-colored green!

Vegetables and fruits are also bothered by high temperatures. A problem known as *stem girdle* is bothersome to vegetables. The high level of heat at the soil surface often scorches the stem of vegetable plants, which causes them to die. A soil mulch such as straw or ground corncobs will reduce the temperature at the soil surface, and will help prevent severe wilting of young vegetable plants. Fruit trees are subject to an injury called *sunscald*. It occurs on very cold but sunny days in winter. The bark temperature on the sunny side of the tree rises during the day, then at night the temperature drops rapidly. The rapid changes in temperature kill the cambium tissue. Often the south sides of the trunks of trees are painted white to prevent injury from temperature changes. Leaves of broadleaf evergreens such as rhododendrons may be severly "burned" by this same midwinter temperature stress. It can be prevented by shading the plants. Vegetables such as newly set tomatoes are affected by sunscald and can be protected by placing a shade such as cheescloth over them for a few days.

Low Temperature. Plants also suffer when the temperature is too cold. This is more of a problem for outdoor plants than for indoor plants. Normally, temperatures inside a house will not be too cold for plants because humans would also be cold. However, precaution should be taken to avoid placing plants too near windows in cold weather, or near air-conditioning outlets in the summer.

Outside plants are affected by frost and freezing weather. Trees and shrubs react to this by going into a dormant condition in the fall. Spring

Figure 11-6. One method to protect tender plants such as tomatoes from cold outdoor temperatures is to cover them with hot caps.

frost is a particular problem for fruit and vegetable growers. Frost will cause the leaves and stems of tomatoes and peppers to turn black and die. Some plants may recover and develop new shoots or leaves, or they may die if the frost is severe. Flowers of fruit trees may be killed by frost which will result in a crop failure.

Several techniques are used to protect outdoors plants from cold temperatures. If the plants are small, as is the case with some vegetables, small plastic covers called "hotcaps" are placed over the entire plant. Many home gardeners cover small plants with milk cartons or newspaper to protect the plants from frost.

Another technique used to protect plants from being damaged by frosts is to continuously spray the entire planting with water all during the time that the air temperature is below freezing (just before to just after dawn, usually). The water on the outside of the plant freezes first. This provides a protective cover that prevents the plant from freezing. The water rather than the plant gives off heat, and the plant tissues are prevented from freezing. This method may seem strange but it works. It is often used with strawberries and fruit crops. A newer practice used for some nursery crops as an all-winter protection from low-temperature injury is to spray crops with a foam. The semipermanent foam insulates the plant. The foam is removed in early spring. Container nursery stock is usually protected over the winter in temporary houses covered with white plastic film, which prevents wide variations in temperature. Portable burners that operate on fuel oil or LP gas are sometimes used in orchards for frost prevention. These create upward-moving air currents that prevent the freezing air from settling around the fruit trees in an orchard or citrus grove. In some orchards or groves, large fans are used to circulate the air for the same purpose.

Certain garden flowers such as roses are protected from freezing temperatures by heaping soil or a mulching material such as tree bark around the base of the plant. This insulates the soil and keeps the crown and roots from freezing and thawing rapidly, which would cause damage to the plants.

Air

Air is essential for proper plant growth. However, even air can sometimes also be a factor in causing plant disorders. Because air is such a common environmental factor involved in plant growth, it is often overlooked as a possible villain in plant disorders. How do you think air causes plant disorders? Most indoor plants do not like an area where the flow of air is either excessive or uneven. If these plants are placed near a door that is opened and closed a great deal, the uneven air flow may cause the plant to lose leaves. The same thing happens if plants are placed near open windows. Part of the time the wind may blow and then the wind may stop.

Plants like an even air flow that is not excessive—they do not like drafty areas. Falling leaves and leaf tips and leaf edges (margins) that turn brown are the main symptoms of drafty areas. Moving the plant or stopping the draft will solve the problem.

If plants that are grown in a home or in a greenhouse, a place where air movement is very slow or nonexistent, are suddenly placed outdoors or in a window in strong wind, the moisture is literally blown out of the leaves and they will be severely "scorched" (dehydrated). Outdoor plants have adapted to air movement, and the same conditions will not damage them.

One way in which plants suffocate from lack of air is through overwatering. If a plant is overwatered the roots will constantly be surrounded by water. The spaces in the soil that are normally occupied by air will be occupied by water. Therefore, the roots will not be able to absorb air or water from the soil. The plant may be wilting while standing in water, because the root hairs have been killed.

You may remember that a plant removes carbon dioxide (CO_2) from the air to be used in photosynthesis. In the winter, plants grown in greenhouses may not receive enough carbon dioxide from the air for maximum growth, because no ventilation to the outside is used. Plants may not show visible signs of this problem, but may simply grow more slowly than they would if more carbon dioxide were available. Roses and carnations have stiffer stems, and chrysanthemums mature faster to produce bigger flowers, when the carbon dioxide level in the greenhouse is increased in winter months. The carbon dioxide level in a greenhouse can be increased through the use of special carbon dioxide generators. These special devices burn various fuels such as kerosene or natural gas to produce more carbon dioxide.

One more way in which air can contribute to plant disorders is by carrying pollution to the plant. The plant disorders caused by pollution will be discussed later in this chapter.

Nutrition

A number of plant disorders are caused by either a lack or an excess of the nutrients needed for plant growth. Some of the symptoms of these disorders are often easily identified, while others are harder to recognize.

Primary Nutrients. Are you ready for another case as a plant detective? Here are your clues. The marigolds and tomatoes you planted several weeks ago seem to be growing fine, but the leaf veins are starting to turn purple. Several days pass and the entire leaf takes on a purplish tint. What is wrong?

This is common and is an indication of a lack of phosphorus. This problem can be controlled by adding superphosphate or a commercial fertilizer high in phosphorus to the crop.

The next case: the leaves on your fruit trees started curling and turning brown or bronze on the leaf margins and at the tips. Later, the leaves appeared scorched. What is the problem?

This time the disorder is probably a potassium deficiency. In houseplants the plant may appear stunted and look "rusty." Normally the leaf margins turn yellow first and then brown. Adding a complete fertilizer containing 5 to 15 percent potassium should solve the problem.

Case number three: your dracaena is stunted and has pale green leaves that are starting to turn yellow, with the lower leaves yellowing and dropping first.

This is a typical case of nitrogen deficiency. Plants that need nitrogen may also be weak and "spindly." For most plants, the yellow color will start in the middle of the leaf blade in the lower leaves. This condition can be corrected by applying a fertilizer containing nitrogen.

Trace Elements. Trace-element deficiencies can also cause plant disorders. Trace elements are elements needed in very small amounts, but if lacking they may cause severe stunting of a plant. A magnesium deficiency results in yel-

A GUIDE FOR DIAGNOSING PLANT DISORDERS

Symptom	Possible Problem	Comment
PLANT IN GENERAL		
Tall and spindly in appearance	Insufficient light Excessive nitrogen Grown too warm	Check the plant management plan for clues
Slow stunted growth	Nutrient deficiency (could be nitrogen or calcium)	If roots are short and stubby calcium could be the problem
	Insufficient level of carbon dioxide	Sometimes a problem in the winter in Northern greenhouses
	Aphids or other sucking insects nematodes or other root pests	Carefully examine roots, soil, and the plant for signs of pests
Plant droops, is wilted	Wilt disease Improper watering Too much light or excessive heat	Examine roots and soil
Pale color	Nutrient deficiency Insufficient light or temperature Various sucking pests	
Plant falls over	Damping off disease Cutworms	Affects seedlings generally Look for evidence of worms
Abnormal white, pink, or black soft masses on plant	Smut disease	
Large tumor-like growth on stems or roots	Gall disease Nematodes	
Dead misshapen shriveled areas on woody stems	Canker disease	
Twigs on ornamentals and fruit trees turn brown or black	Blight disease	
LEAVES		
Yellow streaks (chlorosis)	Mosaic or yellows disease Nutrient deficiency (nitrogen, magnesium, manganese, iron, zinc)	
Tips turn brown	Plant located in a draft Salt spray Potassium deficiency Chemical application improperly done Air pollution	

Figure 11-7. A guide for diagnosing plant disorders.

A GUIDE FOR DIAGNOSING PLANT DISORDERS

Symptom	Possible Problem	Comment
LEAVES (continued)		
Leaves fall off the plant	Drafts Over watering Insufficient light Sucking pests	
Leaves appear scorched	Potassium deficiency Excessive light or heat Improper chemical application	
Leaves rapidly turn black	Frost damage Rot disease	Has there been a killing frost?
Leaves have a white or gray powdery appearance	Mildew disease	
Dark spots on leaves, spot may be surrounded by a darker ring, spots fairly constant in size	Spot disease	
RED, orange spots rapidly turning brown entire leaf becomes brown	Blight disease	
ORANGE, red, light green spots, often rust colored, small BB shaped projections may be present	Rust disease	
LEAVES turn purple	Phosphorous deficiency insufficient temperature	
Bluish tint to leaf	Copper deficiency	
Leaves twist, become crinkly, curled	Improper chemical application	Check plant log
Pale yellow or reddish brown spots, fine webs on leaves, eventually fall	Spider mites	
Parts of leaves eaten	Various pests	
Leaves white and flecked in appearance, undersides spotted with small black spots	Thrips	
FLOWERS		
Buds or flower drops	Low relative humidity various pests	
Blossoms turn black	Blight or rot disease	
Petals chewed	Various pests	

Figure 11-7. (cont.)

lowing between the leaf veins of upper leaves. If the leaf veins are parallel, the yellow color appears as stripes. Plants that are deficient in copper may have leaves with a bluish tint. A manganese deficiency is similar to a magnesium deficiency in that there are yellow areas between leaf veins. Calcium deficiency results in stunted growth; the plant simply stops growing altogether. It may also wilt easily. Roots of plants lacking calcium are short and stubby, and all the root hairs are dead. Even after calcium is applied to these plants, recovery may be very slow. Iron and zinc deficiencies frequently cause yellowing of the leaves between the leaf veins in the upper part of the plant. Applying sprays or powders that contain these trace elements will often solve the problem.

Aids for Diagnosing Nutrition Deficiencies. At times it is difficult to diagnose plant disorders caused by nutrient deficiencies. If you suspect a nutrient deficiency but are not sure, there are two procedures that you can follow. A number of companies sell plant-tissue testing kits. These kits are relatively inexpensive and easy to operate. By applying certain chemicals to plant tissue you will be able to determine whether certain nutrients are lacking. The only problem with this procedure is that some of the kits are not completely accurate.

The second solution is to send a sample of the plant tissue, together with a soil sample, to a laboratory that is equipped with accurate scientific equipment for analysis. Often, state land-grant universities provide this service for a small fee. Your agriculture instructor or county extension agent should be able to give you more information about this service.

Chemicals

Are you ready for another mystery? One day in early spring Pat noticed street workers out sweeping the *grass* that was beside the city streets with a broom. Pat couldn't figure out

Figure 11-8. The improper use of chemicals will damage or even kill plants.

why they were doing this. The grass looked clean. As a matter of fact, winter in this Northern town had just ended, and the grass had been covered with ice and snow most of the winter. Because of this it could not be very dirty. Have you figured out why the workers were sweeping the grass?

This is a tough mystery. In many Northern cities salt is poured onto the streets to melt ice. When the ice melts, the slushy ice and salt is splashed onto the grass next to the street by passing cars. This salt could kill the grass. One symptom of the problem is a browning of the grass near the street.

Salt is also a problem along the coast. The oceans are composed of salt water. Winds blowing in from the ocean may carry a salty mist with them. This salt is deposited on plants inland. The browning of leaf tips is a symptom of salt. Eventually the leaves may fall off the plant on the side next to the ocean. This disorder can be overcome by washing the plants with clear water after a storm. Some plants, such as Japanese black pine, the sycamore maple, and many junipers are highly salt tolerant and are recommended for seaside plantings.

Air pollution is another cause of plant disorders. Plants grown near certain factories that emit pollutants into the air may grow poorly and have leaf discolorations. The manner in which plants react depends upon the type of pollution in the air, since there are many types of pollution. The primary one is sulfur dioxide, which can severely damage greenhouse roses and orchids. Some greenhouses have had to move to pollution-free air because of this problem.

The improper use of chemicals can also result in plant damage. Chemicals such as 2,4-D that are used to kill weeds can also kill horticultural plants if applied accidentally near them. Generally, the plants will become stunted and twisted, with crinkly curled leaves. Fumigants sometimes are used in the greenhouse to kill insects. If the fumigant is applied improperly when the temperature is too high or too low, or the relative humidity is too low, the leaves of many greenhouse plants may be burned (turn brown). These disorders can be prevented by proper and careful use of chemicals.

Mechanical and Cultural Disorders

A number of plant disorders are caused by physical action. When a new home has been built where established trees have been growing for a number of years, sometimes the trees sicken and die. It may be 2 to 3 years before they stop growing altogether. The problem is not in the soil but in how the the soil was moved. If tree roots were exposed for long periods of time, cut, badly scraped, or covered with too much soil during the construction of the house, they may be so badly damaged that the tree will die. The best way to avoid this problem is to not disturb the soil under the spread of the tree branches (most of the heavy tree roots are within this area).

Careless use of lawnmowers can result in plant disorders. Bumping the mower against small trees can damage the cambium layer in the tree. The cambium is a layer of tissue between the wood and the bark. Its function is to produce new cells. Damage to the cambium hinders the formation of new cells and may result in stunted growth. Careful operation of lawnmowers will prevent this problem.

When trees are planted they often have wires attached to them for support. One end of the wire is wrapped around the tree, while the other end is attached to a stake and driven into the ground. This is known as "guying" a tree. Plastic should be wrapped around the tree where the wire is placed. This prevents the wire from girdling or cutting into the tree.

A Management Plan for Maintaining Plant Health

Many plant disorders can be prevented by using common sense, knowledge, and a plant management plan. A plant management plan may be the most important tool you have in maintaining plant health. Whether you work at a greenhouse, floral shop, nursery, golf course, or in fruit and vegetable production, a management plan is needed. Homeowners can also use a management plan for their plants.

A plant management plan should contain two sections. The first section is for planning the care of plants, while the second is used to record actual practices used in caring for the plants.

The Planning Section

Would you go on a long trip to a town or state you had never been to without looking at a map? Probably not. Whenever people go on a trip, they look at a map to plan where to stop at night, see how far to drive each day, and determine when they will arrive at their destination. It would not be wise to start on a long trip without prior planning.

Planning is also important in growing horticultural crops. Growing and caring for plants should not be haphazard.

PLANT MANAGEMENT PLAN

Section A — My Plans

My Goal is: To raise 100 tomato plants and sell the tomatoes at a roadside stand.

Facts about the plant: Tomatoes are warm season plants. It takes about 3 months for a tomato plant to produce marketable tomatoes. Seeds can be planted inside and the resulting seedlings can be planted outside after the last killing frost. The plants should be tied to a wooden stake to keep the tomatoes off the ground. Cutworms, whiteflies, and aphids are pests that will need to be controlled. Growing tomato plants need a lot of water.

Amount of space required: The plants will be planted in 10 rows. Each row will be 3 feet apart and the spacing in the row between plants will be 3 feet. Therefore, a square plot of ground 30 feet by 30 feet will be needed.

What needs to be done	When it should be done	Comments or estimated cost
Buy the seeds	March 10	$5.60
Buy the planting media and flats	March 15	$3.55
Plant the seeds indoors in flats	April 1	
Place the flats in a warm area	April 1	
Water the flats and seedlings	check daily	
Check the seedlings for proper growth	Daily	
Apply fertilizer to seedlings	April 25	Use 12-12-12 ratio ($2.95)
Place seedlings outside during the day to harden them for transplanting	Start May 1	
Prepare the ground for transplanting	May 10	
Mix fertilizer in with the soil	May 10	Use 8-16-16 (cost $5.75)
Buy wooden stakes and drive them in the ground	May 15	$25.00
Transplant tomato plants and tie to stakes	May 20	
Water the plants.	After transplanting, then every 2 days or sooner if needed	
Pull weeds	Every 3 days	
Check for insects or diseases	Daily	
Apply pesticides	As needed	$7.00
Prune unwanted axillary shoots (suckers)	Every 3 days	

Figure 11-9. Good plant management requires a plan.

The planning section of a plant management plan is an outline of what needs to be done and when it should be done to produce a healthy crop. The environmental conditions should be described, along with plans for providing the right environment. Some items that could be included in the planning section are:

1. Type of plants to be grown

PLANT MANAGEMENT PLAN

Section B — My Records

What was done	When it was done	Comments or actual costs
Bought seeds ⟨3 packages – Better Boy / 3 packages – Early Girl⟩	Feb. 27	$4.50
Bought planting media (vermiculite)	Feb. 27	$4.11
Soaked vermiculite and planted seeds	March 28	
Watered flats	April 1	
Watered flats	April 5	
Applied 12-12-12 fertilizer with water	April 11	$1.49 (seedlings coming up)
Watered flats	April 14	
Watered seedlings	April 18	
Placed seedlings outside for the day	April 20	
Watered seedlings	April 21	
Used roto-tiller to prepare ground	April 23	Borrowed from Mr. Jones promised him the first tomato
Applied 12-12-12 fertilizer	April 26	Several plants were turning pale. I think they need more nitrogen or more light. I moved the plants near a south window.
Watered seedlings	April 28	
I placed seedlings outside for the day	April 29	
Bought wooden stakes and drove in ground	May 1	$34.18
Applied "Sevin" in powder form	May 2	Noticed several white flies on the plants. I don't know where they came from. Maybe from sitting outside on the 29th.
Watered plants and set outside for the day	May 3,4,5,6,7	
Transplanted 25 Early Girls	May 8	

Figure 11-10. After establishing the plant management plant, accurate records must be kept.

2. Number of plants to be grown
3. Amount of space needed to grow the plants
4. Method of starting the plants (seeds, cuttings, seedlings)
5. Dates to start or transplant plants
6. Type of soil or planting media required
7. Nutrient requirements of the plants
8. Temperature needed for plant growth
9. Methods for preventing pest infestations
10. The point at which cultural practices, such as pinching buds, should occur
11. Way in which plants will be watered and how often
12. Humidity requirements of the plants
13. Light requirements of the plants

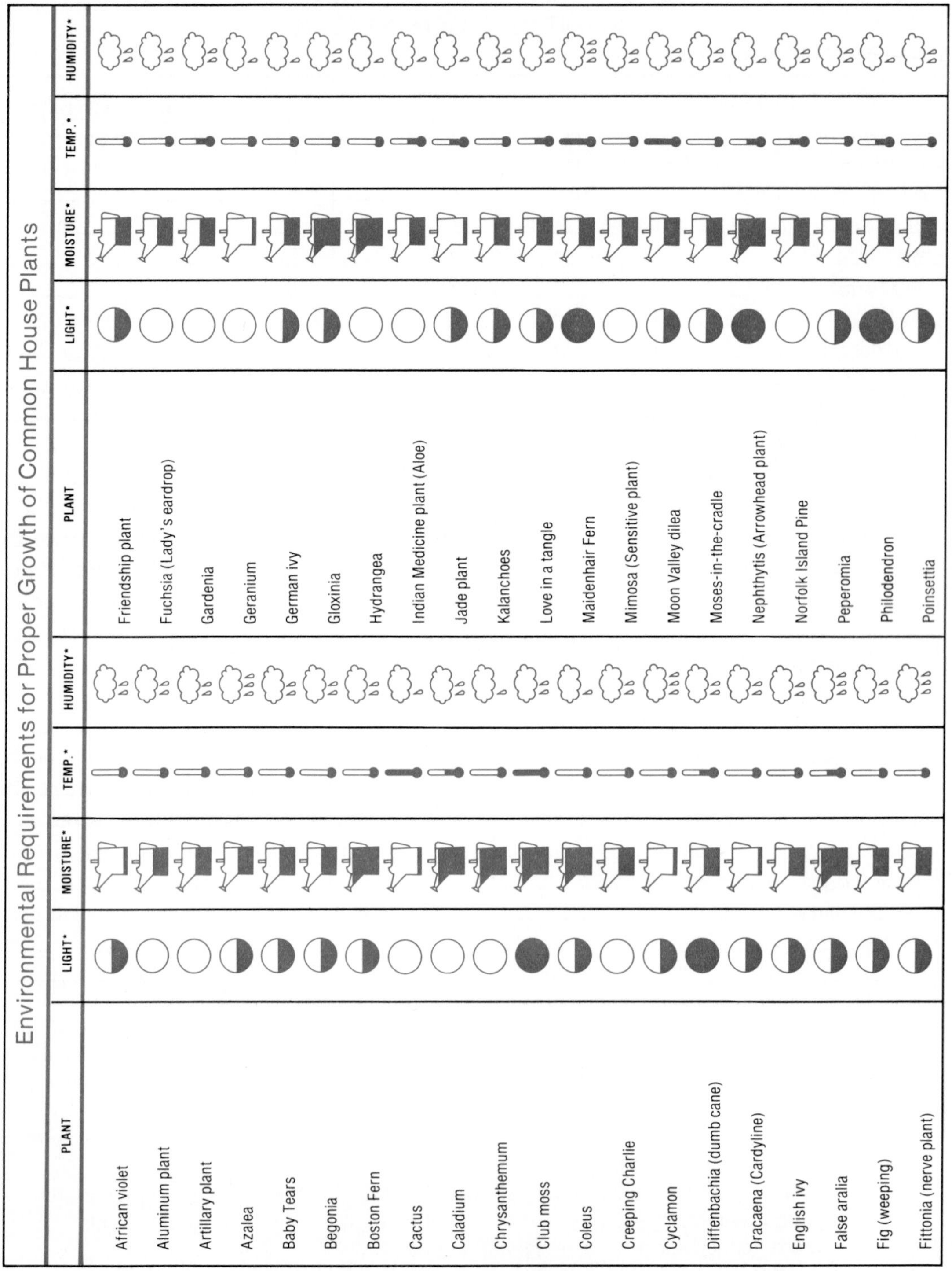

Figure 11-11. Environmental requirements for proper growth of common house plants.

Environmental Requirements for Proper Growth of Common House Plants

PLANT	LIGHT*	MOISTURE*	TEMP.*	HUMIDITY*
Pothos (Devils Ivy)				
Prayer plant (Rabbit Tracks)				
Ribbon plant				
Rubber plant				
Snake plant (Mother-in-Laws tongue)				
Spider plant (Airplane plant)				
String of hearts				
String of pearls				
Swedish ivy				
Umbrella plant (Schefflera)				
Velvet plant				
Vicks vap-o-rub plant				
Waffle plant				
Wandering Jew				
Zebra plant				

Key to Symbols*

Light

High—full sun or bright direct light

Medium—partial sun or diffused light

Low—no direct light, likes shady areas

Moisture

High—likes water, keep soil moist

Medium—soak thoroughly, then allow surface to dry before watering again

Low—soak thoroughly, then allow soil to dry well before watering again

Temperature

Warm—plant prefers 75°F to 85°F (24°C to 30°C)

Medium—plant prefers 65°F to 75°F (18°C to 24°C)

Cool—plant prefers 50°F to 65°F (10°C to 18°C)

Relative Humidity

High—plant prefers high humidity (above 60 percent)

Moderate—plant prefers moderate humidity (around 50 percent)

Low—plant prefers low humidity (30–45 percent, typical of the humidity in most homes)

By writing all this information down in a plan, you will realize exactly what needs to be done to assure healthy plants. This type of planning will help prevent plant disorders. A sample plant management plan is shown in Figure 11-9.

The Record Section

The record section of the plant management plan is a place where you write down dates and procedures that were actually used in caring for plants. You would record when the plants were actually started, watered, transplanted, pinched, fertilized, and pruned. Any applications of pesticides or other chemicals would be recorded. You may wish to record the high and low temperature of each day.

There are several reasons why this type of information is important. One of the most important reasons is to help diagnose plant disorders. Often the symptoms shown by an unhealthy plant could indicate a variety of problems. A plant detective has to carefully sort out all the possibilities. The record section of the plant management plan is extremely important in determining whether the problem could be an environmental factor such as improper watering. If the record shows that the plant has been receiving proper care, then other causes of disorders will be easier to diagnose.

The records are also useful in determining whether fruits and vegetables are safe to eat. After certain pesticides are applied to plants, you must wait anywhere from 24 hours to 2 weeks before the produce can be eaten. By recording the date on which the chemicals were applied, you will know when the vegetables or fruit may be safely eaten. The record section of a plant management plan is shown in Figure 11-10. A plant management plan is a helpful tool in providing the right environment for healthy plants.

Plant Disorders: A Review

As a horticulturist you will be expected to be a plant detective. Many plant disorders are caused by an improper environment. Excesses or deficiencies of water, light, air, temperature, and nutrients can cause plant disorders. Chemicals and mechanical practices can also result in plant disorders. By carefully examining the plant and determining how the plant was cared for, through questions or examination of records, you should be able to determine the cause of the problem. A good plant management plan is helpful in maintaining plant health and diagnosing plant disorders.

THINKING IT THROUGH

1. What characteristics indicate that a plant is healthy?
2. What steps should be taken in determining what is wrong with a plant?
3. How can you determine whether a plant is being overwatered or underwatered?
4. What three symptoms indicate that a plant is not getting enough light?
5. In general, do you think houseplants are kept too cool or too warm? Why?
6. How can "air" cause plant disorders?
7. Can helpful chemicals such as weed and insect killers harm horticultural plants? Explain.
8. How would you go about assessing physical damage to a plant?
9. Why should you have a plant management plan?
10. What information should be contained in a plant management plan?

12

Identifying and Controlling Plant Pests

Dale had started watering the poinsettias when several small white specks appeared on the leaves. "How did this lint get into the greenhouse?" Dale wondered, and bent over to shake the lint off. The moment the leaves started shaking, the air was filled with tiny white specks flying around. Dale jumped back, startled. The white specks were tiny insects, not lint. "Mr. Ramorez should know about this," thought Dale, and headed for the office.

Mr. Ramorez was the owner of the Ramorez Greenhouse and Nursery Company. Dale found Mr. Ramorez in the office and told him that there were tiny white bugs on the poinsettias. "Well, we had better check this out," replied Mr. Ramorez. After examining the plants and the tiny white specks, Mr. Ramorez turned to Dale and said, "What type of pest do you think we have and how should we control it?" Dale didn't know.

Mr. Ramorez went on, "This pest is a fairly common problem in greenhouses. It sucks sap from the poinsettias and other plants. If we don't control it, the plants will grow poorly, wilt, turn yellow, and possibly die. Since you're working here as a part of your school's horticulture program to gain practical experience, I'd like you to determine what the pest is and how it should be controlled. Once you've done that you'll know what to do when you manage or own your own greenhouse!"

Do you know what the pest is? How can it be controlled? As a horticulturist you will need to able to identify and control plant pests. There are many kinds of pests that are destructive to horticultural plants. Pests are responsible for a number of plant disorders. Annually, plant pests cause millions of dollars of damage in the horticulture industry. Controlling plant pests is an important task in horticulture.

CHAPTER GOALS

In this chapter your goals are:

- To distinguish plant disorders caused by pests from those that result from other causes
- To explain why pests are a problem in horticulture
- To demonstrate five general methods of insect control
- To classify pests according to their "eating habits"
- To identify common pests in the horticulture industry

What Is a Pest?

In horticulture, any living organism that competes with or damages the plants we are trying to grow is considered a pest. Using this definition, can you name some horticultural pests?

There are five main categories of plant pests. These are:

1. Insects (and other small insectlike creatures)
2. Snails and slugs
3. Living things with a backbone (vertebrates—animals, birds, reptiles)
4. Weeds
5. Microscopic organisms that cause disease

Insects are probably the most numerous of all plant pests. It is estimated that 3 to 10 million species of insects exist in the world. If insects were allowed to reproduce at their normal reproductive rate for 1 year (and no insects died during this time), insects would form a crust over the entire earth 17 ft (feet) [510 cm (centimeters)] deep. There are also several small creatures that are not scientifically classified as insects that damage horticulture plants. Some

Figure 12-1. Identifying and controlling pets is an important task in horticulture.

of these are mites, ticks, and spiders. The reason why they are not called insects is because they have eight legs. All adult insects have six jointed legs and three body regions.

Snails and slugs are horticultural pests. They belong to a group called *mollusks*. Snails have hard shells and feed on plant foliage. Slugs also feed on foliage and look like snails minus the shells.

Creatures with backbones often damage horticultural crops. Birds eat berries and fruits. Deer and rabbits will eat bark off trees. Moles burrow into the soil, leaving ridges.

Many people do not realize it, but a weed is a horticultural pest, because it is basically a plant out of place. Weeds compete with horticultural crops for nutrients, moisture, and light.

Small microscopic organisms such as bacteria may cause diseases. They could be classified as pests. In this book diseases are discussed separately from pests. In the next chapter you will learn about microscopic organisms and the diseases they cause.

Figure 12-2. Pests damage all types of horticultural crops.

Why Pest Control Is Important

Annually, horticulturalists spend millions of dollars on pest control; if they didn't, they wouldn't be able to make a profit. Pests spread disease, reduce yields, reduce the quality and appearance of plants, and even kill plants.

Plant diseases may be spread by pests. Pests may be carriers of plant disease organisms, just as mosquitoes carry yellow fever to humans. As pests move from one plant to another the disease is spread. Plants weakened by pest damage are more susceptible to diseases. The Dutch elm disease, fruit fire blight, and bud rot of carnations are all spread by pests.

Besides spreading diseases, insects and other pests are harmful in horticulture because they reduce the quality and appearance of horticultural products; reduce yields; and stunt the growth of, or kill, horticultural plants.

Have you ever seen a wormy apple or wormy head of cabbage? Fruit and vegetables damaged by insects are often hard to sell, because of their unattractive, unhealthy appearance. Would you buy a corsage or floral arrangement that was made from plants and flowers whose leaves and petals had been partially eaten by pests or had insects crawling on them? Florists do not use flowers and plants that have been damaged by pests in floral arrangements. The loss of "eye appeal" in many horticultural products as a result of pest damage can be costly.

Not only do pests reduce the quality of horticultural products, but they also reduce plant yields. Many kinds of pests feed on plant leaves. As was pointed out in Chapter 4, leaves are used by the plant in the photosynthesis process. If pests strip the leaves from the plant, then the plant cannot produce the energy needed for the development of fruits, vegetables, and flowers.

Pests often kill plants. If left uncontrolled the San José scale, a small, flat, platey insect that sucks the sap of plants, can kill thousands of acres of orchards simply by sucking up so much plant juice that the plants starve to death. Another sucking pest, the red spider mite, which is microscopic, appears to inject a substance into rose plants that causes them to stop growing. Only a few red spider mites can seriously affect the yield of flowers from a large greenhouse. The Colorado potato beetle simply consumes numerous potato patches annually. Florida chinch bugs can destroy an entire lawn of St. Augustine grass.

If you are employed in horticulture, you will be expected to recognize plant disorders caused by pests. Flower growers must constantly be alert for pests, such as thrips, that will make their flowers unsaleable. Thrips cause leaves of plants such as gladiolus to become whitened and flecked in appearance. Workers who maintain lawns and golf courses should be aware of the Japanese beetle, which is very destructive. The larva of the beetle feeds on grass roots. If the roots are destroyed, the grass will turn brown and die.

No area of horticulture is safe from pests. The loss of horticultural plants and reduction in quality and yields costs the horticulture industry millions of dollars annually. Pest control is a never-ending task performed by those who are engaged in horticulture occupations. As a new horticulture worker, you should be able to recognize an unhealthy plant in the earliest possible stage of a disorder so that control measures can be taken quickly.

Identifying Plant Disorders Caused by Pests

Often the symptoms of plant disorders caused by pests are similar to those of disorders caused by environmental factors or diseases. If Dale had not noticed the tiny white specks on the plant, but instead had noticed the plants growing poorly and turning yellow, the natural conclu-

Figure 12-3. The first step in determining what is wrong with a plant is to examine it. Insects or other pests may be visible.

sion might have been that the problem was a nutrient deficiency or lack of light.

In determining whether a plant disorder is caused by plant pests, you will need a systematic procedure similar to the one outlined in Chapter 11. The first step is to carefully examine the entire plant, including the roots. Are there signs of pests? Can you see the actual pest? Are parts of the plant eaten? Are any parts of the plant an abnormal size or shape? If no signs of pests are present, this does not entirely rule out pests as the problem. At times, pests that cause plant disorders will not be visible. For example, consider the case of a plant that had stopped growing and had started turning yellow, with no visible symptoms of plant pests. When the roots were examined, large abnormal knots were discovered. The knots were caused by a microscopic wormlike pest called a nematode.

In addition to examining the plant, you should ask questions or consult the plant management plan to determine whether pests are the problem. Some questions that could be asked to help verify the problem of pests are:

1. Has the plant or other plants been placed outside recently?
2. Was the plant or other plants moved inside from outside?
3. Has the plant been repotted, using soil from around the house or greenhouse?
4. Have new plants been bought lately?
5. Are there any open doors or windows by which pests could enter?

When plants, soil, or new plants are brought into a greenhouse or home from outside, there is always the possibility that plant pests have "hitchhiked" along. By asking the above questions, one may be able to determine whether plant pests could be a reason for a plant disorder. The same types of information could be gathered from examining the record section of a plant management plan. If you cannot see the pest, the symptom is not one that is caused exclusively by a pest, and the plant has not been exposed to pests as far as you know, try to determine whether the problem is caused by other factors, such as a disease or environmental factor.

Symptoms exhibited by plants whose disorders have been caused by pests vary. Parts of the plant may be discolored or misshapen. Leaves and stems may be stunted. Parts of the plant may be missing. Specific plant disorder symptoms will be discussed later in this chapter, along with the pest that causes the problem.

Methods of Controlling Pests

One of the tasks assigned to Dale by Mr. Ramorez was to determine a way to control the pest. This started Dale thinking: how many different methods of controlling insects are there? How many methods could you name?

Have you ever heard of the "two-blocks-of-wood method"?

Several years ago a radio station in the Southwestern United States advertised a fantastic new insect killer. This insect killer was guaranteed to be 100 percent effective, to kill only harmful insects, and not to harm the environment. Thousands of people who ordered the new insect killer were very surprised when they opened the package. Inside the package were two blocks of wood. The instructions read, "Place the insect to be killed on one block of wood and smash the insect with the other block of wood." Although this method of insect control was quite "effective," it was anything but practical. Today, more desirable methods of insect and pest control are available.

Pest controls can be classified according to the manner in which they destroy or control pests. The most common method of pest control are:

- Quarantine
- Cultural
- Mechanical
- Biological
- Chemical

Quarantine

Quarantine simply means to physically isolate a pest from a population of healthy plants. When greenhouse growers or nursery operators purchase new plant material, they carefully examine it for disorders. If one is found, they isolate the plants and control the pests or disease before placing the plants with disorders near the healthy plants.

Federal and state governments have passed laws that help control pests and diseases. In 1912 Congress passed The Plant Quarantine Act. This law allowed the USDA to prevent the importation of plants and other agricultural products that might contain pests and plant diseases from foreign countries. In 1957 The Federal Plant Pest Act was passed. This replaced the 1912 act and provided the USDA with the authority to stop the entry of pests into the country and the movement of pests within the country.

When plants and other agricultural products are shipped to America, they are carefully inspected before they are allowed into the country. Baggage carried by people may even be inspected for any pests that might have hitchhiked along.

In 1969, a young boy from Florida was vacationing with his family in Hawaii. He picked up an unusual-looking snail and placed it in his pocket. Unfortunately, he got past the quarantine inspectors when he returned home. The snail turned out to be the giant African snail, which feeds on many plants. Shrubs, trees, and lawns in many areas of Florida are now infested with the African snail.

The Japanese beetle, the gypsy moth, and the cereal leaf beetle are three pests that have escaped into this country from other countries. Efforts to control them have not been fully effective, and the loss to woodlands and crops has been very heavy.

If pests can be kept out of the country through the use of quarantines, then horticulturists will not have to be concerned about controlling them. Many states even have quarantine laws that prohibit shipment of selected horticultural products from other states, because of the possibility of spreading pests or diseases to crops not currently bothered by them.

Cultural Control

Cultural control involves the use of recommended agricultural practices to interrupt the life cycle of pests. Fall plowing in place of spring plowing exposes grasshopper eggs and certain grubs to freezing temperatures during the winter and kills them. This is an example of effective cultural control. Keeping the area around greenhouses, orchards, and gardens clear of trash, old lumber, sacks, and dead vegetation will help to

Figure 12-4. This greenhouse operator may have problems because pests can live and breed in the trash and junk surrounding the greenhouse.

control pests. Pests hide in trash piles and can even spend the winter in this trash. If the area is kept clear, pests will not have a home. Crop residues and weeds should also be destroyed, since they can harbor pests. Keeping the grass mowed in orchards discourages mice and rabbits, who often strip the bark from young trees during the winter. Weed-free greenhouses prevent pests from moving from weeds to crop plants.

Another method of cultural control is planting crops at certain times of the growing season, when insects are less of a problem. The egg-laying period of a particular pest can sometimes be avoided by carefully choosing planting times. The Hession fly, a pest of wheat, can be avoided by delaying the planting time of wheat.

Crop rotation is another cultural practice, and is particularly useful to gardeners and nursery operators. Each year the crop is planted in a different spot. This prevents the build-up of pests or disease organisms in the soil.

Cultural control of insects is relatively inexpensive and can easily be incorporated into an overall management plan in horticulture. In the planning section of the plant management plan, you would list when and how you would perform certain cultural practices, such as destroying crop residue and mowing the grass. You would also determine the best planting dates and plan for crop rotation if needed. Often cultural practices are used in combination with other pest control procedures. Whenever a plant management plan is being developed, you should include plans for controlling pests. The manager or operator of a greenhouse, nursery, or garden, or the head greenskeeper at a golf course would determine which cultural practices should be used.

An example of how cultural control can be incorporated into a plant management plan can be seen by looking back to Figure 11-9. In the plant management plan, the student planned on applying pesticides as needed. The only cultural practice mentioned in the plan was to pull weeds. The following items could be added to the plan:

- May 10—Mow area around the garden to eliminate hiding places for pests
- May 20—Plant tomatoes in an area where tomatoes have not grown before
- May 25—Mow every 2 or 3 weeks to keep grass and weeds down

At the end of the growing season, pull up all tomato plants and dispose of them.

Mechanical Control

Mechanical pest control involves the use of specialized equipment. Although not widely used, it is most appropriate in select situations. An example is the electric insect trap, which is effective for certain types of pests in a limited space, such as a greenhouse. This trap has a light or some type of bait used to attract flying insects. Upon entering the trap, the insect passes between wires that are charged with electricity. This creates an electrical arc that electrocutes the insect. The electrical trap does not distinguish between the helpful and the destructive insects, so useful ones may be eliminated along with the pests. Other types of traps contain liquids that kill insects by drowning. Various types of mechanical traps are used to catch moles, rabbits, and mice. Another mechanical device used to control insects, particularly grasshoppers, is the "hopper-catcher" or "hopper-dozer." This device is a slightly curved piece of metal or wood that has a tray at the bottom filled with water or kerosene. It is fastened to the front of a tractor or truck, and driven slowly through orchards where grasshoppers are a problem. When the grasshoppers are disturbed they jump or fly up and hit the back of the machine, then fall into the liquid and drown. Greenhouses in which orchids or snapdragons are grown are usually screened at all openings to keep out bumblebees, which pollinate the flowers and cause them to quickly discolor and drop.

Biological Control

Pests have many natural enemies—birds and animals eat insects; some insects eat other insects; some diseases destroy harmful insects. Using natural enemies to control insects is termed *biological control.*

Growers of citrus trees in California imported the Australian lady beetle to destroy the cottony-cushion scale that was killing citrus trees. The lady beetle is used to control aphids. The mealybug, which harms greenhouse plants and vegetables, has a number of natural insect enemies. The European ground beetle is an enemy of the gypsy moth.

Diseases are also used to control some horticultural insects. Certain fungi have been mixed with water, then sprayed on plants infested with

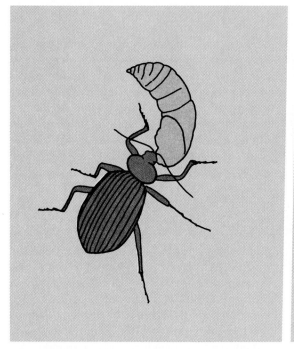

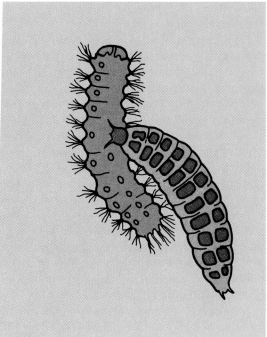

Figure 12-5. (Left) European ground beetle feeding on the pupa of a gypsy moth. (Right) Larva of European ground beetle feeding on the caterpillar.

whiteflies. The fungi live on the bodies of whiteflies and eventually destroy them. This process for controlling whiteflies has been effective in Florida citrus groves and in greenhouse crops. Milky disease, a fungus that may be placed into the soil, has been introduced to control the Japanese beetles.

Radiation has been used in the biological control of insects. Exposure to radiation causes the male of certain insect species to become sterile. When the sterile male mates with a female, the resulting eggs are infertile and will not hatch. Since certain insects, some mosquitoes for example, breed only once, this procedure is effective in reducing the number of insects.

Synthetic sex hormones are used to control certain insects. Fields are sprayed with a synthetic sex hormone. This causes the male insects to become confused. The male cannot find the female insects to breed. This results in a reduction of insects. Sex hormones have also been used to bait insect traps.

Biological insect control is an effective method of controlling insects, but limited to a relatively small number. When biological controls are used, great care must be taken not to destroy the control organism with a chemical application intended for other purposes. Research is constantly being conducted to find new methods of biological insect control.

Chemical Control

The most widespread method of controlling plant pests is through chemicals called *pesticides*. Pesticides can be grouped into three general classes—stomach poisons, contact poisons, and fumigants. The eating habits and life cycle of the pest determine the type of pesticide to be used.

Stomach Poisons. Stomach poisons are sprayed on plants. A poisonous film or powder is left on the plants. When a pest eats the plant, the pesticide goes into the pest's stomach and is spread throughout the digestive system, killing the pest. This type of pesticide is most effective on pests that have chewing mouth parts. Some chewing pests are beetles, cutworms, grasshoppers, and caterpillars.

Contact Poisons. Not all pests have chewing mouth parts. Some have piercing mouth parts that are used to suck sap out of plants. Because they do not chew on plants, stomach poisons will not work on these pests. Instead contact poisons that affect the pest's nervous system are used. The pesticide must come into physical contact with the pest. Some sucking pests are aphids, whiteflies, mealy bugs, and spider mites.

Fumigants. *Fumigants* are poisonous gases that are normally used in tight enclosures such as greenhouses. A poisonous gas is released into the atmosphere. Pests breathe the gas and die.

Pesticides are sometimes the most effective and efficient method of controlling pests. Extreme care must be taken, however, while using pesticides. Many pesticides are dangerous to humans. The safety rules on pesticide labels should be accurately followed. Some pesticides are "restricted" and can only be applied by a licensed applicator.

Pesticides can be applied to plants through a variety of methods. Small hand-held sprayers or dusters are used in the greenhouses. Fogging machines, which distribute the material as a very fine mist, are also used in closed greenhouses. Power sprayers operated by gasoline engines are used in orchards, nurseries, and landscaping. Tractors, airplanes, and helicopters can be equipped with sprayers or dusters to apply chemicals to large plantings. Pesticides

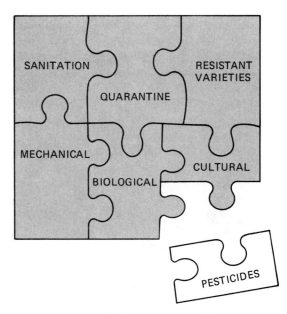

Figure 12-6. Integrated pest control combines a number of control procedures.

and methods of applying them will be discussed in detail in Chapter 14.

Integrated Control

In horticulture, a combination of control methods is often used. In producing one crop the horticulturist may keep the area clean of trash, use electric insect traps, apply fungus spores to kill an insect, and eventually spray the crop with a pesticide if other control methods are not satisfactory. Cultural, mechanical, biological, and chemical controls were all used with the crop. A variety of control methods used together is called *integrated control*. One crop plant may have a number of different pests that are troublesome, and an integrated program for the control of all of them is essential.

Citrus growers often use integrated pest control, including pesticides, cultural practices, mechanical practices, biological practices, and plant quarantine laws to help protect their crops from pests. By adding cultural procedures to the existing plant management plan in Figure 11-9, a grower would develop an integrated pest control plan. The student will use cultural practices such as mowing and destroying crop residue to eliminate the hiding places of pests. Using the rototiller and hoe to get rid of weeds is a mechanical practice. Applying pesticides involves chemical control. If the student were to release ladybug beetles into the garden, then biological control would also be employed.

Horticulturists need an overall management plan and detailed records in order to practice integrated pest control. Growing a healthy crop of oranges, mums, turfgrass, or other horticultural products requires careful planning and management. A record of planting dates, watering schedules, fertilizer application, cultural operations, and procedures used for controlling pests is essential. The time, date, rate, and material used, together with the identity of the pest to be controlled and an indication of how effective the material was, should be recorded. Careful records may give clues to the cause of a sudden disorder in a crop.

Pest Growth and Development

In terms of numbers, insects are our largest group of pests. Insect control is a skill needed by most workers in horticulture. Before a method of controlling insects is selected, one must identify insects that are harmful in horticulture and become familiar with their appearance, eating habits, and life cycles. Knowledge of the life cycle of insects is useful because there is usually one stage in which control is easier than at other stages.

Insect Anatomy

An insect has three distinct body regions: the head, thorax, and abdomen. The mouth parts, eyes, and antenna are located on the head. The central part of the insect body is called the thorax and is where the insect's wings and legs

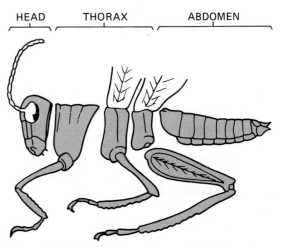

HEAD   THORAX   ABDOMEN

Figure 12-7. Insects have three major body regions.

are located. The rear part of the body is called the abdomen, and tiny openings called spiracles are located there. Insects breathe through the spiracles. All insects have six legs at some stage in their development.

Insect Life Cycles

About half of all insects go through four distinct stages of life. This life cycle is called *complete metamorphosis*. Metamorphosis begins with an egg. The egg hatches into a *larva*, which is often called a worm or grub. In the larval stage, insects are generally very destructive to horticultural plants. Wire worms, white grubs, tent caterpillars, cabbage worms, and tomato hornworms are examples of larvae. After a period of growth the larva changes into a motionless resting stage called a *pupa*. Cocoons, from which butterflies emerge, are examples of the pupal stage. Insects are very difficult to control during the pupal stage, since they neither eat nor move. Insects often spend the winter in the pupal stage. The *adult* insect emerges from the pupa as a beetle, moth, or butterfly, thus completing the life cycle. The adult stage of many insects—beetles, for example—involves eating leaves and stems of plants. The adults lay eggs, and the life cycle begins again.

Not all insects go through complete metamorphosis. A number of insects already resemble the adult when they emerge from eggs. Normally the only difference between the young and the adult is that the young insect does not have wings and is smaller in size. These young insects are called *nymphs*. As they grow they shed their skin periodically and grow a bigger skin. When they reach full-grown size they develop wings and become *adults*. Thus the three stages of their development are egg, nymph, and adult. Grasshoppers develop in this manner. This process of development is called *incomplete* or *gradual metamorphosis*.

A few insects emerge looking exactly like the adult. This is termed *no metamorphosis*. Many species of aphids fall into this category. The three stages of this development are egg, young, and adult. The only difference between the young and the adult is size. Red spider mites (not true insects because they have eight legs instead of six), follow a life cycle similar to that of aphids.

Insect Eating Habits

Understanding the eating habits of insects helps to determine which method of insect control would be the most effective. The mouth parts of most insects can be classified as chewing, rasping-sucking, or piercing-sucking.

Chewing insects actually eat parts of plants. They have teeth (called *mandibles*) and jaws (called *maxillae*). Since they swallow what they eat, stomach poisons are generally effective on insects with chewing mouth parts. Grubs, cutworms, caterpillars, grasshoppers, and most beetles have chewing mouth parts. Grubs often eat the roots of turfgrass. Cutworms devour vegetable and flower seedlings and transplants. Grasshoppers eat grains, fruit, vegetables, and garden flowers. Beetles damage grapes, vegetables, fruit trees, and many ornamental trees

Figure 12-8. Insects can have different kinds of mouthparts: chewing (top); rasping-sucking (middle), and piercing-sucking (bottom).

and shrubs. Slugs and snails, which are not insects, also chew plant parts such as lettuce leaves and strawberry fruits.

A few insects have rasping-sucking mouth parts. These insects bite the plant, causing a break in the surface, then suck the exposed juices. Since such insects do not actually eat the plant, stomach poisons are not effective. Contact poisons are used instead. Thrips, which are small insects, have rasping-sucking mouth parts. They are very destructive of carnation and rose flowers as well as gladiolus flower spikes.

A large number of insects and insectlike pests have piercing-sucking mouth parts. In order to get food they make a microscopic hole in the plant with a sharp, needlelike mouth part called a *stylet* and then suck the sap out of the plant. Insects with piercing-sucking mouth parts include aphids, whiteflies, and leaf bugs. Insectlike pests with piercing-sucking mouth parts include spider mites and nematodes. Aphids and whiteflies suck sap from garden plants, such as tomatoes and most plants grown in the greenhouse. Contact poisons or fumigants are used on insects with piercing-sucking mouth parts.

Problem Pests in Horticulture

Although thousands of pests cause problems in horticulture, a handful of fairly common insects and insectlike pests are particularly harmful in nearly all areas of horticulture. These include aphids, beetles, cutworms, caterpillars, grasshoppers, leaf bugs, scale insects, mealy bugs, whiteflies, thrips, and the fungus gnat. Slugs, spider mites, and nematodes, while not scientifically classified as insects, are also bothersome plant pests. Remember the problem Dale had at the beginning of the chapter? He was assigned the job of identifying the small white flying insect. As we discuss the common horticultural insects, see whether you can identify the insect.

Aphids

Aphids are small, soft-bodied insects about the size of a pinhead. The male aphids have four wings, while the female aphids often have none. Since female aphids greatly outnumber male aphids, most aphids that are observed do not have wings. Aphids can be greenish, yellowish, grayish, pink, or other colors. Root aphids are grayish-white.

Aphids are very common pests with ornamental plants and vegetable crops. They have piercing-sucking mouth parts, which are used to suck juices from plants. The attacked plants have reduced vigor and appear sickly. Aphids secrete a sticky substance called honeydew. One reason why aphids are difficult to control is because ants use this honeydew as food and care for aphids as if they were a dairy herd. Certain species of aphids lay eggs, which are collected by ants in the fall. During the winter the eggs are tenderly cared for by the ants. When spring arrives the ants carry the eggs above ground to stems and leaves where the eggs hatch. Ants will even move their aphid "herds" from one plant to another if the food runs out. A black, sooty mold often grows on the plant where honeydew is secreted.

Aphids are difficult to control for a number of reasons. They reproduce at a very rapid rate. The female aphid normally gives birth to living young. In 6 to 7 days the young are mature and capable of reproduction.

Generally, contact chemicals are used to control aphids. In the greenhouse, fumigants are sometimes used for this purpose. Ladybugs are sometimes used to control aphids in gardens, because they eat aphids.

Beetles

A number of beetles are very bothersome to horticulturists. Mature beetles cause damage by eating the foliage of plants. The larvae of beetles eat plant roots, or bore into the stem of plants. Both the mature beetle and larvae of the beetle are destructive and need to be controlled.

Most beetles have two sets of wings. The forewings are generally hard, and cover the thinner back wings when the beetle is at rest. Beetles have chewing mouth parts. They are

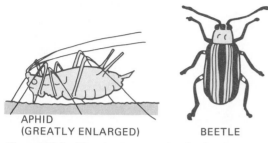

APHID
(GREATLY ENLARGED)   BEETLE

Figure 12-9. Two bothersome horticultural pests—the aphid (left) and the beetle (right).

found in a variety of colors and may be spotted or striped.

Some of the more bothersome beetles are the Japanese beetle, the blister beetle, and a variety of leaf beetles such as the Colorado potato beetle and the cucumber beetle. These beetles can destroy horticultural crops by eating the leaves, stems, flowers, and fruits. The rose chafer, a snout beetle, injects a poison into stems just below a flower and halts development of flowers on roses, chrysanthemums, and marigolds. White grubs are the larvae of May or June beetles. Wireworms are the larvae of click beetles. The larvae feed on roots of plants, which results in stunted growth or the death of plants.

Since beetles and their larvae have chewing mouth parts, stomach poisons are used to control them. Larvae in the soil are controlled with chemicals applied as a soil drench. The adult beetles are usually controlled by means of stomach poisons that are applied to the foliage. Japanese beetle grubs are killed by a fungus growth called milky spore disease. It takes about 3 to 5 years for this fungus to fully occupy a lawn, but it gives permanent control.

Cutworms

Cutworms have a slick, smooth skin (not fuzzy or hairy) and are worm shaped. They are normally 1 to 3 in [2.5 to 7.5 cm] long. Cutworms attach to vegetables, lawns, flowers, trees, and

shrubs during nighttime hours. The "surface cutworm" eats the plant near the surface of the ground. It either cuts the plant down or weakens the stems, causing the plant to fall over. "Climbing cutworms" climb the stems of plants and feed on bulbs, leaves, fruits, and vegetables. The "underground cutworm" remains in the soil and eats underground portions of the stem and the roots.

There are several dozen species of cutworms. The most destructive are the variegated cutworm, spotted cutworm, army worm, glass cutworm, and bronzed cutworm. Cutworms occur in a variety of colors and color combinations.

The cutworm is the larval stage of certain species of moths. Cutworms have chewing mouth parts and are thus susceptible to stomach poisons. They are also subject to attacks by other insects. Chemical control of cutworms is often used commercially.

Caterpillars

Caterpillars form a large group of plant pests, and are similar to the cutworms except that they have fuzzy or hairy bodies. Caterpillars are the larval stage of moths and butterflies. Caterpillars

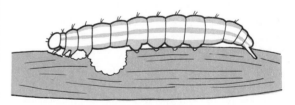

CATERPILLAR

CUTWORM

Figure 12-10. Two worm-like pests—the cutworm (bottom) and the caterpillar (top).

have chewing mouth parts and feed on the foliage of fruit trees, flowers, and vegetables. The gypsy moth stage has caused serious losses in timber trees, and there is currently no effective control for this pest. Dusting or spraying plants with a stomach or contact poison will control many kinds of caterpillars.

Grasshoppers

Since the beginning of history, grasshoppers have probably been the most destructive insect known to humans. Grasshoppers, at times called "locusts," can damage any type of crop, including such horticultural crops as vegetables, flowers, and ornamentals.

Grasshoppers are identified by a pair of large powerful back legs. They also have wings and a long, slender body. They are capable of hopping and flying. There are many species of grasshoppers that are harmful in agriculture and horticulture.

Grasshoppers have chewing mouth parts and eat on all parts of plants. A swarm of grasshoppers can completely devour a crop. There is a famous seagull monument in Salt Lake City, Utah, which commemorates an occasion when a large flock of seagulls consumed a swarm of grasshoppers that seriously threatened crops in that vicinity. Poisoned baits in which stomach poisons, bran, and water are mixed have been used to control grasshoppers. Certain stomach poisons and contact poisons in spray form may be used successfully on plants infested with grasshoppers. A number of mechanical devices such as the hopper catcher have been used to greatly reduce populations of grasshoppers in pastures and hayfields.

Leaf Bugs

One group of insects are known as "true bugs" and include a variety of leaf bugs, squash bugs, and chinch bugs. These bugs are closely related; if they were human we would call them "cousins" to each other. These insects are very

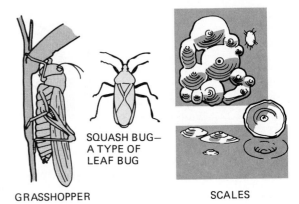

SQUASH BUG—
A TYPE OF
LEAF BUG

GRASSHOPPER

SCALES

Figure 12-11. Three horticultural pests: the grasshopper (top), the squash bug (middle), and scales (bottom).

similar in size and shape. They are three to four times longer than wide, have two pairs of wings, and have antennae. The prevailing colors for these bugs are green, black, and red. They are often flecked, spotted, or striped.

This group of bugs sucks sap from leaves, vines, and stems of plants. Affected plants will look unhealthy, turn yellow, and wilt. A variety of vegetable crops and ornamentals are affected. Chinch bugs can severely damage a lawn in a few days.

A number of control methods can be used. Cultural control should be used because these bugs hibernate under the protection of dead vines, leaves, and trash. The squash bug has been battled mechanically by small pieces of boards placed in the garden. At night, squash bugs like to gather under the boards. In the morning the gardener destroys the bugs by stepping on them. This procedure is not used commercially but contact poisons are used commercially with this group of bugs.

Scale Insects

Scales are small insects that secrete a waxy material that forms a shell over them. For a few days after they hatch, they move about on plants and have no protective cover. At this stage they are more easily controlled with spray materials, but the timing of application is very important. The scales are often circular in shape and resemble the back of a turtle. Scales are found primarily on fruit trees, but can be found on greenhouse plants and many woody ornamentals as well.

Scales damage the plant by sucking sap. Infected plants will grow poorly, turn yellow, and eventually die if the infestation is severe. Only certain chemicals are effective when sprayed on plants to control these insects. A very thorough spraying job is needed, since the shells tend to protect the insects. Special penetrating materials are sometimes added to improve control. In Northern states, dormant oil sprays, which are used while the plants are dormant (or leaves will be severely burned), control scales on fruit trees and evergreens. Biological control has also been used on certain types of scales. One example is the importation of the Australian lady beetle to kill the cottony-cushion scale, which was a serious pest of citrus in California.

Mealybugs

Mealybugs are small insects that have small ridges on their backs and very small white spines radiating out from their bodies. They have a thick, waxy coating that gives them a bright white appearance. Only certain pesticide formulations can penetrate this coating. Mealybugs are particularly bothersome in citrus groves and on ornamental and greenhouse plants. Often mealybugs will mass together and appear to be very slow-moving white blotches on plants. Mealybugs suck sap from leaves and fruits, which causes the plant to grow slowly. On mature plants, such as fruit trees, fruit production is severely reduced. Biological control methods have been very successful in citrus orchards. Coccinellid beetles are released in orchards to feed on mealybugs. In greenhouses, fumigants and contact poisons can be used successfully.

Whiteflies

The whitefly is a small, snow-white, four-winged fly. There are several species of whiteflies that attack citrus trees, vegetables, flowers, and greenhouse plants. Whiteflies are not readily visible on plants, since the early stages are green and have no wings. They feed on the underside of the leaves. To determine whether whiteflies are a problem, shake several plants. Flying little white specks indicate a whitefly problem. This was the insect that Dale was asked to identify.

Whiteflies suck sap from plants, which causes plants to grow poorly, wilt, turn yellow, and possibly die. Whiteflies go through gradual metamorphosis. The nymph is wingless and has a greenish tint. Both contact poisons and fumigants are used to control whiteflies. Until recently they were very difficult to control in the nymphal stage. Spores of a certain fungus are now sprayed on the plants. This fungus attacks the whiteflies, giving excellent biological control.

Since the whiteflies were in an enclosed area (the greenhouse), Dale decided that a fumigant might be a good way to control the whiteflies.

Thrips

Thrips are minute, slender insects that normally have four very slender fringed wings. They are often difficult to see, simply because they can move very quickly. Thrips have a rasping-sucking type of mouth. They "chew" on leaves and stems to break the surface, and then suck up the plant juices. As symptoms of thrip damage, leaves become whitened and flecked in appearance. The undersides of leaves will be spotted with small black spots. Tips of leaves may eventually wither, curl up, and die. Thrips are especially bothersome in the greenhouse and on flowers, making carnations and gladioli unsaleable. They also attack vegetables and fruit crops. Thrips undergo gradual metamorphosis and can be controlled with contact poisons or fumigants.

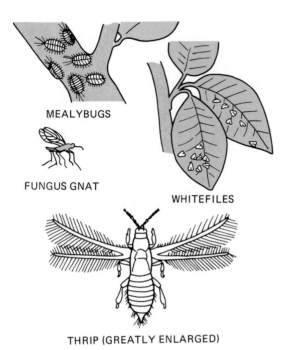

MEALYBUGS

FUNGUS GNAT

WHITEFILES

THRIP (GREATLY ENLARGED)

Figure 12-12. Four additional horticultural pests are: mealybugs, white flies, thrip and fungus gnats.

Fungus Gnats

Several kinds of small, long-legged flies with dark gray or black bodies are known as fungus gnats. The name comes from the fact that these gnats often lay their eggs in fungi or other dark, damp places (such as moist soil). Within a few days, the eggs hatch into larvae that have nearly transparent bodies with dark heads. The larvae feed on the roots of many greenhouse plants.

Plants being attacked by the fungus gnat larva will grow poorly and leaves may turn yellow. No visible signs of insect damage are available, because the damage is done to the roots.

Even if there are so few gnats that no damage is evident to the plant, customers react unfavorably when the plant they purchased releases little black gnats into their home.

To determine whether fungus gnat larvae are present, you should remove the soil from around the plant roots and examine it. Small brown scars on the roots or eaten root hairs indicate that the larvae have been feeding on the plant. The larva changes into the pupa stage and finally emerges as an adult fungus gnat. Several chemicals, applied as soil drenches, are used to kill the larva. Steaming of soil prior to planting plants will destroy the fungus gnat eggs and prevent infestation.

Insectlike Pests

Horticultural crops are often damaged or destroyed by three other pests that are not scientifically classified as insects: spider mites, slugs, and nematodes.

Spider Mites. Spider mites are extremely small (nearly microscopic) members of the spider family, and are by far the most troublesome pest of horticultural crops. They have eight legs instead of the six legs that insects have. Spider mites are generally greenish, reddish, or white, with two distinct dark spots on the abdomen. They appear as small specks on plants. If spider mites are suspected, place a piece of white paper under a stem of the plant and tap the stem. If present, some spider mites will drop onto the paper, and look like tiny, moving specks of black pepper. A hand lens will make them clearly visible. Spider mites suck sap from leaves. Leaves infested with spider mites become blotched with pale yellow or reddish-brown spots, and the plant will stop growing, develop a sickly pale appearance, and gradually die. Often the degree of stunted growth seems all out of proportion to the few numbers of spider mites present. They appear to inject a substance that causes the plant to stop growing. Vegetables, fruit trees, ornamentals, and greenhouse plants are injured by spider mites. Both cultural and chemical control can be used with spider mites. They often spend the winter in weeds, so keeping the area clear of weeds will reduce the number of spider mites. Contact insecticides and fumigants are also used on spider mites. Spider mites have the ability to become resistant to a control material that is overused. After a period of several years, some growers rotate control materials in order to avoid this problem. They also try new materials as they are introduced in hopes of gaining complete control.

Slugs. Slugs are slimy, legless, soft-bodied creatures that feed on a variety of horticultural plants. They are really snails without shells. True snails are controlled in the same manner as slugs if they become a problem on a crop. Symptoms of slug or snail damage include leaves or fruits that have been partially eaten. Often a shiny, varnishlike film will be visible on the plant. Slugs are from ½ to 4 in [1.3 to 10 cm] long, and are a grayish to tannish color. They feed on plant foliage and are found in damp areas, such as under boards and trash. Therefore, the greenhouse or garden area should be kept clean. Poisoned baits and stomach poisons have been used to control slugs, but they are sometimes difficult to control with chemicals.

Nematodes. Nematodes are microscopic wormlike pests. The root-knot nematode burrows into the roots of plants and causes the plant roots to swell. Roots of the affected plants appear knotty and swollen, and look like small, misformed sweet potatoes. The only above-ground visible symptom of nematodes is an unhealthy plant. The root-knot nematode is one of the most troublesome kinds. Nematodes attack all kinds of horticultural plants. Many kinds suck juices from plant roots without actually going inside the roots. Preventative measures, such as steaming or fumigating soil prior to planting, give satisfactory control. Several chemicals can be applied as a drench to the soil to control nematodes.

Problem Animals in Horticulture

Animals and birds can damage horticultural crops. Can you name any animals that are plant

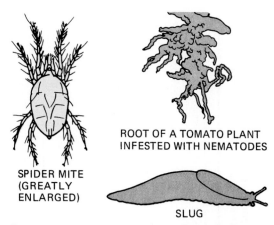

ROOT OF A TOMATO PLANT
INFESTED WITH NEMATODES

SPIDER MITE
(GREATLY
ENLARGED)

SLUG

Figure 12-13. These horticultural pests are not insects: spider mites, slugs, and nematodes.

pests? The more common ones are rabbits, mice, groundhogs, chipmunks, moles, and deer.

Rabbits eat lettuce, cabbage, beans, peas, and of course carrots. They also eat grass, berries, and the bark from fruit trees. Rabbits can be kept out of gardens by means of a rabbit-proof fence.

Mice eat fruits, nuts, vegetables, and bark from young trees. They can be controlled by mechanical traps or poisoned baits. Keeping the area clean will also help control mice.

Chipmunks and groundhogs have been known to like sweet corn, beans, and other vegetables. The best method of controlling these pests is to locate their burrows and place a fumigant gas in the burrow.

Moles generally do not feed on horticultural plants but cause damage to gardens and lawns by digging burrows near the surface of the ground. Moles are carniverous, and eat both grubs and earthworms. The burrowing creates ugly ridges in the lawn or garden. The roots of plants may be cut off or dried out, because they are not in contact with the soil. Treating the soil to eliminate grubs and earthworms removes the food for moles, and is the most effective control method for moles.

Deer can cause serious damage to fruit trees and ornamentals by chewing the bark and buds.

Pests In Horticulture—A Summary

Pest	Description	Type of Metamorphosis	Metamorphic stage in which pest is most destructive	Type of Mouth part	Horticultural Crop Affected	Type of Damage
Aphids	Small soft bodied insects. About the size of a pinhead. May have four wings or none. Color varies. Most common colors are greenish, yellowish, pink, and gray.	No metamorphosis	Young and Adult	Piercing-Sucking	Ornamentals Vegetables Greenhouse plants	Most species suck juices from plants and deposit honeydew on the plant which supports a black sooty mold.
Beetles	Relatively large bodies. A hard fore wing rests over smaller back wings. Found in a variety of colors and color combinations.	Complete	Larva and Adult	Chewing	Lawns Vegetables Ornamentals Orchards	Eats foliage and roots
Cutworms	Worm like insects with smooth skin. Occur in a wide variety of colors and color combinations. Active at night.	Complete	Cutworms are the larval stage of certain moths.	Chewing	Vegetables Flowers Lawns Orchards Shrubs	Cuts down plants, eats buds, leaves, fruits, roots, vegetables.
Caterpillars	Fuzzy or hairy worm like insects. About 1 to 3 inches long	Complete	Caterpillars are the larval stage of certain moths and butterflies.	Chewing	Flowers Vegetables Orchards	Eats foliage
Grasshoppers	Winged insects with long slender bodies about 1 to 2 inches long. Large powerful back legs.	Incomplete	Nymph and Adult	Chewing	All horticultural crops	Eats foliage, stems, fruit, vegetables.
Leaf Bugs Squash Bug Chinch Bug	Insects with 2 pairs of wings, antenae, about 3 or 4 times longer than wide. Found in a variety of colors and combinations.	Incomplete	Nymph and Adult	Piercing-Sucking	Vegetables Ornamentals Lawns	Suck sap from leaves, vines, stems.
Scales	Small insects with flat, roundish hard waxy shells.	Incomplete	Nymph and Adult	Piercing-Sucking	Fruit trees Greenhouse crops Lawns	Sucks sap from plants.

Figure 12-14. A summary of pests.

Pests In Horticulture—A Summary

Pest	Description	Type of Metamorphosis	Metamorphic stage in which pest is most destructive	Type of Mouth part	Horticultural Crop Affected	Type of Damage
Mealybugs	Small white wingless insects with small spines radiating out from the body.	Incomplete	Nymph and Adult	Piercing-Sucking	Citrus trees Ornamentals Greenhouse plants	Sucks sap from plants
Whiteflies	Small snow-white, four winged flies.	Incomplete	Nymph and Adult	Piercing-Sucking	Ornamentals Greenhouse plants Vegetables Citrus	Sucks sap from plants
Thrips	Minute slender fly-type insect which normally has four slender fringed wings.	Incomplete	Nymph and Adult	Rasping-Sucking	Greenhouse Crops Flowers Vegetables Fruits	Sucks sap from plants
Fungus gnats	Small long-legged flies with dark gray or black bodies	Complete	Larva	Larva-Chewing	Most greenhouse plants	Feed on plant roots
Spider mites	Very small spiders, greenish, reddish, or white with two dark spots on the abdomen.	No metamorphosis	Young and Adult	Piercing-Sucking	Most greenhouse plants	Sucks sap from plant and injects a harmful substance into the plant.
Snails and Slugs	Snails—small creatures with a curving shell Slugs—slimy, legless soft body creatures one-half to four inches long, gray or tan color.	No metamorphosis	Young and Adult	Chewing	Ornamentals and vegetables	Eats foliage, flowers and produce
Nematodes	Thin microscopic worm-like pests.	No metamorphosis	Adult	Piercing-Sucking	Greenhouse crops Flowers	Eats from the inside of plants or sucks juices from outside the plant
Animals and Birds	——	——	Adults	Chewing	Vegetables Ornamentals Flowers	Eat foliage, roots, produce

Figure 12-14 (cont.)

In winters with deep snows the problem is most severe. Certain repellant spray materials are effective preventatives.

Generally, birds are beneficial because they eat insects. However, at certain times of the year birds may eat grapes, berries, and other types of fruit. They sometimes cause considerable losses in such field crops as corn, wheat, and oats. Several methods of controlling birds have been tried with limited success. Nets, cannons that make a loud boom, and recordings of "distress" bird calls have been played in crop areas. Research is being conducted to find effective and economical ways to control birds.

Weeds

In horticulture, weeds are a pest. Weeds compete with crops for moisture, light, and nutrients. If weeds are not controlled, they can outgrow the desirable crop in some cases and eventually kill it. Weeds also provide a home for insects and other pests.

Weeds can be controlled in several ways. Chemicals, called herbicides, are used to kill weeds. One problem is that herbicides can also kill plants. Caution must be used in selecting a herbicide. More details about herbicides are given in Chapter 14.

Weeds are controlled by mowing, plowing, hoeing, and pulling the weed by hand. When pulling or hoeing is used, care should be taken to remove the roots along with the top of the weed. Mowing only cuts the top of the weed off, but through regular mowing the weeds are prevented from maturing and producing seeds.

Pests in Horticulture: A Review

Controlling pests is important in horticulture. Greenhouse workers, nursery operators, gardeners, fruit producers, golf course greenskeepers, commercial flower growers, and landscape employees all need to be able to recognize plant disorders and determine the cause of the disorder. If pests are the trouble, the horticulturist will be expected to identify the problem and control it.

In horticulture, pests are responsible for decreased yields, reduced quality, spread of diseases, poor growth, and even death of plants. Insects, insectlike creatures, snails and slugs, animals (including birds), and weeds are often pests.

Aphids, beetles, cutworms, caterpillars, thrips, grasshoppers, scales, leaf bugs, mealybugs, and whiteflies are particularly destructive insects in horticulture. These insects have either complete or gradual metamorphosis. Their type of mouth parts can be classified as chewing, rasping-sucking, or piercing-sucking. Insects and pests can be controlled by quarantine, or by mechanical, cultural, biological, or chemical means.

THINKING IT THROUGH

1. What is a pest?
2. How are pests harmful in horticulture?
3. What procedures should be followed to determine whether pests are causing a plant to be unhealthy?
4. Name five general methods for controlling pests, and give an example of each.
5. What is the difference between complete and incomplete metamorphosis?
6. How does the type of mouth part a pest have influence the type of insecticide that is used?
7. Which pests are problems in fruit and vegetable production?
8. Which pests are found in greenhouses?
9. What pests cause damage to lawns and ornamentals?
10. List six horticultural pests that are not insects. What type of damage does each pest do?

13

Identifying and Controlling Plant Diseases

The week of vacation with the family had been fun, but Chris was looking forward to arriving home and getting ready for the annual all-city floral show. She was going to exhibit roses in the big show. Earlier in the spring, Chris had selected and planted three All-American rosebushes in the backyard. The roses were part of her supervised occupational experience program in horticulture. Many hours of tender care had been invested in the rosebushes.

As the family car pulled into the driveway, Chris jumped out and ran to the backyard. Chris was shocked at the sight of the rosebushes. The leaves on each rosebush were covered with black spots. Chris didn't know what the problem was. Earlier, the roses had been in good health. After the initial shock had worn off, Chris sat down to figure out what the problem was.

No signs of insects were

found on the rosebushes. Chris had been watering, fertilizing, and caring for the roses properly. The record book for the roses showed nothing unusual. Chris asked the neighbors on each side whether they had used any chemicals lately, but neither of the neighbors had. After considering all the facts, Chris decided that the roses had a disease. Do you agree?

If you agreed, you are correct. Do you know the name of the disease? Do you know what causes plant diseases, or how plant diseases are spread? In this chapter, you will learn how to identify and control common horticultural diseases.

The identification and control of plant diseases is important in the horticulture industry. In the United States, the loss caused by plant diseases in horticultural crops amounts to over a billion dollars annually. Besides causing monetary losses, diseases reduce yields and detract from the appearance of plants. Fortunately, many horticultural diseases can be prevented and controlled.

Greenhouse growers must constantly be alert for diseases that could spread rapidly through the greenhouse. Arboriculturists treat diseases and other disorders of trees. Groundskeepers on golf courses guard against turf diseases. Fruit and vegetable growers are continually protecting their produce against disease. Some firms specialize in providing disease-and-pest-control services for horticulture businesses. Employees in all areas of horticulture need to be able to identify and control diseases.

CHAPTER GOALS

Your goals in this chapter are:

- To identify the symptoms of 10 common horticultural diseases
- To describe the causes of diseases in horticultural plants
- To determine how diseases in horticultural plants are spread
- To recommend methods of controlling disease in horticultural plants

Causes of Diseases

After deciding that the roses had a disease, Chris phoned Ms. Franklin, the horticulture teacher, and asked her to come over. Ms. Franklin promised to drop by first thing in the morning.

"I'm sure glad to see you," stated Chris when Ms. Franklin arrived the next morning. "I think my roses have some type of disease." After examining the roses, Ms. Franklin said, "It looks like your roses have a disease known as black spot—that's a fairly common disease with

roses." Chris thought about this for a moment before asking, "What caused the disease?" Do you know what caused the disease?

The causes of most plant diseases can be

Figure 13-1. Disease control is an important task performed by horticulturists.

placed into two groups: *improper growing environment* and *microorganisms*. (Microorganisms are tiny animals or plants that can be seen only with a microscope.)

Improper Growing Environment

Horticultural plants need moisture, light, air, nutrients, and the proper temperature for growth. If these environmental factors are out of balance, plants will not grow properly. Some plant diseases (or disorders) are caused by too much or too little of these environmental factors. These were discussed in Chapter 11.

Microorganisms

Three types of microorganisms are responsible for most diseases of horticultural plants: bacteria, viruses, and fungi. Each type of organism has distinct characteristics. You will be better able to control diseases if you are aware of the characteristics of each microorganism.

Bacteria. *Bacteria* are small, one-celled living things. Scientists classify them as plants, but they differ from most plants because they contain only one cell and cannot produce their own food. (This is because they do not contain chlorophyll.)

Bacteria can be seen only through a microscope. Twenty thousand (20,000) bacteria placed end to end would measure about 1 inch (in) [2.5 cm (centimeters)].

There are thousands of different types of bacteria. Many are beneficial to humans; some kinds aid in the production of nitrogen in plants, in waste disposal, in the making of cheese and butter, and in tanning leather. Over 900 types of bacteria are found in the stomach (rumen) of a cow. They help the cow digest feed. Only about 1 percent of all bacteria are harmful. However, harmful bacteria cause a number of diseases in both humans and plants.

Bacteria multiply by dividing in half. Under ideal conditions, some bacteria can divide every

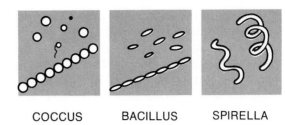

COCCUS   BACILLUS   SPIRELLA

Figure 13-2. Bacteria are responsible for many horticultural diseases.

20 to 30 minutes. Once a bacterial disease gets started it may be difficult to control, since bacteria reproduce so quickly.

Bacteria can enter plants only through natural openings (such as stomata) or through wounds. Once inside the plant, bacteria multiply rapidly, destroy cell walls, and move throughout the plant.

If you were to look at bacteria under a microscope you would discover that they come in three shapes—round (called *coccus*), rod-shaped or oval (called *bacillus*), and spiral-shaped (called *spirella*). Some bacteria have little hairlike attachments called *flagella* which are used for moving. Rod-shaped bacteria are responsible for most of the bacterial horticultural diseases. Horticultural diseases caused by bacteria include rots, leaf spots, and blights.

Viruses. The second group of microorganisms that cause plant diseases are known as *viruses*. Scientists do not fully understand viruses. They are extremely small; about 250,000 viruses might measure 1 in [2.5 cm]. They are so small they cannot be seen through ordinary microscopes, so electron microscopes are used to study viruses.

Viruses reproduce only when they are in the cells of plants or animals. When isolated in a laboratory, viruses do not move or reproduce. Generally, viruses do not kill plants but do reduce production and affect the appearance. Two fairly common diseases caused by viruses are mosaics and yellows. Asters often contract a virus disease known as "aster yellows," which causes asters to be stunted and yellow.

Fungi. The great majority of plant diseases are caused by fungi. *Fungi* are classified as plants. They are similar to bacteria in the fact they contain no chlorophyll and must obtain their energy from some other source, often horticultural plants. Fungi differ from bacteria in two respects. Often fungi will contain more than one cell, and some fungi are large enough to be seen without a microscope. Molds, mildews, and even mushrooms are all fungi.

Fungi reproduce sexually and asexually. Many fungi reproduce through spores. Spores are microscopic, one-celled bodies. They are similar to the seeds produced by a plant. Spores are extremely small and lightweight. They are easily spread by wind, water, insects, humans, equipment, and animals. Spores may lie inactive in soil for a number of years before they start growing. When conditions are right the spores will start growing and develop into fungi. Generally, fungi and spores develop more rapidly in moist climates.

A number of common horticultural diseases are caused by fungi; all rusts, mildews, and smuts for example. Many of the rots, blights, and spots are also fungus diseases. The black spot on Chris's roses was caused by a fungus.

Transmission, Prevention, and Control of Disease

After learning that black spot disease of roses was caused by a fungus, Chris was curious as to how the disease could be controlled. How do you think the disease can be prevented or controlled? If one understands how bacteria, viruses, and fungi are transported, then methods for controlling disease can be planned and implemented. Disease prevention should be incorporated into a management plan.

Insects

The spread of disease is aided greatly by insects. Insects with piercing-sucking mouth parts make holes in the plant, which allows bacteria, viruses, and fungi to enter the plant. Microorganisms may be carried from one plant to another by "hitchhiking" on the body of insects. As insects feed, moving from plant to plant, they carry the disease organism with them. Some viruses even enter the body of insects and move between plants inside the insect.

Aphids, thrips, and leafhoppers are known for transporting virus diseases. The Dutch elm disease, a fungus, is transported by bark beetles. Bees and flies are guilty of carrying the bacteria that causes fire blight in apple and pear trees.

Since a number of diseases are spread by insects, one method of controlling disease would be to control insects. While controlling insects will not completely eliminate diseases, it will help reduce disease. Methods of controlling insects are discussed in Chapter 12.

Soil

Many bacteria and fungi live in the soil. Root rot and damping off are two diseases caused by microorganisms that live in the soil.

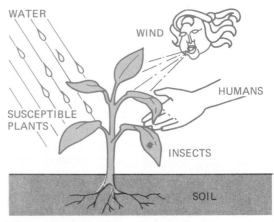

Figure 13-3. Diseases can be spread in a variety of ways.

Have you ever planted seeds that did not grow? If you dug them up you might have found that they had rotted. This was caused by disease organisms in the soil.

Soil pasteurization can be used to help control soil-borne diseases. The soil used to grow plants in greenhouses should be sterilized before it is used. This will kill many of the bacteria and fungi in the soil that cause disease.

Seeds are often coated with fungicides before they are planted. A *fungicide* is a chemical that protects against damage from fungi. Seeds that are coated with a fungicide are often pink or purple in color. These seeds should never be eaten by humans or fed to animals.

Wind

The spores produced by fungi are very lightweight and may be carried many miles by the wind. Some spores have been known to blow over 1,000 miles. They then land on a plant and develop into fungi. A number of the rust diseases are spread through the wind.

Diseases spread through the wind are difficult to control. A number of chemicals are available and can be sprayed on plants. When fungi spores are deposited on the plant by the wind, the chemical kills the spore.

Humans

Humans are responsible for transporting a number of diseases. Often they do so without realizing what they are doing. For example, disease may be spread by pruning. After being used to prune the limbs of a tree or shrub that is infested with a disease, the same pruning shear may be used on nondiseased trees or shrubs. The bacteria causing the disease are carried on the pruning shears to noninfested plants. Diseases may also be spread through grafting or taking plant cuttings with the same knife.

Humans may spread disease while performing common tasks such as watering or hoeing. They or their equipment may brush up against a diseased plant and thereby come into contact with the disease organism. The disease is then carried by the person or on the equipment. When the person or equipment comes into contact with nondiseased plants, the organism may be transferred to that plant. Disease organisms may even be carried on soil that sticks to the bottom of people's shoes.

Common sense and basic sanitation are needed to prevent the spread of disease by humans. Disinfecting equipment after working with diseased plants is a good practice.

After handling diseased plants, you should wash your hands well before handling other plants. You should also minimize the amount of physical contact you have with a diseased plant, since disease can be carried on clothing.

Water

Water is responsible for spreading disease in a variety of ways. Bacteria are often transported by water. Many bacteria can "swim" or move about in water. Therefore, bacteria may move from plant to plant if water is in contact with the plants. The splashing of water during rain or during watering can splash bacteria up onto a plant.

Fungi generally grow best under moist conditions. They often need moisture to release spores. These spores need moisture in order to germinate. You have probably noticed more mold and mildew in damp areas than in dry. Fungus caused diseases are not as bad during a dry growing season as in a wet growing season.

Careful watering of plants will help to reduce the spread of disease. Avoid splashing water and soil onto plants. Plants should not sit in pools of water for long periods of time.

Variety Selection

Certain varieties of plants catch diseases more easily than others. In tomatoes, the Better Boy resists fusarium wilt (a bacterial disease), while Marglobes and Ponderosas are susceptible. Planting resistant varieties will help prevent the spread of disease. However, note that some plants that have excellent disease resistance may be subject to other problems, such as lack of winter hardiness or low yields.

Quarantines

Just as there are quarantines to prevent the spread of insects from one area to another, there are quarantines to prevent certain plants from being moved from a diseased area to a nondiseased area. Citrus nursery stock from India or China cannot be imported into California, Florida, South Texas, or Arizona because it may carry a disease called citrus canker.

Sanitation

General sanitation procedures will help reduce the spread of disease. Burning plants that are diseased will destroy the disease-causing organism. Keeping the area clean of weeds and debris will help, since microorganisms may live in trashy areas when they are not attacking horticultural plants. Removing and burning diseased limbs from trees is a sound sanitation practice.

Common Disease of Horticultural Crops

After Ms. Franklin explained how diseases were transmitted, Chris wanted to know whether there were common diseases that could be expected to attack plants. Ms. Franklin explained that there were hundreds of horticultural diseases, but many of them were closely related and had similar symptoms. These diseases were often grouped together and discussed as one disease. "Let's walk around the block and I'll try to point out some of these common diseases," said Ms. Franklin, "and let's start here with your roses."

Leaf Spots

Many diseases are characterized by spots on the leaves of plants, which are generally caused by bacteria or fungi. They are more common in rainy seasons and when the humidity is high. The spots may be small or large circles. Often the outer edge of the spot may be a different shade from the inside of the spot. In some cases, the inner part of the circle may fall out and cause small holes. You can tell that this is not insect damage, because each hole will be surrounded by a spot that is a different color from the leaf.

Black spot of rose is one common spot disease. Dark brown to tan spots on the upper half of iris leaves is known as iris leaf spot. If the same spots were on gladiolus, the disease would be called gladiolus leaf and flower spot. Spots on chrysanthemums that change from yellow to black are symptoms of chrysanthemum leaf spot. Round gray spots, surrounded by a yellow ring, on watermelons, cucumbers, and squash is characteristic of cucurbit septoria leaf spot. Some of the diseases that have dark brown or black spots are Boston ivy leaf spot, English ivy bacterial leaf spot, delphinium black spot, and maple tar spot. The plant that is damaged is included in the name of the disease.

Leaf spot diseases are even found on grasses. The spots may be white, tan, or brown. Small white spots in lawns and on golf courses are characteristic of a disease know as dollar spot. Turf workers and golf course groundskeepers must be able to identify and treat dollar spot disease.

Control of leaf spot diseases is aided by several practices. Removing infected leaves from plants and burning them helps to destroy the disease-producing bacteria or fungi. Several sprays are available to help protect plants from fungi and bacteria. These sprays normally contain sulfur or copper. Ask your vocational horticulture teachers or county agricultural agent for local recommendations.

Blights

After talking about the spot diseases, Ms. Franklin and Chris started down the street. They stopped at Mr. Camp's house. Mr. Camp had several fruit trees beside his house. Fruit trees, along with a number of other horticultural plants, may have blight diseases. There is no one single characteristic that is common to all blight diseases. *Blights* normally result in rapid and noticeable detoriation of the plant. Young growing tissue such as shoots, leaves, and twigs are killed. Leaves may develop spots that rapidly expand into large patches that cover most of the leaves. Blossoms and twigs may die.

Fire blight is particularly harmful to apple and pear trees. It first affects the blossoms, which wilt rapidly and turn a dark brown. The leaves then turn brown or black. The leaves and twigs look as if they had been burned in a fire. Only certain portions of the tree may be affected. Cutting out diseased portions of the tree and burning them is one method of controlling fire blight. Selecting tree varieties that show resistance to fire blight is another way to control the disease. Orchard owners and employees must be able to identify blight and treat the problem.

Ornamentals are susceptible to several types of blight. In humid areas of the South, azaleas may lose all their flowers to the azalea petal blight. Lilies may contract the lily botrytis blight. Orange or red spots develop on the lower leaves and spread upward, turning black until the entire plant is affected. The lilac phytophthora blight causes the flowers and growing tips of lilacs to turn brown and die. Spraying the plants with a bordeaux mixture before plants become infected will prevent the blight. *Bordeaux* is a mixture of copper sulfate, lime, and water. You should ask your instructor for additional information about bordeaux.

Rots

Chris and Ms. Franklin continued down the street. The sight of cannas with irregular brown areas on the leaves, blackened stalks, and rotten flowers in Mrs. Johnson's yard caught Ms. Franklin's eye. The plants had canna bud rot.

A number of diseases called *rots* affect horticultural plants. Symptoms of the disease are death and decay of parts of the plants. The infected parts will turn brown or black and are often wet or soggy. As the name implies, the plant rots. Often a smelly odor is given off by plants with a rot. The rot may affect leaves, buds, stems, fruits, roots, rhizomes, and bulbs.

The disease may be hard to diagnose if the roots are infected, because the roots are not readily visible. If plants lose vigor and become sick or stunted, root rot may be the culprit. Examine the roots and if they are moldly or brown or gray, then root rot is the probable cause.

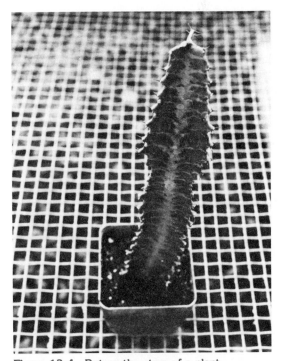

Figure 13-4. Rot on the stem of a plant.

Rots are caused by fungi and bacteria. Excess moisture contributes to the development of rots.

Narcissus bulbs may decay into a brownish blob because of narcissus basal rot. Stem rot affects many horticultural crops such as dahlias, mums, peonies, zinnias, and tomatoes. The stems of these plants will start rotting. The rot will look like a snowy-white mold with black spots the size of BB's in the mold. The stem rapidly turns black. Peaches and plums are affected by brown rot. Small round brown spots develop on the fruit and spread rapidly, covering the entire fruit. The fruit then shrivels up.

Removing rotten fruit or parts of the plant helps in the control of rots. Watering properly to avoid an excess of water will also aid in the control of rot.

Rusts

As Chris and Ms. Franklin came to the corner of the block, Ms. Franklin suggested that they stop at Terry's Flower Shop, which was across the street. Terry had graduated from the school's horticulture program several years earlier, and had recently purchased the flower shop. A greenhouse was connected to the flower shop.

After Ms. Franklin explained that she and Chris were looking at plant diseases, Terry suggested that they look at the carnations in the greenhouse, since they had rust. Chris thought, "Someone is trying to play a joke. Flowers don't rust." Do you think Terry was joking? If you said no, you are right.

Rust is a fungus disease found in a variety of horticultural plants. Normally, the disease is characterized by small spots on the leaves of plants. The spots are often yellow, orange, red, or brown, colors that often resemble rust. These spots may be on both the top and bottom of the leaf. A small projection shaped like a tube or a round projection about the size of a BB often develops on the leaves. These projections are called *pustules*. They release red to brownish spores into the air infecting other plants.

Rust diseases have been a problem for centuries. The Romans had a god called Robigus who was named after the disease. On April 25 of each year, the Romans conducted a big festival to please Robigus. They thought that if the god was pleased, their crops would not be damaged by rust.

Carnations and chrysanthemums are two greenhouse plants that are particularly bothered by rust. On carnations and chrysanthemums, spots that are light green or yellow develop into pustules, then break open to release thousands of brown spores. The plants become stunted and have leaves that become distorted.

Greenhouse operators and workers can perform several tasks that will aid in the control of rust. The best control is to maintain a well-ventilated greenhouse, take care not to splash water on the leaves, and remove and destroy infected plants.

Rusts attack a number of garden plants, including beans and asparagus. In asparagus, the leaves and stems have a reddish or brownish tint. When they are brushed against, a cloud of spores is released. Planting varieties of asparagus that are resistant to rust is the best control.

Some rusts spend their life on one plant, while other types of rust need two different plants in order to reproduce. The cedar-apple rust needs two different plants to survive—cedar (or juniper) and apple. The fungus develops on red cedars or junipers. Small round globes up to the size of ping-pong balls develop. These are called galls. They are green first, but change to a dark brown later.

Orange-colored tubes then protrude from each ball. Spores are released from the tube and carried through the air. If they land on an apple or crabapple tree, the spores will produce yellow spots on the leaves, which gradually turn reddish-orange. A black dot will be in the center of the spot. Small BB-sized pustules will appear on the underside of the leaves. In late summer, the pustules will release spores. The spores will be carried through the air and will reinfect junipers or red cedars. Here the spores will develop into the galls mentioned earlier.

This disease causes deformed or small apples. Little damage is done to the juniper or cedar. The best method of controlling this disease is to remove one of the hosts—either the cedar or the apple. A *host* is an organism (such as a plant) that provides a home for another organism (such as a disease). The galls also can be picked off the cedar. Some of the other rusts are: rose leaf rust, snapdragon rust, hawthorn rust, and gladiolus rust.

Smuts

After leaving Terry's Floral Shop, Chris and Ms. Franklin stopped at the Morales home. Mr. Morales always planted a patch of sweet corn in the backyard. The tips of several ears of corn were covered with a big black mass. The black mass was nearly fist sized and was fluffy like cotton candy. This is *smut.*

Smut is a fungus disease that often affects garden crops. In sweet corn, pale green blisters develop on the stalks, leaves, and ears. These blisters will enlarge until they burst open, releasing large numbers of black spores. The blisters will then turn black. Plants affected by smut are often misformed.

The spores can survive in old stalks or in the soil. They may lie inactive in the soil for 6 to 7 years before they develop. The best control is good sanitation. Parts of the corn plant that have the green blisters should be removed and burned. The old stalks that are infected should be burned after the harvest.

Other plants that are susceptible to smut are onions, Bermuda grass, bluegrass, carnations, delphiniums, dianthius, gladioli, and violets.

Mildew

As Chris and Ms. Franklin continued their walk around the block, Chris noticed some zinnias with a whitish, powdery look to the leaves.

Figure 13-5. Smut, a fungus disease, on an ear of corn.

"That," exclaimed Ms. Franklin, "looks like powdery mildew."

Mildews are a fungus disease of horticultural plants. The two main types are the powdery mildews and the downy mildews. A wide variety of plants such as apples, melons, grapes, lilacs, phlox, dahlias, roses, asters, and chrysanthemums are affected by mildew.

Powdery mildews grow on the surface of plants. They form white or grayish splotches that are velvety to the touch. They are found on leaves, stems, and buds. Powdery mildew develops best during periods of humid weather, warm days, and cold nights.

Downy mildews are generally white, gray, or purple. They usually develop in velvetlike patches on the undersides of leaves. They like wet weather.

Several chemical sprays are available to help in the control of mildews. Good ventilation also helps to control mildew.

Wilts

Chris noticed that several tomato plants in the garden were droopy. They looked as if they needed to be watered. Ms. Franklin explained that underwatering was one possibility, but there was also a disease called wilt that could be the problem.

Wilt disease is caused by fungi, bacteria, or viruses. Generally, the plants show no external damage such as spots or decay. The plant becomes wilted and there may be a change in the color of the leaves and stems. The wilting is caused by microorganisms in the water-conducting tissue (xylem) of the plant. These microorganisms produce wastes that help to clog up or block the cells. Water cannot be transported in the plant. This causes the plant to wilt. This disease is not easily controlled.

Carnation fusarium wilt is characterized by the plant's wilting, turning pale green or yellow, and becoming stunted. If you were to take a knife and slice into the carnation stem lengthwise, you would find brown streaks. This is caused by the disease microorganism. Soil pasteurization and destruction of diseased plants help to control the disease. Gladioli, strawberries, melons, celery, cabbage, and asters are several plants that are affected by wilt diseases.

Figure 13-6. Plants exhibiting symptoms of wilt.

Galls

Next, Chris and Ms. Franklin stopped at the Hamiltons'. Their home had several large oak trees in the front yard. One had large tumorlike growths on it. These are known as galls. *Galls* are found on a variety of plants. They are large, tumorlike growths that may attack roots, stems, or leaves. Some are caused by insects, but many are caused by bacteria and fungi.

Many fruit trees and roses are susceptible to crown gall. Big round swellings are found on the plant trunk or stem, near the ground. This disease is caused by bacteria in the soil. The best prevention of this disease is to plant varieties that are less susceptible to crown gall.

Azaleas, rhododendrons, and camellias are affected by leaf gall. A fluffy white or pink growth develops on the leaves. The best control is to remove infected leaves.

Cankers

One of the oak trees in the Hamiltons' yard had a dead area that was discolored and looked like an unhealed wound. This malady is a disease known as canker.

Canker is a fungus disease found primarily on woody plants. Cankers attack the stem and result in shriveled, misformed areas on the trunk or stem. Often, cankers crack open and expose the wood. This reduces the vigor of the plant and can result in the death of that area.

Cankers affect many horticultural crops, including apples, grapes, roses, gardenias, blueberries, azaleas, walnuts, and pyracantha. Certain plant varieties exhibit a resistance to cankers. The primary control method is to cut the diseased area out of the plant.

Mosaic and Yellows

The asters growing near Chris's home looked strange. They were yellow and stunted, and they had a disease known as aster yellows.

Figure 13-7. A canker on a birch tree.

Plants taking a yellow appearance are often termed "yellows." The name of the plant precedes the word yellows—as in aster yellows and peach yellows. Yellows is caused by a virus.

A closely related disease is called *mosaic*. Leaves on affected plants turn yellow in areas. The yellow color often is streaked, with a pattern of light and dark areas. This disease is also caused by a virus.

In later stages of both yellows and mosaic, the leaves may become distorted and malformed; flowers may not open properly, plant growth will be stunted. It is often difficult to tell the difference between yellows or mosaic and the disorders caused by environmental or nutritional factors. This is when keeping a plant log comes in handy. After studying the records and determining that the environmental and nutritional factors have been favorable, mosaic or yellows may be the problem.

These two diseases are controlled to an extent by using resistant varieties, controlling insects, and keeping the area clean of weeds.

Damping Off

Soon Chris and Ms. Franklin had completed the tour. "I have one more question," said Chris.

"I have problems with small seedlings started from seeds. Do you know why they are dying?" Ms. Franklin explained that the problem could be a disease known as *damping off*, common in young seedlings. People who specialize in growing bedding plants must take special precautions to prevent damping off.

It is caused by fungi found in the soil. Young seedlings are somewhat weak. The fungi grow in the stem tissues, which causes them to become weaker, until eventually the plants fall over and die. This disease is aided by keeping the soil wet. The best prevention is to avoid overwatering and to plant seeds only in pasteurized soil. Seeds may be chemically treated to prevent damping off. The type of plant will determine which chemical should be used.

After answering this question, Ms. Franklin left. Chris decided to make notes in the record book to help prevent future disease problems.

Disease Control and a Plant Management Plan

After a plant gets a disease there is usually no easy cure. In many cases, parts of the plant will have to be removed and destroyed. The best "cure" is to prevent it in the first place.

Prevention of disease and other plant disorders is enhanced through a comprehensive plant management plan. In the plant management plan, methods of watering plants can be outlined to minimize the spread of disease. Provision for pasteurizing soil and implementing sanitation procedures would be included. Special plant-growing practices such as fertilizing, shading plants, and applying pesticides should be detailed. This will aid not only in the control and prevention of disease but also in the prevention of other plant disorders, such as environmental and nutritional deficiencies or insect damage.

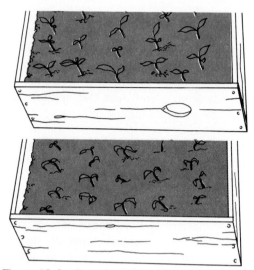

Figure 13-8. (Top) Seedlings planted in pasteurized soil are healthy. (Bottom) Seedlings planted in unpasteurized soil. Some have been killed by damping off.

The record section of the plant management plan (refer to Figure 11-10) is an invaluable aid in diagnosing plant diseases. The symptoms of certain nutritional deficiencies are very similar to those of certain diseases. By consulting the records it may be possible for you to determine whether the problem is a disease or a nutrient deficiency.

The overall plant management plan for a greenhouse, nursery, golf course, or other horticultural firm is generally developed by the manager or head groundskeeper, who determines when things should be done and records most of the facts and figures. If the employees notice any unusual conditions, such as insect infestations, disease symptoms, or other signs of plant disorders, they should notify the supervisor. Workers may make daily entries, such as watering and mowing, in the record.

Identifying and Controlling Plant Diseases: A Review

Billions of dollars are lost annually in the horticulture industry because of diseases. Many plant diseases are caused by bacteria, fungi, and viruses. Bacteria are single-celled microorganisms that reproduce rapidly, contain no chlorophyll, and can enter plants only through natural openings and wounds. Fungi are plants without chlorophyll. They vary in size and shape and often reproduce through spores. Viruses can be seen only through electron microscopes.

Diseases are spread by insects, water, wind, soil, and humans. Disease can be partially controlled through sanitation and soil pasteurization, chemicals, and the planting of resistant varieties.

Common horticultural diseases are leaf spots, blights, rots, rusts, smuts, mildews, wilts, galls, cankers, mosaic, yellows, and damping off.

A comprehensive plant management plan is essential in horticulture. By using a comprehensive plant management plan the horticulturist is able to manage the environment to maximize plant growth and minimize plant disorders.

THINKING IT THROUGH

1. What causes diseases?
2. How are diseases spread?
3. What general steps should be taken to prevent the spread of diseases?
4. What are the differences among fungi, bacteria, and viruses?
5. What is the difference between leaf spots and rots?
6. How do cankers and galls differ?
7. Why do some rusts need two different species of plants to reproduce?
8. How will a plant management plan aid in controlling plant disease?

14

Using Pesticides Safely

"Bret, I am going to take a vacation next week," stated Mr. Wheeler. "Do you think you can take good care of the greenhouse until I return?" "Yes, sir!" Bret replied. Bret had worked for almost a year with Mr. Wheeler, growing floral plants in a small greenhouse. Bret was happy that Mr. Wheeler was giving him the responsibility of operating the greenhouse.

On the first day that Mr. Wheeler was gone, Bret noticed that some of the poinsettias' leaves appeared to be turning a light brown. "They probably need a little fertilizer," thought Bret as he picked up a plant to examine it. A careful examination of the plant revealed a very fine webbing, like a spider web. "This may be more serious than I thought," said Bret as he went into the office to get a sheet of white paper and a magnifying glass. He returned, took the plant, and shook it over the white paper.

Several small dark specks landed on the paper. Bret examined the specks with the magnifying glass. It was just as he thought—spider mites. As he looked at several other plants, Bret discovered there was a severe infestation of spider mites.

"Now what am I going to do?" wondered Bret. "I guess I should apply some type of pesticide, but I'm not sure what should be used or how it should be applied. Mr. Wheeler always took care of applying the pesticides." Would you know what to do in this situation? Do you think a pesticide is the solution? If a pesticide is the solution, are there any restrictions on the use of pesticides? Would Bret be allowed to apply the pesticide?

Applying pesticides is a task that requires extreme care. Without a proper knowledge of pesticides, an inexperienced applicator could cause serious damage to the environment or to people. Death could even be the result. Bret should not attempt to apply pesticides without proper supervision or instruction.

Applying pesticides is a task often performed by workers in horticulture. In this chapter you will learn about the different types of pesticides, regulations concerning them, and ways to apply them safely.

CHAPTER GOALS

In this chapter your goals are:

- To identify different types of pesticides
- To interpret pesticide labels
- To apply pesticides safely

What Are Pesticides?

Have you ever sprayed on some mosquito repellent during the spring and summer? If you have, you were using a pesticide. The moth balls placed in stored clothes is a pesticide. The substance applied to your lawn to kill crabgrass or dandelions is a pesticide. The fly strip hanging in the garage is a pesticide. The can of Raid® is a pesticide. Rat poison is also a pesticide. Any material used to kill or repel pests is a pesticide.

Pesticides are used to control weeds, insects, rodents, bacteria, fungi, and other horticultural pests. Some pesticides are relatively safe to use, such as the household or garden types, while others are very dangerous. Many of the pesticides used in horticulture are dangerous.

It should be noted that there are a number of ways to control pests other than with pesticides. However, using pesticides is often the most efficient and effective method of controlling certain pests.

Categories of Pesticides

Since the term pesticide is so broad, scientists have developed a system of classifying pesticides according to the type of pest controlled. Horticulturists generally use the more specific group name, such as insecticide or herbicide, instead of the broader term pesticide.

Insecticides. Any material used to control insects is called an *insecticide*. Whiteflies, aphids, and beetles are some of the insects controlled with insecticides. Mites, spiders, and other insectlike creatures are often controlled with insecticides. Some groups of pesticides control more than one type of pest.

Herbicides. Undesirable plants (weeds) are controlled with *herbicides*. Some herbicides are selective and will kill only certain types of plants, while other herbicides will kill any plant to which they are applied.

Fungicides. A number of horticultural diseases are caused by fungi. The substance used to

Figure 14-1. Pesticides are used to control weeds, insects, rodents, bacteria, fungi, and other horticultural pests.

control fungi is called a *fungicide*. Fungicides are used widely on golf courses and other turf operations.

Rodenticides. Rats, mice, and other small animals that damage horticultural crops are called rodents. Therefore, the material used to control these rodents is labeled a *rodenticide*. Rat poison is a rodenticide.

Miticides. The substance used to control mites is called a *miticide*. It could be used on the spider mites discovered by Bret.

Acaricides. Spiders, ticks, and some mites are controlled with substances know as *acaricides*.

Molluscicides. Snails and slugs are bothersome pests to many horticulturists. *Molluscicides* are used to control snails and slugs. The slug bait sold in many garden stores is a molluscicide.

Bactericides. Bacteria are controlled with *bactericides*. Bacteria are responsible for a number of horticultural diseases.

Nematocides. Small microscopic wormlike organisms called nematodes are responsible for a number of plant disorders. The pesticides used to control nematodes are called *nematocides*.

How Pesticides Work

Pesticides may be classified according to the way they control the pest. Some pesticides have to be eaten by the pest, others have to come into physical contact with the pest, and still others have to be inhaled into the body by breathing.

Stomach Poisons

A number of pesticides have to be eaten by the pests in order to work. This type of pesticide is often sprayed or dusted on horticultural crops. The pest feeds on the plant that has been sprayed and takes the pesticide into its body, which kills the pest. An example would be a caterpillar eating on the leaf of a tomato plant that has been treated with carbaryl (Sevin™). Stomach poisons are generally most effective on pests with chewing mouth parts.

Systemics

A type of pesticide that is very similar to a stomach poison is called a *systemic*. Systemics are applied to the plant either by being sprayed with the material or by being mixed with the water. The plant then absorbs the pesticide. The systemic does not harm the plant. The pesticide is inside the plant, in the plant sap. When a pest eats on the plant or sucks sap from the plant, it is taking the systemic pesticide into its body. This kills the pest in the same manner as a stomach poison would. Systemics are very good for controlling pests with piercing-sucking mouth parts, and are widely used in horticulture. Some of the more common systemics are Thimet™, Di-Syston™, Systox™, and Cygon™.

Contact Poisons

One group of pesticides kills pests whenever they come into contact with the pest. For example, malathion could be applied to the leaves of roses to control aphids. When aphids come into contact with the malathion they are killed. Contact poisons do not have to be eaten by the insect.

Fumigants

Some pesticides are in a gaseous form. The vapors or fumes are breathed by the pest, which then kills the pest. These fumigants work best in enclosed structures such as a greenhouse. Fumigants are especially good at killing pests in hard-to-reach areas, such as cracks, crevices, and rolled-up leaves. The spider mites that Bret discovered earlier in this chapter could be controlled by means of a fumigant. Malathion and lindane are two pesticides that are available as fumigants.

Other Types of Pesticides

There are several other ways in which pesticides may affect pests. Recently a new kind of pesticide called *sterilants* has been introduced. These pesticides prevent pests from reproducing. *Pheromones* are a group of pesticides that cause pests to act unnaturally. Some pheromones have caused insects to run around in circles until they were exhausted. *Protectants* are materials that are placed on or near plants. They either smell bad or taste bad. One example is moth balls. Protectants do not kill the pest but succeed in keeping pests away. *Anticoagulants* prevent blood from clotting. This is often used in rat poisons. After eating the material, rats start

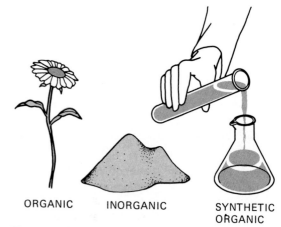

ORGANIC   INORGANIC   SYNTHETIC ORGANIC

Figure 14-2. Pesticides are developed from minerals (inorganic), plants (organic), and chemicals (synthetic organic).

bleeding internally. The blood will not clot, so they bleed to death internally.

Sources of Pesticides

Have you ever wondered where pesticides come from? The earliest pesticides were made from parts of plants or existing minerals. For example, nicotine is a contact poison that is made from tobacco. The mineral sulfur is a pesticide. Today many pesticides are developed from the combination of various chemicals. All pesticides can be placed into three groups according to the source of origin.

Inorganic Pesticides

Some pesticides are made from minerals such as sulfur, lead, and arsenic. These pesticides were popular in the late 1800s and early 1900s. They kill insects primarily as a stomach poison. They were used on horticultural crops such as grapes, fruit trees, and vegetables. Some examples of inorganic pesticides are lead arsenate, calcium sulfate, and plain sulfur. Inorganic pesticides are not used to a great extent today because they are harmful to humans and to other animals that are beneficial to humans.

Organic Pesticides

Several pesticides are made from plants. These pesticides have been in existence for hundreds of years. Rotenone, which is made from the roots of a South American tree, and pyrethrum, which is made from certain flowers, are two of the more common organic pesticides. Organic pesticides usually act either as a stomach poisons or as contact poisons. They are relatively safe to humans but effective on pests. The high cost of producing organic pesticides, however, has limited their use.

Synthetic Organic Pesticides

This group of pesticides is manufactured rather than natural. Such pesticides are made by combining various chemicals such as carbon, hydrogen, chlorine, and phosphorus. Most of the synthetic organic pesticides are only about 30 to 40 years old. They were developed primarily during World War II for use in warfare, and are very effective in controlling insects but may also be dangerous to humans.

Synthetic organic pesticides can be divided into three main groups. Horticulturists must be aware of each group, because each group affects pests and humans differently.

Chlorinated Hydrocarbons. Lindane, endrin, toxaphene, BHC, methoxychlor, dicosol, TDE, and DDT are all chlorinated hydrocarbons. They are very effective in controlling pests, but have several disadvantages. One problem is that the pesticide remains long after it is needed, and when accumulated it causes problems for many animals and birds. This is because the pesticide can be stored up in their bodies. Because of this many of the chlorinated hydrocarbons such as DDT can no longer be used in the United States.

Organophosphates. Parathion, malathion, EPN, and TEPP are organophosphates. They are very effective in killing many insect pests, but can also kill warm-blooded animals and humans unless extreme care is taken. One advantage of the organophosphates is that they break down after about 2 weeks to 1 month and do not leave residues.

Carbamates. Some of the safest pesticides available are found in the carbamate group. Carbamates break down quickly and leave no harmful residues. Most of the carbamates are not very harmful to humans or warm-blooded animals. Some carbamates used in horticulture are carbaryl (Sevin™), and aldicarb (Temik™).

Pesticide Formulations

If you were to walk into a store to select a pesticide, it would be easy to become confused by all the different forms in which pesticides come. Some pesticides come as liquids, and some as dusts, granules, or powders. There is a reason why they come in various forms.

Often, the ingredients that are used to kill pests cannot be used in their pure form. They must be changed or mixed with something else. They are often changed or mixed in order to make them more convenient or safer to use. These mixtures are called *formulations*. Some formulations are ready for use, while others have to be diluted with water or some other liquid such as oil.

Liquid Formulations

Many pesticides come in liquid form. There are five different liquid formulations, and each liquid except one has a letter that stands for it.

Emulsifiable Concentrates (EC or C). If you see a pesticide with an E or EC on the label, this means that the pesticide is a liquid but has to be mixed with water. The mixture (*emulsion*) of pesticide and water is then sprayed on the plant. You should read the label carefully to determine how strong to make the mixture.

Solutions (S). A *solution* is a liquid that is normally ready to be used straight from the container. It does not have to be diluted or mixed with another liquid.

Flowables (F or L). Some pesticide ingredients are normally found as solids. These ingredients are ground up into a fine powder and put into a liquid, which creates a thick liquid. The name *flowable* means that it is a flowable solid. Flowables are often mixed with water and applied to plants by means of a sprayer.

Aerosols (A). Some liquid pesticides are packaged in pressurized cans. The pesticide is placed in the can in liquid form along with a propellant (gas), and then released as a fine mist. Raid™ is an example of an aerosol pesticide.

Liquefied Gases (Fumigants). Several pesticides are liquids until they are released into the soil or atmosphere, at which time they change to a gas or vapor. These *liquified gas* pesticides are brought in containers, ready for use.

Dry Formulations

Like liquid pesticide formulations, there are five types of dry pesticide formulations.

Dusts (D). The pesticide ingredient is often ground into a very fine powder and combined with an inactive powder such as ground clay or ash. This *dust* is placed on plants in powder form. It is not mixed with water. Dusts are easy to use but can be blown by the wind.

Granules (G). Small pellets of pesticides are called *granules*. The granules will not stick to

NO-DISEASE

BENOMYL FUNGICIDE

Wettable Powder (WP)

ACTIVE INGREDIENT
Benomyl [Methyl 1-(butylcarbamoyl)-
2-benzimidazolecarabamate]   50%
INERT INGREDIENTS   50%
U.S. Pats 3-541,213 & 3-631,176   EPA Est 352-WV-1
EPA Reg. No. 352-354-AA

Keep out of reach of children

CAUTION! MAY IRRITATE EYES, NOSE, THROAT AND SKIN
Avoid breathing dust or spray mist.
Avoid contact with skin, eyes and clothing.
Wash thoroughly after using. Keep away from fire or sparks.

In case of contact, flush skin or eyes with plenty of water; for eyes, get medical attention.

Figure 14-3. Letters on the pesticide label indicate the form of pesticide such as flowables, solutions, granules, and emulsifiable concentrates.

a plant, so they are placed on the ground around plants. When they become wet they dissolve and release the pesticide.

Wettable Powders (WP or W). *Wettable powders* look like dust but must be mixed with water before they are applied. They are applied to plants by means of a sprayer. The mixture has to be stirred. Wettable powders do not *dissolve* (form a solution) when they are mixed with water.

Soluble Powders (SP). *Soluble powders* look like dust and are mixed with water. When mixed with water, the soluble powder dissolves and creates a solution. After the mixture is completely dissolved no further stirring is needed. Soluble powder solutions are applied to plants by means of a sprayer.

Baits (B). A *bait* is a pesticide that is mixed with an edible or attractive material. The bait attracts pests that eat the material. Baits are often used to control rats, slugs, and grasshoppers.

Understanding the Pesticide Label

Before a pesticide is used, the applicator should examine the label carefully. Much valuable information is printed on the label, if you know how to interpret it. Horticulture employees should be able to read a pesticide label. If you do not follow the instructions on the label you are subject to either civil or criminal penalties. The following information is contained on a pesticide label.

Name

Each pesticide has three different names—a brand name, a common name, and a chemical name. The brand name is a name developed by the manufacturer to identify and advertise the product. Sevin™ and Systox™ are examples of brand names. You can tell when a name is a brand name because the symbol™ will appear next to the name, indicating a trademark.

The Environmental Protection Agency (EPA) has approved a number of common names that can be used to identify pesticides. The common name is often used to identify a pesticide, because two or three companies could be manufacturing the same pesticide but each would call it a different brand name. For example, Cygon™, De-Fend™, and Rogor™ are all brand names for the same pesticide—dimethoate is its common name. Carbaryl is the common name for Sevin™, and demeton is the common name for Systox™.

The third type of name required on each pesticide label is the chemical name. This name is developed by chemists and is often very long and complex. A systemic fungicide often used in horticulture has the brand name of Benlate™. The common name is benonyl. What do you think the chemical name is? It is Methyl 1-(butylcarbamoyl)-2-benzimidazolecarbamate. It would be difficult to use this name every day.

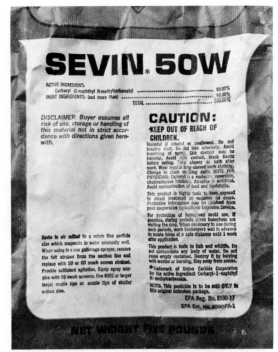

Figure 14-4. The label on a pesticide should be read carefully.

That is why brand names and common names are used instead.

Use Classification

All pesticides are classified by the EPA as either a general-use pesticide or a restricted-use pesticide—the label must specify whether the pesticide is for general or restricted use.

General-use pesticides may be purchased by anyone. They will do little, if any, harm to the applicator or environment if they are applied according to directions. The label on the pesticide will say "general classification."

A *restricted-use pesticide* can be purchased and used only by certified applicators or by people under the direct supervision of a certified applicator. These pesticides are very dangerous and must be used with extreme care. They can harm humans or the environment.

Figure 14-5. (Left) General-use pesticides may be used by anyone. (Right) Restricted-use pesticides may only be used by certified applicators.

Signal Words and Symbols

Each pesticide label must have the word DANGER, WARNING, or CAUTION on the label. Each word has a special meaning. The word DANGER means that the pesticide is highly toxic (poisonous). A taste to a teaspoonful of the pesticide could kill an average person. TEPP, parathion, endrin, and demeton (Systox™) are examples of highly toxic pesticides.

The word WARNING means that the pesticide is moderately toxic. A teaspoon to a tablespoon of the pesticide would kill the average person. Lindane, rotenone, and dimethoate (Cygon™) are examples of moderately toxic pesticides.

The word CAUTION is used when the pesticide has low toxicity. It would take from an ounce to more than a pint of the material to kill an average person. Carbaryl (Sevin™),

malathion, atrazine (Aatrex™), and simizine (Princep™), are pesticides with low toxicity.

If a pesticide is highly toxic, the skull and crossbone symbol, accompanied by the word POISON, must be on the label—in addition to the word DANGER.

Ingredients and Net Content

Each label must list all the toxic ingredients found in the pesticide according to the percentages of each ingredient you know exactly what you are getting. Often a pesticide will contain several ingredients. The chemical name must be listed along with the common number if there is one. The percentage of inactive (inert) ingredients must be listed but the inert ingredients do not have to be identified by name.

PESTICIDE LABEL SIGNAL WORDS		
DANGER	WARNING	CAUTION
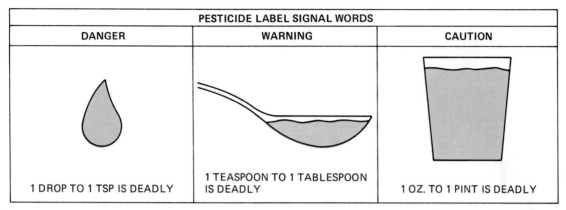 1 DROP TO 1 TSP IS DEADLY	1 TEASPOON TO 1 TABLESPOON IS DEADLY	1 OZ. TO 1 PINT IS DEADLY

Figure 14-6. The signal word on a pesticide label indicates how deadly the pesticide is.

The total amount of pesticide in the container must also be listed on the label. This could be in ounces, grams, pounds, kilograms, gallons, liters, or some other type of measure.

Pesticide Formulation

The formulation of the pesticide is listed on the label. Often only the letters such as D, G, WP, or EC, are listed. You have to know what the letters stand for. In this case the corresponding formulas are dusts, granules, wettable powders, and emulsifiable concentrates.

Directions for Use

A pesticide label contains much useful information. One bit of information relates to using the pesticide. This includes a list of the pests it can control, the types of equipment to use to apply the pesticide, the way to store and dispose of the container, and restrictions on the use of the pesticide.

Other Label Information

The pesticide label has to have seven more items of information in addition to those already discussed. These are:

Child Hazard Statement. The child hazard statement is simply a statement that says "Keep out of reach of children." This statement must appear on all pesticides.

Statement of Practical Treatment. The statement of practical treatment tells how to administer first aid to a person who has been poisoned or who has come into contact with the pesticide. If a person has to go to a doctor, the container with the label should also be taken to the doctor.

Reentry Statement. Some pesticides require people to wait for a number of hours or days before they reenter an area where the pesticide has been used. The reentry statement tells how long you must wait.

Precaution Statement. Precautions must be used with a number of different pesticides. The precaution statement tells what precautions must be taken. Examples include: keep away from heat or flame, do not use on windy days, this product is highly toxic to bees, and avoid contact with the skin.

Name and Address of Manufacturer. The name of the company that manufactures the pesticide, along with its address, must be on the label.

Figure 14-7. Protective gear should be worn when using pesticides.

EPA Identification Number. A number assigned by the EPA must be on each pesticide label. This shows that the pesticide has been registered with the government.

Misuse Statement. It is against federal law to use pesticides in an improper manner. Pesticides should not be used at more than the recommended rate, and should be used only for the pests specified on the label. If you do not follow the directions properly you are breaking the law. The misuse statement warns you of this.

Safety Precautions

Before applying a pesticide, you should follow certain safety precautions. You cannot be too cautious or too careful when using pesticides. Pesticides can be harmful to you if they get on your skin, are breathed in, or are swallowed. To prevent this from happening, common sense and protective equipment are needed.

Protective Clothing

Directions on the pesticide label will tell you what type of protective clothing should be worn. If the pesticide is fairly safe (CAUTION on the label) you may need to wear only a long-sleeve shirt and long pants (or coveralls). However, if the pesticide is highly toxic, special clothing must be worn.

Gloves. Rubber (neoprene) gloves are often worn when highly toxic or concentrated pesticides are used. The gloves should extend over the wrist and be worn inside the sleeves of your shirt. The reason why the sleeves are on the outside of the gloves is to keep the pesticides from running into the gloves in the event that it is spilled or begins to run down a shirt sleeve. The gloves should not be lined with cloth, because the cloth could absorb the pesticide. The pesticide label will specify the type of gloves to wear if they are needed.

Head Covering. A wide-brimmed waterproof hat should be worn on the head when you are using toxic pesticides. This will help protect your neck, mouth, eyes, and face. Plastic hard hats are good for this purpose, because they can be washed. Cloth or felt hats may absorb the pesticide and should not be used.

Waterproof Clothing. A waterproof raincoat or apron may be needed when you are handling highly toxic materials. Unlined neoprene boots may be worn with the waterproof clothing. Trousers should not be tucked into boots, because the pesticide could run down the leg into the boot. Again, check the pesticide label for specific directions.

Eye Protection. It is a good idea to wear safety goggles or a face shield when you are using *any* type of pesticide. Your eyes can absorb many pesticides, but certain pesticides, such as parathion, are so toxic that one drop of concentrated parathion in the eye will kill the average person.

Respiratory Protective Equipment

Many pesticides can accidentally be ''breathed'' into the body. To prevent this, a specialized piece of equipment called a respirator is placed over the mouth and nose. The respirator filters out the harmful substances but still allows you to breathe. Some respirators are hooked to separate oxygen supplies. Respirators are used with highly toxic pesticides or in cases where the applicator will either be exposed to the pesticide for long periods of time or be working in an enclosed area such as a greenhouse. The label on the pesticide container will state when a respirator is needed.

First Aid

Extreme care should be taken while you are using pesticides to avoid contact with the substance. If you do come into contact with the material, appropriate action must be taken. The label on the pesticide will tell what type of action to take.

If you get pesticide on your skin, you should wash the affected area immediately with soap and water. Quick washing may prevent any serious damage from occurring. If the pesticide is highly toxic, a doctor should be contacted. Any clothing that has come into contact with the pesticide should be removed.

Pesticide in the eyes can be very serious. The eyes should be immediately flushed with running water for at least 15 minutes. You should also call a doctor.

If someone swallows a pesticide, it is very important that the label on the pesticide container be consulted. With some pesticides the recommended procedure is to get the victim to vomit. However, with other pesticides it is dangerous to cause vomiting. Therefore, it is

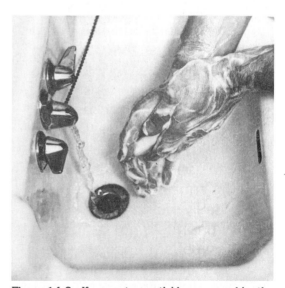

Figure 14-8. If you get a pesticide on your skin, the area should be washed thoroughly with soap and water.

important that the label be read carefully. In all cases the victim should be taken to a doctor immediately.

Inhaling pesticides is also dangerous. Someone who has inhaled a pesticide should move to an area with fresh air and lie down. Loosen clothing and keep the person warm and quiet. A doctor should be contacted.

After using pesticides, you should take a bath or shower. The hair should be washed and the fingernails should be scrubbed to remove any hidden pesticide residues. Clothing that was worn while applying pesticides should be washed before it is worn again. The clothes should not be washed with other clothes.

Storing and Disposing of Pesticides

Proper storage and disposal of pesticides is an important safety practice. Pesticides should be stored in their original containers, in a cabinet or room that is locked. Children and animals should not be allowed near pesticides. The storage area should be well lighted so that the labels can be read. The label will provide additional storage information.

There are several ways to dispose of empty pesticide containers. Directions for disposing of the container will be found on the label. Some containers have to be buried in the ground at least 18 in (inches) [46 cm (centimeters)] below the surface. Other containers may be burnt. Some counties have specially designed landfills just for empty pesticide containers. Leaving empty pesticide containers lying around is not safe. They should be disposed of properly.

Pesticide Application

Pesticides may be applied to horticultural crops in a variety of ways and with a variety of equipment. The equipment used to apply pesticides can be grouped into sprayers, dusters, and spreaders.

Figure 14-9. Pesticides should be stored in a locked cabinet or room.

Spray Equipment

A number of pesticides come in liquid form and are applied to plants as sprays. Sprayers can be very simple or very elaborate.

Aerosol. An aerosol is a sealed container in which a pesticide and a compressed gas are placed. Many mosquito repellants and household pest sprays are sold in aerosol cans. The nozzle is held down to release the pesticide as a fine spray. Aerosols are used for small jobs.

Trigger-Pump. Plastic bottles with a nozzle and trigger are called trigger-pump sprayers. They operate like a water gun. The pesticide is placed into the bottle and squirted onto the plant. A trigger-pump sprayer can be bought for less than $2. It does not hold much pesticide,

TRIGGER-PUMP SPRAYER

HAND-PUMP SPRAYER

AEROSOL

SPRAY COMES
OUT HOLES

HIGH-PRESSURE SPRAYER

COMPRESSED AIR
SPRAYER

GARDEN HOSE
SPRAYER

Figure 14-10. Different types of sprayers.

however, and is used primarily by people who grow horticultural plants as a very small-scale hobby.

Hand-Pump. A hand-pump sprayer looks like a bicycle tire pump attached to a small can. The pesticide is placed into the can, and the handle of the pump is pushed and pulled to spray the pesticide. This type of pump is sold for $2 to $3 in most garden stores or variety shops. It is used a great deal for spraying ants and flies around many homes.

Garden Hose. Several spray attachments screw onto the garden hose. Often a clear plastic

container is screwed onto the end of the garden hose. The pesticide is placed into the container. The pressure of the water pulls the pesticide out of the container and it is mixed with the water. The pesticide is then applied to the plants as they are watered.

Compressed Air Pump. Compressed air pumps are commonly used for spraying lawns, shrubs, gardens, and greenhouse plants. A compressed air pump is a round cylinder with a handle in the top. The handle is pushed up and down to build up air pressure inside the cylinder. A spray gun on the end of a hose attached to the cylinder is used to apply the pesticide. Most compressed air pumps will hold 1 to 3 gallons (gal) [3.8 to 11.4 liters (l)] of liquid.

Power-Driven Sprayers. Many of the sprayers used in horticulture are powered by gasoline or diesel engines or by tractors. The sprayers are normally mounted on or pulled by tractors. These sprayers are capable of covering large areas of land at moderate speeds.

On some sprayers, the pesticide is applied through nozzles that are evenly spaced on a long bar or pipe (boom) that is parallel to the ground. Other sprayers use a fanlike device to blow the pesticide onto the plants. This type of sprayer is often used in orchards. A hose with a spray gun on the end is used on sprayers where the pesticide has to be aimed at a target, such as the top of tall trees. Helicopters and airplanes are also used to apply pesticides.

Dusters

There are several types of dusters for applying pesticides that come in dust form. Some dusts are sold in containers with holes in the end, similar to salt shakers. The pesticide is then applied by shaking the container. Some dusters have a hand crank that operates a fanlike device. The dust is thrown onto plants by turning the crank. Power dusters with fans or blowers to blow the dust onto the plant are available.

Spreaders

Granules are applied to horticultural crops by means of a spreader. Portable hand spreaders can be used. They have a hand crank which operates a revolving metal or plastic disc. The disc has fins which throw the granules out as the disc revolves.

Large-granule spreaders are pulled behind tractors. The large spreaders resemble grain drills or corn planters, and operate on the same principle. They are used to cover large amounts of land. The amount of granules applied can be regulated.

Deciding When to Use a Pesticide

The best way to handle pest problems in horticulture is to prevent them before they occur. A plant management plan is helpful in doing this. Procedures for reducing pest problems are identified in the plant management plan. Some of the procedures listed may include the planting of disease-resistant plant varieties, general sanitation procedures, and adoption of approved cultural practices. Prevention is the best strategy for controlling horticultural pests.

No matter how well a person plans, there is always the possibility of a pest problem. When a pest problem occurs the horticulturist should first identify the problem. Is it a disease, insect, nutrient deficiency? After the problem has been identified, the horticulturist must determine whether the problem is serious enough to warrant action. Sometimes the problem is so small or so late in the season that the best thing to do is nothing. If the problem is serious enough to need correcting, then the different methods for controlling the problem should be compared.

If the problem is a pest, the grower may want to compare the advantages and disadvantages of each control method. Some questions that should be asked are:

1. How effective is this method?
2. How much will it cost?
3. Will it have bad side effects?
4. How safe is the procedure?
5. How much time will it take to produce results?
6. Is specialized equipment needed?

After answering these questions, the grower should decide upon the best method of control and implement it. The control method may be mechanical, cultural, or biological, or it may be the use of a pesticide. Whatever method is chosen, care should be used before making the decisions. The vocational horticulture teacher or county agricultural extension agent should

Frequently Used Pesticides

Common Name	Brand Name	Pesticide Use	Label Signal Word
Amitrole	Weed Azol	Herbicide	CAUTION
Atrazine	Aatrex	Herbicide	CAUTION
Azinphosmethyl	Guthion	Insecticide	DANGER
Benefin	Balan	Herbicide	CAUTION
Benomyl	Benlate	Fungicide	CAUTION
Bromacil	Hyvar X	Herbicide	CAUTION
Captan	------	Fungicide	CAUTION
Carbaryl	Sevin	Insecticide	CAUTION
Carbophenothion	Trithion	Insecticide	DANGER
Chloramben	Ambien	Herbicide	CAUTION
Chlordane	Ortho-Klor	Insecticide	WARNING
Chloroxuran	Tenoran	Herbicide	CAUTION
Crotoxyphos	Ciodrin	Insecticide	WARNING
Dalapon	Dowpon	Herbicide	CAUTION
DCPA	Dacthal	Herbicide	CAUTION
Demeton	Systox	Insecticide	DANGER
Diazinon	------	Insecticide	WARNING
Dicamba	Banvel	Herbicide	CAUTION
Dimethoate	Cygor, De-Fend, Rogor	Insecticide	WARNING
Diuron	Karmex	Herbicide	CAUTION
DNBP	Premerge	Herbicide	DANGER
DSMA	Ansar, Sodar	Herbicide	CAUTION
Endrin	------	Insecticide	DANGER
EPTC	Eptam	Herbicide	CAUTION
Ferbam	Fermate	Fungicide	CAUTION
Fenuron	Dybar	Herbicide	CAUTION
Lead arsenate	------	Insecticide	WARNING
Lindane	------	Insecticide	WARNING
Linuron	Lorox	Herbicide	CAUTION
Malathion	Cython	Insecticide	CAUTION
Maneb	Manzate	Fungicide	CAUTION
Methyl parathion	------	Insecticide	DANGER
Methoxychlor	------	Insecticide	CAUTION
Monuron	Telvar	Herbicide	CAUTION
Nabam	Dithane	Fungicide	WARNING
------	Nemagon, Fumazone	Nematicide	WARNING
Nicotine	------	Insecticide	DANGER
Norea	Herban	Herbicide	CAUTION
Paraquat	Weedol	Herbicide	WARNING
Parathion	------	Insecticide	DANGER
Phorate	Thimet	Insecticide	DANGER
Picloram	Tordon	Herbicide	CAUTION
Prometon	Pramitol	Herbicide	CAUTION
Prometryn	Caparol	Herbicide	CAUTION
Rotenone	------	Insecticide	WARNING
Siduron	Tupersan	Herbicide	CAUTION
Silvex	Kuron, Weedone-TP	Herbicide	CAUTION
Simazine	Princep	Herbicide	CAUTION
------	Slugit	Molluscicide	CAUTION
Sodium arsenite	Atlas A	Herbicide	DANGER
Sodium chlorate	Atratal	Herbicide	CAUTION
Sodium fluoride	------	Insecticide	WARNING
Strychnine	------	Rodenticide	DANGER
Trifluralin	Treflan	Herbicide	CAUTION
Warfarin	------	Rodenticide	DANGER
------	Zectran	Molluscicide	CAUTION

Figure 14-11. Table of commonly used pesticides.

be contacted if help is needed in making a decision.

That is what Bret did at the beginning of this chapter. After finding the spider mites on the poinsettias, Bret contacted his vocational horticulture teacher. The teacher examined the situation with Bret and determined that a fumigant used properly would kill the spider mites, and would not harm the plants. The teacher showed Bret how to use the fumigant safely and helped fumigate the greenhouse. The spider mites were killed. When Mr. Wheeler returned at the end of his vacation, he was pleased with the action taken by Bret.

Using Pesticides Safely: A Review

Applying pesticides is a task often performed by horticulture employees. Pesticides can be very toxic and must be used carefully. Protective clothing should be worn while using pesticides. Pesticides should be kept away from children and stored in a locked area.

Pesticides may control pests by killing or repelling them. Some pesticides work as stomach poisons, while others are contact poisons or fumigants. Pesticides are derived from minerals (inorganic), plants (organic), or chemicals (synthetic organic). Pesticides come in liquid or dry formulations. Sprayers, dusters, and spreaders are used to apply pesticides.

The label on a pesticide container contains much valuable information. Before using a pesticide, read the label carefully. All methods of controlling pests should be carefully analyzed before you select a method.

THINKING IT THROUGH

1. What is a pesticide?
2. What factors should be considered before you use a pesticide?
3. List eight groups of pesticides, and the type of pest that is controlled by each.
4. How do pesticides control pests?
5. What is a systemic pesticide?
6. Name the three main sources of pesticides.
7. If you were to see the letters EC, S, F, D, G, and WP on a pesticide label, what would it mean?
8. Why does each pesticide have three names?
9. What is the difference between a general-use pesticide and a restricted-use pesticide?
10. What do the words DANGER, CAUTION, and WARNING mean on a pesticide label?
11. What type of safety precautions should be taken when you are applying pesticides?
12. How should pesticides be stored?
13. What are the different types of sprayers used in applying pesticides?

FIVE

Horticultural Plant Growing Structures and Equipment

What is the weather like outside? The chances of its being 68–75° F [20°–24° C], sunny, with light breezes and an ideal humidity for growing plants, is remote. Of course, it depends on where you live and the time of year. Weather is a factor that affects both humans and plants, and there is little we can do to control it. Or are there ways to improve the weather? In a limited sense, plant-growing structures control the "weather." At least these structures modify the plant environment to achieve a more ideal condition. These plant-growing structures vary in type and construction. A shade or lathhouse may be used to produce foliage plants in the deep South. Glass, fiber glass, or vinyl materials may be used for greenhouses in the temperate midcontinent region. In central Minnesota, a totally enclosed and insulated growing structure uses artificial light for plant production. As a horticulture worker, you will need to be familiar with the types of structures and ways of maintaining them.

A person who works in a flower or greenhouse business will use many skills in replacing and repairing benches, maintaining the structure, and often even installing irrigation systems. The vegetable and bedding plant producer and the plant propagator may use more plastics for greenhouses and may use hothouses or cold-frames.

As you gain experience, you will recognize the tremendous amount of labor that is involved in producing quality plants. In the past, the drudgery of hand or stoop labor may have kept people from choosing a career in horticulture. Today, much of the manual labor has been replaced by machines. These machines do more of the physical labor, so people can accomplish more. The small engine provides power for many of these machines. The lightweight 2-cycle engines are used as backpack sprayers, vacuums, and post-hole diggers. The larger 4-cycle engine powers rotary and reel mowers, lawn tractors, and other equipment. The golf course superintendent and the assistant superintendent understand and direct maintenance programs for both small engines and greens. The landscape maintenance supervisor can troubleshoot engine problems and make minor repairs and adjustments. The riding reel-mower has taken the hard work out of a beautiful lawn.

As the horticulture industry grows, it uses more tools and equipment. As a result, it needs more workers who understand sophisticated machines. Machines may also be powered by electric motors. Greenhouses use motors for ventilation fans, soil shredders, power carts, and other equipment. Many of these are driven by belts or chains, which require periodic maintenance. The tree service owner uses chainsaws and hydraulic lifts to prune or repair large trees. Many operations use lawn tractors and other small industrial tractors for power. The park or cemetery worker uses a small tractor and a gang-mower to keep the grass clipped. The greens maintenance crew uses hydraulic driven reel-mowers to keep the putting surfaces in tournament condition. Of course, there are still hand tools and they must be redressed. The hand shear and loopers need to be sharpened and all tools require good maintenance. After all, tools are expensive and you want them to do a good job and last as long as possible. In this unit, you will learn to develop many of the mechanical skills that future horticulture careers demand.

15

Horticultural Plant Growing Structures

The students in the horticulture class were excited as they loaded onto the bus. It was the first field trip of the year. After everyone had been seated, Ms. Tucker, the teacher, stood and said, "Now I want everyone to remember the purpose of this field trip. We are going to see the types of structures used to grow horticulture crops commercially. The company we are going to visit, National Plant Producers, Inc., has its headquarters here but also has plant producing facilities in Florida and California. We will learn what types of equipment and buildings are used to grow plants in all these locations. Are there any questions?" There were no questions. The bus pulled out of the school parking lot toward the headquarters of National Plant Producers, Inc.

What do you think the students will learn on the field trip? Do you know what types of structures and equipment are used for growing horticultural crops? As a horticulturist you will be expected to know the types of equipment and facilities used in producing horticultural crops. You may be expected to repair, maintain, and assemble equipment and facilities used in pro-

ducing horticultural crops. In the future you might own a horticulture business and will make decisions concerning which plant-growing structures are needed in your business.

CHAPTER GOALS

Your goals in this chapter are:

- To identify the structures and equipment used to produce horticultural crops
- To describe the purpose of greenhouses, hotbeds, cold frames, wintering houses, and shade houses
- To identify six types of greenhouses
- To compare the advantage and disadvantages of the materials used in constructing greenhouses
- To explain how greenhouses are heated and cooled
- To determine the mechanical skills needed to maintain plant-growing structures properly

After a short trip the bus arrived at National Plant Producers, Inc. The students entered the main building and were shown to a large meeting room where they were introduced to Mr. Weaver, the vice president of the company. "Welcome to National Plant Producers. We are always glad to have visitors. Today we will be showing you through our facilities and explaining them to you. In order to help you understand our operations, it might be wise to give you a brief overview of the different structures used in the various areas of horticulture."

Plant-Growing Structures in the Horticulture Industry

Plant-growing structures are used to provide a more favorable growing environment for plants. The temperature and moisture are often con-

trolled in plant-growing structures. Because of this, plants can be grown in many locations and at times of the year when plants could not normally be grown. Without plant-growing structures, horticultural operations would be limited to a few areas of the United States. As a result, horticultural products would be expensive to buy if they were available.

Several different types of structures are used in producing horticultural crops. The type of structure used depends upon the type of horticultural operation and the geographic location of the business.

Nursery operators use a variety of plant-growing structures, depending upon where they are located. In the northern part of the United States the term "nursery" refers to a place for the production of woody outdoor ornamental plants. Trees, shrubs, and rosebushes are often grown directly in the soil in rows, or in containers on specially prepared surfaces. In the winter these container-grown plants are often placed in a structure known as a wintering house to protect them from the cold weather. Cold frames may be used for "hardening" transplants. Greenhouses are used for plant propagation. Some nursery stock may be placed in cold storage facilities until early spring or fall when it is shipped.

In some Southern and Western states, such as Florida, California, and Hawaii, the term "nursery" refers to a business that grows woody outdoor plants, such as trees, shrubs, and rosebushes; foliage plants; flowering plants; and bedding plants. Many plants are grown outdoors. In Florida, mums are grown outdoors under woven nylon structures. In California,

Figure 15-1. Various structures are used to produce horticultural crops. Among the most common structures are: the greenhouse (top left); the plant growing surface (top right); the shade house (bottom left); and the cold frame (bottom right).

many crops such as daisies are grown outdoors in full sun. Lathhouses and other types of shading structures are used to protect nursery crops from excessive heat or sun in these areas. Greenhouses may also be used for propagation purposes in the South and West.

Vegetable and fruit producers require several types of specialized structures. They need packing sheds and storage sheds for the produce. Many vegetables and fruits are kept in cold storage for short or long periods of time. Apple producers often place apples in cold storage for extended lengths of time. Greenhouses, cold frames, and hotbeds are used to start vegetable transplants. The production of out-of-season vegetables occurs in greenhouses. Tomatoes are grown in greenhouses during the winter in many parts of the country.

Florists and commercial flower growers use specialized facilities. If you have been in a floral shop you have probably noticed a big walk-in cooler. This cold storage facility is used to prolong the life of cut flowers. Various bulbs are also stored in cold facilities. Greenhouses and shading structures are used by commercial flower growers.

After the brief overview of the types of structures used in horticulture, Mr. Weaver was ready to start the tour. "Since greenhouses are used in all areas of horticulture, we'll start our tour in the greenhouses," stated Mr. Weaver. Soon Mr. Weaver and the students were standing in the middle of a greenhouse.

The Greenhouse

Most people think of greenhouses when they think of a plant-growing structure. A greenhouse is a building with roof and walls that are made largely of glass or other transparent materials. The environment inside a greenhouse is controlled. The temperature and relative humidity can be controlled in most greenhouses. Thus an ideal growing environment can be created.

The first greenhouse was built by the Roman Emperor Tiberius Claudius Nero. He was sick and was ordered by his doctor to eat only cucumbers. His gardeners built a structure that resembled a greenhouse, called a specularium. The specularium was built from thin sheets of stone that were translucent. Nero's gardeners were able to grow cucumbers all year long in the specularium.

Greenhouses are used for a variety of purposes. Plant propagation is a major activity conducted in many greenhouses. Many plants are started from seeds or cuttings in the greenhouse and then transplanted outside into the soil when they become large enough. Plants that are propagated asexually, such as sansevieria (mother-in-law's tongue or snake plant) from leaf cuttings, are started in the greenhouse.

Plants that are sold at specific seasons of the year, such as poinsettias at Christmas and Easter lilies at Easter, are grown in greenhouses because the weather will not allow the plants to be grown outside at that time of the year in most parts of the country. Certain popular vegetables, such as tomatoes, are grown in greenhouses during the winter.

Today there is a variety of styles and types of greenhouses. When more than one greenhouse is in one location under one operation, it is referred to as a "greenhouse range"—whether or not the greenhouses are actually joined together. Greenhouses are usually erected on gentle slopes so that there is good air drainage. In Northern states, special care must be taken to locate greenhouses away from structures that might shade them in midwinter. A building 20 ft [6 m (meters)] tall casts a shadow 40 ft [12 m] long in the Chicago area in December.

Parts of a Greenhouse

Like the human body, a greenhouse has specific parts. Horticulture workers should be familiar with the parts of the greenhouse because they may be asked to repair certain parts of the greenhouse or may need to communicate with others. If you were employed in a greenhouse and told your employer that the "thingamajigs" on the side of the greenhouse were stuck, the employer would not know what you were talking about. If instead you said that the side ventilators were stuck, the employer would know what you meant. A worker in a greenhouse should be able to identify the parts of a greenhouse, just as a mechanic has to identify the parts in a car.

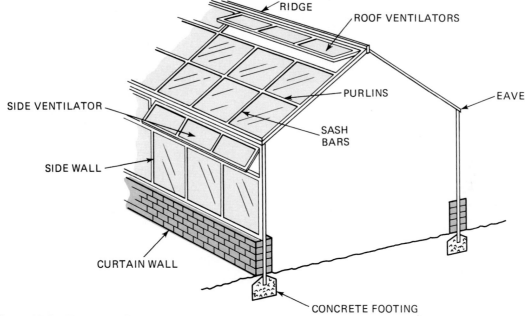

Figure 15-2. The parts of a greenhouse.

Sidewall. As the name implies, the *sidewall* is the wall on the side of the greenhouse. Often load-bearing posts called side posts are located in the sidewall. The bottom 2 or 3 ft [60 or 90 cm] of the sidewall is often made of concrete block or corrugated asbestos. This section of the sidewall is called the curtain wall.

Concrete Footers. When a building is built, it should have a firm foundation. The entire weight of a greenhouse rests on the side posts or sidewall. Therefore, each side post or sidewall is set on a separate *concrete footer* that extends well below the frost line (3 to 4 ft [90 to 120 cm] below the surface in Northern states). The concrete footers support the weight of the structure. Temporary greenhouses may not have concrete footers.

Side Vetilator. Some, but not all, greenhouses may have a portion of the sidewall hinged so that portions of the wall can be opened. These

areas swing out and are called *side ventilators*. They are used for ventilating the greenhouse.

Eave. The area of the greenhouse where the top of the sidewall joins the roof is called the *eave*. Glass greenhouses have an L-shaped metal strip where the roof and sidewalls join. This is known as the *eave plate*. This eave plate is heated by the warm air inside the greenhouse, and prevents ice from accumulating at the eave, which would cause breakage from its weight. In ridge and furrow greenhouses, this eave plate is replaced by a metal gutter that is heated in cold weather to prevent ice from forming. Some greenhouses have curved sidewalls that blend into the roof. These types of greenhouses do not have eaves.

Roof. The *roof* is composed of several parts. Bars of metal or wood running from the eave to the ridge (peak) of the roof are used to hold glass in place. These pieces are called *sash bars*. Bars running the length of the roof (under and

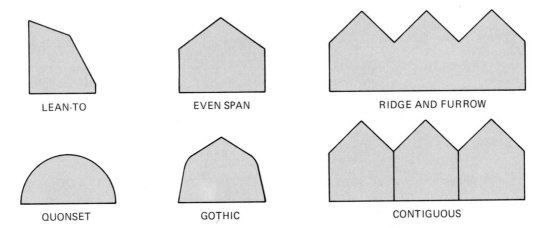

LEAN-TO   EVEN SPAN   RIDGE AND FURROW

QUONSET   GOTHIC   CONTIGUOUS

Figure 15-3. Various greenhouse styles.

at a right angle to the sash bars) are called *purlins*. They help support the sash bars. The peak of the greenhouse roof is called the *ridge*. On each side of the ridge the first row of glass may be set into a ventilator that is hinged at the top. These ridge ventilators allow ventilation even during rain, provided that there is no wind.

After pointing out the parts of a greenhouse, Mr. Weaver was ready to show the students several different types of greenhouses.

Types of Greenhouses

Greenhouses come in various sizes and shapes. In choosing a greenhouse, one must consider many factors. Each type of greenhouse has advantages and disadvantages.

Even-Span. Many commercial greenhouses are even-span greenhouses. This type of greenhouse has two ends, two sidewalls, and a roof that slopes in two directions. The ridge is the highest point in the center of the greenhouse. The distance from the ridge to the eave is the same on both sides of the greenhouse.

The even-span greenhouse normally stands by itself, but could be attached to a building. In a commercial range such greenhouses are usually arranged so that many are placed at right angles to a long corridor greenhouse. The space

between them is wide enough so that the shadow of one greenhouse does not fall into the next one. One advantage of the even-span greenhouse is the uniform light exposure to all areas of the greenhouse.

Ridge-and-Furrow or Contiguous. In many large commercial operations, a number of even-span greenhouses are connected to each other at the eaves. There may or may not be walls between the greenhouses. If walls have been erected under the adjoining eaves, the greenhouses are referred to as *contiguous*. If no walls are present and there is open space throughout the area covered by the greenhouse, the structure is referred to as a *ridge-and-furrow* greenhouse.

Large areas can be economically covered under one roof with a ridge-and-furrow or contiguous greenhouse. This arrangement is more economical to operate than the same amount of space in a number of separate even-span greenhouses. A ridge-and-furrow greenhouse lends itself to the production of horticultural crops that require the same growing environment. Since there are no walls within a ridge-and-furrow greenhouse, it is difficult to have different plant-growing environments. However, a ridge-and-furrow greenhouse may be quickly adapted as a contiguous greenhouse by simply installing walls to separate the sections for crops requiring different temperatures. Polyethylene film can be used for temporary walls.

In areas of heavy snowfall, the ridge-and-furrow type is impractical because snow may bridge over the adjoining gutters and become so heavy that it breaks the glass or causes the structure to collapse.

Quonset. A greenhouse that looks like an upside down half-circle is called a *quonset*. It is a relatively inexpensive structure. The semicircular frames of bent pipe or laminated wood support the greenhouse. These frames are frequently covered with corrugated fiber glass to provide a permanent inexpensive greenhouse. At times the frame may be covered with a flexible plastic and used seasonally.

Gothic. The *gothic* greenhouse resembles the quonset greenhouse except that the roof is not perfectly rounded but is slightly pointed. Such greenhouses are usually covered with corrugated fiber glass. The main advantage of this type of greenhouse is that snow more readily slides off the roof than it will with a quonset type.

Lean-To. The *lean-to* greenhouse is attached to a building. It looks like an even-span greenhouse that was split lengthwise and attached to a building. One advantage of a lean-to greenhouse is that it can be attached to an existing building. This reduces the cost of the greenhouse. Since the lean-to is attached to a building, it may be in the shade part of the time, which is a disadvantage. They should be located on the south side of the building to which they are attached. This type of greenhouse is popular with people who enjoy horticulture as a hobby. Few lean-to greenhouses are used commercially.

Selecting a Greenhouse Type

If you decide to start a horticulture business and need a greenhouse, care should be taken in choosing the type to build. The ridge-and-furrow might not be a wise choice in areas of heavy snows. The quonset greenhouse normally is an inexpensive frame but requires a more expensive covering material. Each type of greenhouse has advantages and disadvantages that should be carefully considered before you select a greenhouse type.

Greenhouse Coverings

Most greenhouse coverings can be grouped into three categories—glass, rigid plastics, and flexible plastics. As is true with types of greenhouses, each covering has advantages and disadvantages. As an employee in a horticulture operation, you may be required to repair or replace the covering. If you build your own greenhouse you will have to make the choice of what to cover it with. You will need to be familiar with the characteristics of each material.

Glass. The standard greenhouse covering used for many years is glass. Under normal conditions glass will outlast other covering materials, and transmits more light than any other covering material. After many years of use, glass still transmits light easily. Many other greenhouse materials cannot make this claim.

Double-strength or tempered glass is often used in greenhouses. These types of glass are stronger than regular glass. Even though they are stronger, they can still be broken by heavy hail or hard objects accidentally or intentionally thrown at them. Glass becomes brittle with age, but usually gives 30 to 50 years of good service. It is interesting to note that larger panes of glass (called "lites") are less easily broken than smaller ones—with the explanation being that they are more flexible. The possibility of glass breakage and the high cost of glass are the main disadvantages of using glass as a greenhouse covering material.

Replacing Glass Lites. Employees in greenhouses that are covered with glass may be expected to replace lites. The process of replacing glass lites is termed *glazing*. There are two reasons why lites have to be reglazed. Putty used in glazing may dry out and cause the lite to become loose. At times lites may be broken and will need to be replaced.

The procedure for installing new glass panes is relatively simple. The broken glass, old putty, and glazier points should be removed from the sash bar where the pane is to be replaced. Putty or other types of bedding compound are placed in the grooves where the glass goes. *Glazier points* (small pointed pieces of metal) are then used to hold the glass in place while another strip of bedding compound is placed on the top of the glass alongside the sash bar. Aluminum bar caps may be screwed to the sash bars to further hold the glass in place.

Rigid Plastics. Fiber glass is a rigid plastic used to cover greenhouses. It is a plastic material into which glass fibers are embedded. *Fiber glass* is not as clear as glass and transmits about 85 percent of the light that glass does. However, careful research has shown that plants grown under fiber glass are identical to plants grown under glass. As fiber glass ages, it has a tendency to become cloudy and dirty, which may result in less light transmission. (Tedlar-coated fiber glass will not become dirty, because the special coating excludes the ultraviolet rays that cause the clouding.) One other disadvantage of fiber glass is that it will burn rapidly if exposed to a high temperature or open flame.

On the plus side, fiber glass is two to four times more resistant to breaking than glass. It may be more expensive to buy than glass, but it requires a less expensive supporting structure. Fiber glass is not as heavy as glass. The supporting sash bars are spaced about 3 ft [0.915 m (meters)] apart for fiber glass. Sash bars on glass greenhouses are spaced about 21 in [53 cm] apart.

Flexible Plastics. Plastic film is being used to cover "temporary" greenhouses in a number of horticulture operations. Most of the plastic is polyethylene. *Polyethylene* is the lowest-cost covering of all materials. It comes in large rolls and can be installed easily on a greenhouse frame. It is often used to cover a wood frame. Polyethylene transmits about 85 percent of the available light. Two layers of plastic are often placed on a greenhouse frame. Air is blown between the two layers to insulate and reduce heat loss.

The major drawback of polyethylene plastic film is its short life. Regular polyethylene will last about 9 to 11 months. Ultraviolet (uv)-resistant polyethylene may last 12 to 18 months. Bright sun causes the film to become brittle. The plastic has to be replaced every year or two.

Plastic films are often used on seasonal greenhouses. They are designed primarily to protect plants from an occasional freeze during the winter and to provide a more favorable growing environment for starting plants.

Horticulture employees are often expected to help replace plastic on the temporary green-

Figure 15-4. Horticulture employees replace plastic or cloth on greenhouses, usually each year.

Figure 15-5. Most wooden greenhouses should be painted every three to five years to protect them from decay.

houses. The old plastic and wooden strips are removed from the greenhouse. The new plastic is placed over the frame and is normally held in place by wooden strips that are nailed to the frame. You may have to saw new wooden strips and you will be expected to use a hammer to nail them in place.

The Greenhouse Frame

In a human body, the skeleton is the frame. In a greenhouse the side posts, purlins, and sash bars serve as the skeleton. They, along with rafters or trusses, give the greenhouse strength. The greenhouse frame can be all wood, all aluminum, a combination of wood and steel, or a combination of aluminum and steeel. Each material has advantages and disadvantages. Horticulture workers often perform maintenance on the greenhouse frame. The type of frame will determine how much maintenance is required. If you are considering building a greenhouse, the material should be selected carefully; it will determine the cost and life of the greenhouse.

All Wood. Greenhouses with a wood frame are the cheapest to build. The sash bars in a

wooden greenhouse are usually cypress, cedar, or redwoood. Other parts of the frame may be made of redwood or other wood that has been treated with a wood preservative.

The primary disadvantage of a wooden structure in a greenhouse is that the pieces must be about four times as large as metal pieces to be of the same strength. This cuts down on the amount of midwinter light reaching the plants, resulting in measurable differences in the vigor of plants grown under metal and wood.

Another disadvantage of wood is that it may warp or decay. The wood should be painted every 3 to 5 years to protect it from decay. The color of the paint is important to assure maximum light reflection. White or aluminum paint is preferred. Only paint especially manufactured for greenhouses should be used. A few wooden parts of the greenhouse may have to be replaced every year after the first 5 years or so. Painting skills are important to greenhouse workers.

Painting involves more than just a paint brush. A knowledge of paints and painting is important because there are many different types of paints. Before paint is applied, the surface to be painted should be prepared. A steel brush and scraper is used to remove loose bits of old paint. The surface is then sanded lightly so that it will be smooth when the paint is applied.

Plants and equipment should be covered or moved to prevent paint from falling on them. Paint should be thoroughly mixed before it is used, and then applied to the wood in even, long strokes. Care should be taken not to get paint on the glass or in other places where it does not belong. After painting, the brushes should be cleaned and stored.

With proper maintenance and painting, a wood greenhouse can last 20 to 50 years. However, it should be noted that regular painting and maintenance are time consuming and expensive.

All-Aluminum. Many of the newer greenhouses are made of an aluminum alloy. An aluminum greenhouse does not require painting, does not rot or rust, and should last for an indefinite period of time. Aluminum can be molded into a variety of shapes, which helps reduce the bulk of the frame. One aluminum piece can take the place of two or three wood pieces.

Aluminum greenhouses may be ordered in prefabricated kits and assembled with a little skilled labor. The primary disadvantage of the aluminum-frame greenhouse is the high initial cost.

Wood-and-Steel. A wood-and-steel combination is found in many older commercial greenhouses. The steel provides most of the structural support in the form of trusses, side posts, and purlins, while the wooden sash bars hold the glass in place. The steel parts of the greenhouse may be in the form of angle iron and/or pipe.

The wood-and-steel greenhouse is more expensive than all-wood but is not as expensive as the all-aluminum greenhouse. Steel can rust. Therefore, the steel and the wood both should be painted, each with a different type of paint.

The steel-and-wood greenhouse will last longer than the all-wood greenhouse, but does require maintenance.

Aluminum-and-Steel. An aluminum-and-steel combination is found in many large commercial greenhouses. Steel is used to build the basic frame providing strength with minimum bulk. The aluminum is used for supporting the glass. This combination provides strength, is durable, and requires little maintenance. If steel and aluminum are in direct contact, it results in a reaction that causes disintegration of the aluminum, so felt pads or other material must be used between the two metals at contact points. The aluminum-and-steel greenhouse is expensive to build, which is the main disadvantage.

After discussing the construction of greenhouses, Mr. Weaver asked whether there were questions. Several students raised their hands. One student wanted to know the purpose of the big pipes running the length of the greenhouse. Another student desired to know why big fans were located on the end of the greenhouse. Do you know the purpose of the pipes and fan? They are used for heating, cooling, and ventilating the greenhouse.

Heating, Cooling, and Ventilating the Greenhouse

An important factor to consider in constructing greenhouses is how the greenhouses will be heated and cooled. Nearly all kinds of plants grow best when the temperature is between 45 and 90° F [7 to 32° C]. Each plant has a more specific range within which it grows best. For example, carnations require 55° F [14° C] at night and 60° F [16° C] during the day; while poinsettias need 65° F [18° C] at night and 75 to 80° F [24 to 27° C] during the days. There are few places in the United States where the temperature constantly remains between 45 and 90° F [7 to 32° C]. In some areas of the country, especially in the North, special attention is given to heating the greenhouse. In the South, cooling the greenhouse is a major problem. Employees in greenhouses often perform maintenance on cooling and heating systems.

Greenhouse Heating Systems

Most commercial greenhouses use one of two systems for heating the greenhouse, either hot water heat or steam heat. Another system widely used commercially in Central, Northern, and Sunbelt states is forced hot air.

Hot Water Heat. Water is heated and then circulated through the greenhouse in pipes. The pipes become hot (because of the hot water inside) and give off heat. The circulation of the hot water is regulated by a thermostat.

Heat given off by the hot water system is uniform throughout the greenhouse. The temperature of the water can be adjusted in mild weather. The water may be heated only to 20° F [12° C] above the outdoor temperature in mild winters, but in very cold weather it may be heated to nearly 200° F [93° C]. This saves on energy cost in mild weather.

The pipes for hot water heat may run along the sides of the greenhouses or under the benches. In large greenhouses, large pipes are needed to carry enough hot water to heat the structure. Hot water heat is found in both large and small greenhouses. Because larger pipes are needed for circulating hot water than for circulating steam, the installation cost is greater than for a steam system; however, the pipes generally have a longer life than the pipes in a steam system.

Steam Heat. In the steam heat system, water is heated until it changes into steam. The steam is then circulated through pipes throughout the greenhouse. Heat is given off by the pipes, which are heated by the steam.

Because of the higher temperature of steam as compared to hot water, greenhouses can be heated more quickly. Also, smaller pipes can be used. Steam can be transported longer dis-

tances more efficiently than hot water can. For this reason, steam systems are usually used in very large greenhouse ranges.

Steam heat systems do not have the flexibility of the hot water system in regulating the temperature of the steam. Water must be heated to boiling in order to turn to steam. The heat of steam is fairly constant.

Boilers are used to heat the water that turns into steam. In some states people who operate boilers have to be licensed, which could be a disadvantage of the steam heating system in those states.

Forced Hot Air. In the forced hot air system a heater burns a fuel such as natural gas or oil to produce heated air. One or several such heaters are installed in each greenhouse, and the heated air is distributed throughout the greenhouse in a long 18 in [45.72 cm] diameter plastic tube called a *convection tube* that is suspended just under the ridge. Small holes about the size of oranges in the sides of this tube (or sleeve) allow for even distribution of this heated air throughout the greenhouse.

In cold climates this system must be supplemented with hot water or steam lines along the lower walls of the greenhouse to overcome a cold zone that always develops there.

Heating System Maintenance. The pipes used for heating the greenhouse require periodic maintenance. Dirt, rust, or chemical residues may build up in pipes in which water is circulated. The pipes are taken apart for cleaning. A chain attached to a wire is pulled through a pipe to knock rust and other residues loose. Another method of cleaning pipes is to use commercial muriatic acid. The acid loosens the residue. Employees in horticulture may be expected to disassemble and assemble pipes. Special wrenches called pipe wrenches are used. At times pipes may need to be replaced. New pipe is cut to the correct length and then threaded. A pipe cutter and pipe threader are used. Green-

Figure 15-6. Greenhouse employees may be expected to possess basic plumbing skills.

house workers who possess basic plumbing skills are valued employees.

Greenhouse Cooling Systems

During spring and summer, the greenhouse can become extremely hot because of the intensity of the sun. If the greenhouse is not cooled, the plants may wilt and die. Most commercial greenhouses are cooled by the fan-and-pad system. Shading is a supplemental technique for reducing the heat in greenhouses.

Fan and Pad. The most commonly used greenhouse cooling system is the fan-and-pad method. It will reduce air temperature within the greenhouse by about 20 to 25° F [12 to 15° C]. Large exhaust fans are mounted on one

side of the greenhouse. Pads of aspen wood, excelsior, or composition fiber are placed in frames in the ventilator space on the opposite side of the greenhouse. (These pads resemble flat shredded wheat.) The pads are kept wet by a system that circulates water through them. The ridge ventilators and ventilators in the fan side of the greenhouse are closed, which means that all air entering the greenhouse has to pass through the water-soaked pads. The exhaust fans are operated to pull the cooled air through the greenhouse. The pads are replaced at periodic intervals. This task is performed by greenhouse workers. This method of cooling the greenhouse is relatively inexpensive.

Since this cooling system increases the humidity in the greenhouse, precautions should be taken to prevent diseases. Mildews and rots thrive in moist areas. To help prevent diseases, water that circulates through the pads is turned off while the exhaust fans are operated in the evening after the outdoor temperature cools down. This will decrease the relative humidity.

Shading. Greenhouses are often shaded in addition to being cooled by a cooling system. A white paintlike substance called a *shading compound* is applied to the roof of the greenhouse. This reflects some sunlight, thereby reducing air temperatures within the greenhouse. Greenhouse workers assist in applying the shading compound and in removing it in the fall of the year. Some greenhouse operators also place muslin cloth above the plants to shade them in the case of plants such as African violets and many orchids which are damaged by strong summer sunlight.

Ventilation Systems in the Greenhouse

For maximum plant growth, a method of ventilating the greenhouse is needed. In addition to helping regulate the temperature, the ventilation

Figure 15-7. Greenhouses are shaded in the summer to reduce heat.

the next chapter.) Some newer greenhouses do not have ventilators, but have intake louvers and exhaust fans instead.

Exhaust Fans. Some greenhouses have exhaust fans mounted in the end or side walls. These fans may be manually operated or connected to a thermostat. Louvers in walls opposite the fans open automatically when the fans operate so that new air is drawn into the greenhouse.

Convention Tubes. Many modern greenhouses have an air circulation system that uses a long tube called a *convection tube*. The tube is suspended a few inches below the ridge of the roof. This tube is 1–2 ft [0.305–0.61 m] in diameter and runs the length of the greenhouse. One end of the tube is attached to a fan, while the other end of the tube is sealed. Small holes about the size of an orange are spaced evenly every 2 or 3 ft [0.61 or 0.915 m] along the tube. The fan blows air into the tube. The air is forced out the holes and creates a number of small air

system is needed to circulate air and maintain the proper relative humidity. This helps reduce diseases and is necessary for maximum plant growth. Gentle air movement also results in better plant growth because it permits a more rapid exchange of gases between the leaves and the atmosphere.

Ventilators. Ventilators along the ridge and on the sides are found in many greenhouses. In older greenhouses, these ventilators were opened and closed by hand. Ventilators in more recent greenhouses are operated by electric motors that are controlled by thermostats. Ventilators open and close as the temperature in the greenhouse changes. Often employees in greenhouses have to service the electric motors that operate the ventilators. (This is discussed in

Figure 15-8. To provide uniform ventilation in the greenhouse a fan blows air through a convection tube. Evenly spaced holes in the tube let air escape.

currents, which circulate air through the greenhouse. The fan can be adjusted to use air from the outside, air from the inside, or heated air from a hot air heater.

After explaining how the heating, cooling, and ventilating system worked, Mr. Weaver asked, "How do you think we water the plants in our greenhouses? If we used a hose or watering can, it would take a long time." Do you know how plants are watered in commercial greenhouses?

Watering Systems for the Greenhouse

Plants need water to grow properly. In commercial greenhouses, watering is usually done by a semiautomated system of some kind. Watering by hand may be done in the case of plants requiring special attention. If hose watering is done by hand, a special device called a "water breaker" is attached to the hose in order to allow a large volume of water to be applied at very low pressure. This prevents washing of the soil. Four different systems of watering are typically used in commercial greenhouses.

Subirrigation.   Potted plants in pots up to 6 in [15 cm] in height may be watered from below. A level, water-tight bench is needed. The surface of the bench is covered with a special water-retaining thin fiber mat, or with a "bubble"-type plastic sheet. These mats or sheets are simply soaked with water several times a day, and the water is drawn up into the soil in the pots. Water is then drawn up into the plant by capillary action.

The Spaghetti System.   The spaghetti system, also called the Chapin System, uses many small plastic tubes. A plastic tube (about the size of a spaghetti strand) with a small weight on the end is placed on the soil in the pot to be watered. The other end of the tube is connected to a plastic water pipe. Hundreds of small tubes are connected to the water pipe. When the

Figure 15-9. Orchid plants are usually watered using the spaghetti or Chapin system.

water is turned on, each pot receives water. The water supply can be regulated automatically with a time switch; the usual length of time for watering is about 7 minutes. This is a time-saving way to water hanging baskets.

Spray System.   The spray system is used primarily for cut flower crops such as carnations and chrysanthemums. Water pipes are placed on the soil surface on the greenhouse bench. Small spray nozzles, spaced evenly along the water pipe, distribute water in a flat spray over the soil surface. An important point is that the leaves of the plants are kept dry to avoid foliage diseases.

Mist System.   The mist system is used for the propagation of plants from cuttings or seeds. Nozzles are mounted about 2 to 3 ft [0.61 to 0.915 m] over the propagation bench. When the water is turned on, a fine mist is sprayed over the bench. The length of time of each spraying and intervals between sprayings are regulated by a time clock.

Occasionally the nozzles may become clogged with dirt or chemical deposits from the water. Small wires or chemicals are used to clean the nozzle. As a greenhouse employee, you may be required to perform this task.

After Mr. Weaver had explained how watering systems work, he and the class walked through several more greenhouses. One of the students noticed that the benches in the greenhouses were not like those at school and were arranged differently. The student asked Mr. Weaver how many different types of benches are used and how they should be arranged. Do you know the different types of benches commonly found in horticulture and how they are arranged?

Benches and Beds in the Greenhouse

Plants grown in the greenhouse are placed in beds or on benches. One task often performed by a greenhouse worker is assembling benches.

Bench and Bed Construction

Benches may be off or on the ground. They are often called *raised beds* if they are off the ground. *Raised benches* are used for potted plants and require less bending by people who are working with the plants. Air circulates better around raised benches. Raised benches are generally 30 in [76.2 cm] off the ground and 4 ft [120 cm] wide.

The soil surface in raised beds may be 6 to 12 in [15 to 30 cm] above the floor of the greenhouse. The raised beds are used for growing cut flower crops such as roses, carnations, chrysanthemums, and snapdragons. Several materials are used in the construction of beds and benches.

Wood and Wire Mesh. A frame of wood is covered with 1 in² [2.54 cm²] wire mesh. This type of bench is used for growing potted plants. The wire mesh supports the pots, allows for good air circulation, and provides good drainage.

Corrugated Asbestos. Hard, wavy sheets of concrete and asbestos are often used for green-

Figure 15-10. Several types of benches are used in the greenhouse. On the left a wood and wire mesh bench. On the right a corrugated asbestos bench.

house benches. This material does not decay. A frame of wood or metal is used to support the material. The benches should be sloped slightly to allow for drainage, and should also have drainage holes every few inches.

Concrete.

Raised beds or benches are often constructed of reinforced concrete. The botton of the bench is V-shaped, with holes in the bottom of the V at intervals to allow for drainage. Concrete is long lasting but may need to be patched periodically as it chips or cracks. Expansion joints must be provided at 12-ft [3.66-m] intervals so that the expansion and contraction when the soil is steamed for pest control will not crack the concrete.

Concrete Blocks and Snow Fence.

Temporary benches are often constructed of concrete blocks and snow fence. The concrete blocks are used as legs, with one laid flat on the ground and another block placed upright on top of it. The legs are spaced about 6 ft [1.83 m] apart in two rows. Each row is 4 ft [1.22 m] apart. Wooden 2 × 4 in [5 × 10 cm] boards are laid on top of the blocks in the two rows. Snow fence is then laid on top of the wooden boards. Potted plants and bedding plants are often grown on this type of bench.

Arranging the Benches.

Benches within a greenhouse can be arranged in a variety of ways. The two most common arrangements are *peninsular* and *longitudinal*. In the peninsula arrangement, an aisle about 4 ft [1.22 m] wide runs down the center of the greenhouse lengthwise. The benches are placed at right angles to the aisle, about 18 in [45 cm] apart. The peninsula is an efficient arrangement. It provides a large amount of usable bench space. All the benches are accessible from the one center aisle.

In the longitudinal arrangement, the benches are placed in rows the length of the greenhouse. An aisle about 18 in [45 cm] wide is between each row of benches. This arrangement requires more walking, so less bench space is available. However, water systems may be eaiser to install because of the long, continuous bench arrangement.

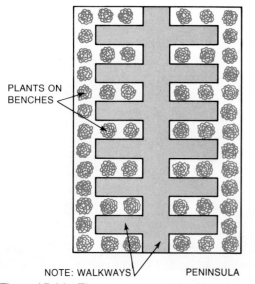

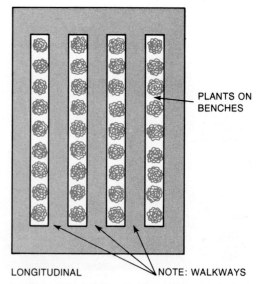

PLANTS ON BENCHES

PLANTS ON BENCHES

NOTE: WALKWAYS   PENINSULA

LONGITUDINAL   NOTE: WALKWAYS

Figure 15-11. The two most common bench arrangements within the greenhouse are the pennisula and longitudinal.

The Floor

The floor of the greenhouse should be composed of small stones, porous concrete, or other water-absorbing material. There are several advantages to this type of floor. Water will be retained and relative humidity will be easier to control. An entire floor of solid concrete retains heat and does not provide moisture to maintain the needed humidity during winter months.

It may be desirable to construct the aisles and walkways of concrete. Often equipment will be moved between the benches in the greenhouse. It is easier to move equipment or concrete. A concrete aisle will not be muddy. It is easy to keep clean. Gravel or stone can be used in place of concrete if it is well packed. The space under benches is usually filled to a 4-in [10-cm] depth with ½-in [1.25-cm] stones to provide needed humidity and to prevent weeds from growing. The aisles should be just wide enough to accommodate the equipment, because greenhouse space is valuable.

Headhouse

A workroom called the *headhouse* should be located near the greenhouse or should be connected to the greenhouse. Often headhouses are located on the north side of the greenhouse to avoid shading the greenhouse. The headhouse contains work benches and bins for storing soils, fertilizers, and other planting media. Cabinets with locks for storing pesticides are usually located in the headhouse. The heating plant is often located in the headhouse.

"Let's go outside now and I'll show you a number of other plant-growing structures," stated Mr. Weaver.

Other Plant-Growing Structures

There are a number of other plant-growing structures used in the horticulture industry be-sides greenhouses. These structures are not as complex as the greenhouse, but are just as important in many horticulture operations. Some of these structures are hotbeds, cold frames, shade houses, wintering houses, and container plant surfaces.

Hotbeds

A *hotbed* is used primarily to start plants in the early spring. It functions as a miniature greenhouse. Hotbeds are not used as much as they once were, because they are difficult to work in and require a lot of labor.

A square or rectangular pit is dug into the ground 12 to 18 in [30 to 46 cm] deep. A frame is constructed around the edge of the pit. The frame may be concrete blocks, concrete, brick, or wood. The frame extends above the ground 1 to 2 ft [30 to 61 cm]. The frame is sloped so that the south or east side is about 1 foot [30 cm] lower than the north side.

Glass sashes are then placed on top of the frame. Since standard size glass sashes are 3 ft × 6 ft [91 cm × 183 cm] the cold frame should be 6 ft [183 cm] wide and a multiple of 3 ft [91 cm] in length (examples: 6, 9, 12 ft or 1.8, 2.8, 3.7 m). The sun is used to heat the hotbed, along with another source of heat. Electrical cables are often buried beneath the soil and are used to heat the hotbed. The buried electric cables give off heat just as an electric blanket would. Hot water or steam lines can be used to heat the hotbed. Horse manure has been used to heat hotbeds. The pit is filled to a depth of about 12 in [30 cm] with horse manure. On top of the manure is placed 4 to 6 in [10 to 15 cm] of soil. As the manure decays, heat is given off.

Cold Frames

A *cold frame* resembles the hotbed in appearance. The main difference is that the cold frame does not use additional heat. The only heat it gets is from the sun.

A cold frame serves several useful functions. In the spring it can be the halfway point between the greenhouse and the outdoors. Plants grown in a greenhouse should not be removed from the greenhouse and planted outside immediately. They need a transition or hardening period in which to become adjusted to wind, higher light intensities, and fluctuating temperatures. Plants can be removed from the greenhouse and placed in the cold frame. The cold frame environment will not be as warm as the greenhouse or as cool as night temperatures outside.

In the fall, the cold frame can be used to store potted plants such as roses and hydrangeas before bringing them into the greenhouse for the winter.

Shade House and Lathhouse

In the summer months, many plants require shade. A *lathhouse* is made of wood poles and rolls of snow fence or other similar materials. The wooden poles are erected in the ground about 10 to 12 ft [3.05 to 3.66 m] apart. The

Figure 15-13. Container-grown plants are placed in wintering houses in Northern states during the winter.

poles are connected at the top with lumber. A snow fence is rolled out on top of this frame. This reduces the sunlight by as much as 50 percent.

The shade house or cloth house makes use of a frame similar to that of a lathhouse. The frame is covered with a material such as saran or plastic screen. Shade houses are used extensively for growing cut flowers in Florida. Foliage plants grown in the South are often placed under shade houses to acclimate them to a lower light intensity before they are shipped to Northern states for sale.

Shade houses and lathhouses are used to protect potted plants and container-grown plants. Often you will see plants for sale at nursery or garden centers being displayed under a lathhouse. This protects the plants from the sun. In the South and in many areas of California, plants such as chrysanthemums and azaleas may be grown year round in shade houses. In the North, asters, snapdragons, and chrysanthemums may be grown in the shade house during

Figure 15-12. Lathhouses or shade houses are used to protect plants from excessive sun.

the summer. Nurseries use lathhouses for growing Taxus and azaleas for 2 to 3 years before lining them up out in fields.

Wintering Houses

In some parts of the country, especially in the North, a plastic tunnel is built over plants growing outdoors. The tunnel is built with polyethylene or a similar plastic, stretched over a metal or wooden frame. This is called a *wintering house*. No heat is used in the wintering house. The temperature in the tunnel may drop below freezing, but it will not be as severe as the outside temperature. Wintering houses are used on small fruit trees and other nursery plants. Plants grown in containers are placed in a wintering house during the winter in Northern states. A wintering house closely resembles a plastic greenhouse.

Container Plant Surfaces

More and more plants are being grown outdoors in containers. The containers are made of metal, plastic, or peat. Container growing started in

Figure 15-14. Many nurseries are growing plants in containers on specially prepared surfaces.

California and Florida because many plants there have a very short dormant season and can only be readily moved in containers. It is now widely used throughout the United States.

Special surfaces are used to produce container-grown plants. An asphalt pavement similar to that used on roads is being used. Gravel surfaces are also very common. This type of surface is much easier to work on than ordinary soil. The surface does not erode easily and can be sloped to provide drainage. The main reasons for growing plants on a special surface are so that the roots of the plant will not grow out of the container and into the soil below and to provide excellent drainage.

Cold Storage Facilities

Cold storage facilities are found in many horticulture operations. Florists prolong the life of cut flowers by placing them in refrigerated storage. A carnation can be kept for 1 week if it is stored at 33° F [0.5° C]. Narcissus and hyacinth can be stored for 2 weeks at 33° F [0.5° C].

Fruit and vegetable producers use cold storage facilities for extended periods of time to preserve their produce. Apples can be stored for up to 8 months at 32° F [0° C]. Oranges can be kept at 32° F [0° C] for 2 to 3 months. Late-crop potatoes can be stored at 40° F [5° C] for 5 to 8 months.

Cold storage facilities are used by nurseries to store bare-root nursery stock. Deciduous fruit trees and shrubs are stored at 32° F to 35° F [0° C to 2° C] during the fall and winter. They are then sold during the spring. Many bulbs such as amaryllis and gladiolus are stored over the winter at a temperature of about 45° F [8° C].

The type of cold storage facility needed will depend upon the type of horiticulture business. Florists need coolers with glass fronts to display the cut flowers. Fruit and vegetable growers, along with nurseries, require large walk-in facilities. Some storage facilities have been built into sides of hills or underground because soil serves to insulate the facility and has a tendency to be cooler than above-ground storage. Your grandparents may have had an underground root cellar to store potatoes, apples, and other crops if they lived in the country. Cold storage facilities built above ground need to be well insulated to retain the cool air.

A cold storage facility is like a giant refrigerator. The temperature is maintained with a thermostat. Many cold storage facilities have benches or shelves inside.

Specialized Equipment

Many types of specialized equipment are used to produce horticultural crops. In some fruit production operations, large machines called hydrocoolers are used to rapidly cool the fruits after they are harvested. Special machines may be used to sort and grade vegetables and fruits. Buildings are often needed for grading and packaging produce. Mechanical harvesters may be used with such crops as beans, apples, and tomatoes. Most of this equipment has gasoline or diesel engines or electric motors. In the next chapter, general principles of machinery maintenance will be discussed.

Horticultural Plant-Growing Structures: A Review

A number of different plant-growing structures are used in horticulture. Greenhouses are commonly used, and are found in a variety of shapes with the even-span, ridge-and-furrow, and quonset found generally in commercial horticulture operations. The frame of the greenhouse can be wood, aluminum, steel, or a combination of these materials. The covering is glass, rigid plastic, or plastic film. Greenhouses are heated with hot water, steam, or forced air. Exhaust fans and pads are used to cool the greenhouse. Ridge and side ventilators or convection tubes are used to ventilate the greenhouse.

Other structures used to grow and maintain plants are hotbeds, cold frames, lathhouses, shade houses, wintering houses, container plant surfaces, and cold storage facilities.

Employees in horticulture are expected to maintain the plant-growing structures. These tasks include replacing glass, painting, applying shading compound, cleaning spray nozzles and pipe, covering greenhouses with plastic, and assembling benches.

THINKING IT THROUGH

1. Of the six types of greenhouses discussed in this chapter, which would you choose for a permanent commercial greenhouse? Why?
2. What are the advantages and disadvantages of wood, steel, and aluminum in the greenhouse frame?
3. What are the advantages and disadvantages of glass, fiberglass, and polyethylene as greenhouse coverings?
4. What are the three types of greenhouse heating systems? How does each operate?
5. How can a greenhouse be cooled?
6. Why is good ventilation needed in a greenhouse?
7. How does the spaghetti watering system work?
8. What material would you use for building permanent greenhouse benches? Temporary benches?
9. What is the difference between a hotbed and cold frame?
10. What function does a lathhouse serve?
11. What are the purposes of a plant-growing surface?
12. Why are cold storage facilities needed?
13. What is a wintering house?
14. What types of mechanical skills are needed by horticultural employees?

16

The Care, Operation, and Maintenance of Small Engines

With a trend toward reducing the amount of labor in horticulture, the use of small engines has increased dramatically in the past 25 years. Today, many tasks such as mowing, trimming, and tilling are almost exclusively mechanical operations. The person using a rototiller can perform more work than several people who are using spades or hoes, and much more easily. The portable nature of the small engine allows the worker to use the machine to do many jobs. And new machines like the small-engine-powered vacuum sweeper and the "Weed-eater"® are being developed each year.

Workers employed in the landscape industry use small engines every day. The reel-type mower and the rotary mower are used for grass cutting and trimming. The lawn and garden tractor provides power for many operations like dethatching lawns, aerating greens, trimming golf fairways and roughs, and removing leaves or clippings. The arborist and arboriculturist use small engines to power the chainsaw, pruning saws, wood chippers, and many types of chemical applicators and sprayers. Nursery workers find both the lawn and garden tractors useful for many jobs. Field-grown stock can be planted in a growing area that has been plowed, disked, and bedded with the tractor. The smaller tractor provides the power for cultivation and spray application. A high-clearance nursery tractor may be used to root-prune and to ball the plants. People who work in greenhouses can reduce the amount of manual labor by using small engines to drive the mulcher and the soil shredder. They may even find that a utility tractor and trailer can move plant materials and medium more easily than a wheelbarrow. Also, many greenhouse owners use the small engine to power a stand-by electric generator in the event of a power outage. Horsepower is replacing human power. Since much of the horsepower comes from economical small engines, the worker of the future will benefit by knowing the basic principles of operation, service, and adjustment of the small engine.

CHAPTER GOALS

Your goals in this chapter are:

- To identify 4-cycle engines and describe their characteristics
- To identify 2-cycle engines and describe their characteristics
- To perform preventive maintenance on small engines
- To service the ignition system
- To service the carburetor system
- To service the rewind starter
- To troubleshoot small engines and diagnose engine problems.

Basic Engine Principles

The small engine used today is an internal combustion engine. This means that the fuel is mixed with air and is compressed inside the cylinder of the engine. At the proper time, this compressed mixture is ignited by a spark. This "explosion" releases a large amount of energy, which drives a sliding cylinder or piston down and transfers power to a crankshaft and the flywheel.

In 1838, an Englishman named Barnett developed the idea of a 2-stroke cycle engine. His engine worked in the laboratory but it was not successful in general. Scientists used many strategies but they were largely unsuccessful, until a German inventor by the name of Otto built the first successful small engine in 1876. The first successful engine was a 4-stroke cycle and was often called the Otto-cycle engine. It was not until the later 1940s and early 1950s that small engines became popular. Considering the evolution of humanity, the small engine is a very recent development.

Figure 16-1. The small engine can reduce the drudgery of physical labor required only a few years ago. These engines can be operated and serviced by horticulture employees who understand basic engine principles.

Figure 16-2. The small engine has application in almost any horticulture job. New tools are being developed each year to make these jobs easier.

The 4-Stroke Cycle Engine

Engines are described by the number of strokes required by the piston to complete one cycle or series of events. A *stroke* is the movement of the piston from one end of the cylinder to the other.

Engine Operation

The 4-stroke cycle engine, generally called the 4-cycle engine, completes a cycle for every four strokes of the piston. These occur in a sequence of the intake, compression, power, and exhaust strokes. During each stroke, very different processes are occurring as shown in Figure 16-3.

The Intake Stroke. Initially, the piston is at the position closest to the cylinder head. As the piston moves away from the sealed cylinder head, a vacuum is created. The intake valve is opened, allowing an air-gasoline mixture from the carburetor to fill the cylinder. The intake valve closes as the piston reaches the point farthest from the cylinder head. This results in a sealed cylinder filled with a combustible mixture. If the small engine uses diesel as a fuel, only air is drawn into the cylinder initially. The diesel fuel will be injected into the cylinder under high pressure during the compression stroke.

The Compression Stroke. As the piston begins the stroke, the air-gasoline mixture in the cylinder is compressed as the piston travels toward the top of the cylinder. As any gas is compressed, the temperature of the gas rises and the gas becomes easier to ignite. Just before the piston reaches the top of the cylinder, called "top dead center (TDC)," the mixture is ignited. In diesel engines, the fuel is injected just before TDC. The ignition of the mixture does not result in an actual explosion but has a very carefully controlled burning rate. As the flame travels across the piston, heat and pressure are released and produce power for the engine.

The Power Stroke. The pressure against the top of the piston causes it to move away from the cylinder head as shown in Figure 16-3. The piston is connected to an offset shaft called a *crankshaft.* The up-and-down motion of the piston is converted to rotary motion by the crankshaft. The crankshaft of the engine transfers this rotary motion to a machine. By the time the piston reaches the end of this stroke, the majority of the energy from the fuel has been released as mechanical energy and heat. The remaining exhaust gases cannot be used to produce power.

The Exhaust Stroke. When the piston reaches the bottom of the stroke, called "bottom dead center (BDC)," the exhaust valve opens. As the piston moves toward the cylinder head, the exhaust gases are forced out through the muffler. The exhaust valve closes at the end of the stroke and the engine is ready to repeat the cycle. The crankshaft has turned two complete revolutions, with power exerted on the power takeoff shaft of the engine.

Engine Characteristics

Just as legumes are different from grasses, engines have different operating requirements. The fuel and lubrication for a 4-stroke cycle engine is different from that used in a 2-stroke cycle. Not all engines use the 4-stroke cycle principle, but there are several ways to identify those that do. The best technique is to become familiar with the owner's manual that is furnished with *every* new engine. This small reference book not only gives the basic information about the engine, but also provides information concerning the operation, service, and adjustments that should be made. Following these simple directions will add years to the life of any engine. Another technique is to read the manufacturer's nameplate, which is attached directly to the engine. This nameplate will provide much useful information, such as the model, type, and serial numbers, in the event you need new engine

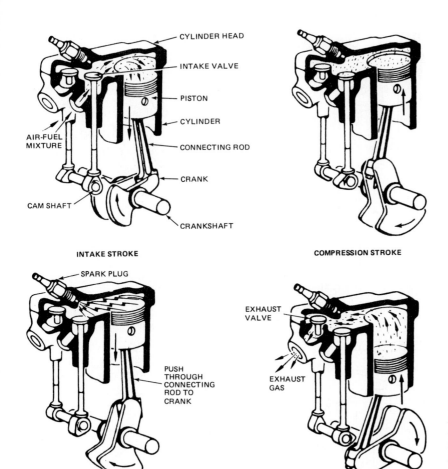

Figure 16-3. The four strokes are required to complete one cycle. Many engines use this piston and valve sequence. They are commonly referred to as "4-cycle engines."

parts. The 4-cycle engine always has an oil sump, or reservoir, with a filler plug, an oil dipstick, or a visual oil level indicator. This is essential for engine lubrication. If you cannot find these, the engine is a 2-stroke cycle. A third check is to locate the muffler. A 4-cycle muffler connects at the cylinder head end of the engine. If the muffler appears to be in the middle of the engine, it is a 2-stroke cycle engine. A final check should be made by identifying the carburetor. The 4-cycle engine carburetor feeds the air-fuel mixture into the top of the cylinder near the cylinder head, while a 2-cycle engine car-

buretor may be attached to the opposite end of the cylinder. If you are still unsure whether the engine is a 4-cycle, ASK! Remember, many very costly mistakes can be prevented by asking your instructor or supervisor.

Applications of the 4-Cycle Engine

By design, the 4-cycle engine has some advantages and some limitations. A larger volume of oil is available for lubrication, which extends the service life. It also usually operates at lower speeds to produce the same horsepower. By

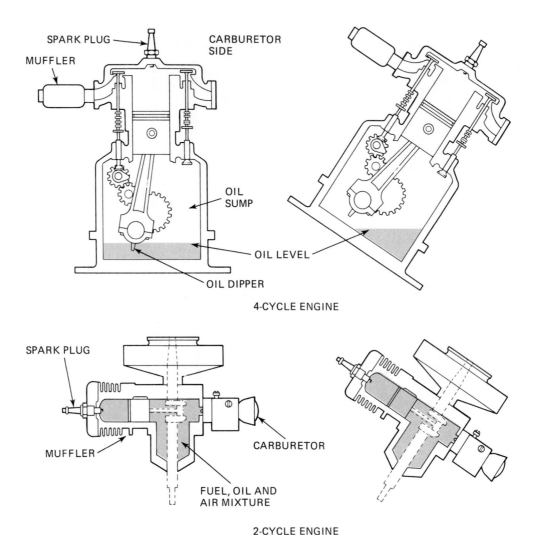

SPARK PLUG

MUFFLER

CARBURETOR SIDE

OIL SUMP

OIL LEVEL

OIL DIPPER

4-CYCLE ENGINE

SPARK PLUG

MUFFLER

CARBURETOR

FUEL, OIL AND AIR MIXTURE

2-CYCLE ENGINE

Figure 16-4. The 4-cycle engine can be distinguished from the 2-cycle engine by using several simple checks. The 4-cycle engine is usually limited to a level or nearly level application.

design, it is very suitable for multicylinder engines. It operates on regular gasoline or diesel fuel, depending on the type of engine. Because of the oil sump, these engines must be operated in a near-level position. On slopes of 15 degrees and steeper, the lubrication system may not work properly, thus causing engine failure. This is a limitation for chainsaws, power edgers, Weed-eaters,® and other very portable tools.

The 2-Stroke Cycle Engine

The 2-stroke cycle engine is often referred to simply as a 2-cycle engine. It is popular for many applications, because it produces a high horsepower output in relation to the engine weight. This is especially important if the nursery worker is required to carry the engine as a part of a backpack sprayer. Also, the 2-cycle engine such as those on chainsaws may be used in many different positions without affecting the operation.

Engine Operation

In the 2-cycle engine, the piston controls both the intake and the exhaust gases in order to produce a power stroke with each revolution of the crankshaft. The top of the piston is cast in a wedge shape, causing a turbulence in the air-fuel mixture. Also, a leaf-type valve, sometimes referred to as a reed valve, controls the flow of the air-fuel mixture into the engine. In the 2-cycle engine, the intake and compression sequence occur during one stroke of the piston, and the power and exhaust sequence occur during the following stroke.

The Intake-Compression Stroke. As the piston moves from the bottom of the cylinder to the top, two separate events take place. Because of a vacuum, an air-fuel mixture is drawn through a reed valve into the crankcase. At the same time, the piston travels past the intake-exhaust port. These are passages that open into the cylinder from the crankcase and the muffler. Now the piston has sealed the cylinder and begins the compression sequence; when the piston reaches top dead center (TDC), the maximum air-fuel charge is drawn into the engine crankcase and the second charge has been compressed and is ready for ignition.

The Power-Exhaust Stroke. As the piston reaches TDC, a high-voltage spark ignites the air-fuel mixture in the cylinder. This energy drives the piston to the bottom of the cylinder, transferring power to the crankshaft and the power take-off shaft. At the same time, the air-fuel mixture in the crankcase is put under moderate pressure by the downward movement of the piston. As the piston moves into the lower part of the cylinder, the intake and exhaust ports are opened. The exhaust gases escape through the muffler, and the fresh air-fuel charge is drawn into the cylinder. The wedge-shaped top on the piston serves as a deflector for both the exhaust and intake gases, but there is some mixing of the two. When the fresh air-fuel charge is diluted with exhaust gases, energy is reduced. Thus the 2-cycle engine is not as efficient as the 4-cycle engine.

Applications of the 2-Cycle Engine

The 2-cycle engine is lightweight and compact. It is available as fractional horsepower model airplane engines or as 400+ horsepower diesel truck engines. The engines used in horticulture usually range from 2 to 5 horsepower. These engines are cast from aluminum or magnesium alloys to reduce the weight and improve the cooling. The crankcase must remain sealed for proper operation, so the engines should not be dropped or struck. These alloys will crack or break with rough use or abuse. Common applications of the 2-cycle engine include chainsaws, lawn mowers, edgers, and sprayers. Because the operator either pushes or carries the tool, weight is important. Also, because they have fewer parts, 2-cycle engines usually have a lower initial cost.

Advantages and Limitations. As pointed out, the high ratio of horsepower to weight makes the engine a good selection for many applications. The engine can be operated in any position—upright, on its side, or even upside down! Because of this, almost all chainsaws are powered by 2-cycle engines. The lower initial cost also makes this engine a good choice.

Engine Lubrication. Because the 2-cycle engine does not have oil in the crankcase, the lubrication system is designed differently. The oil used to lubricate the engine is mixed with the fuel. As the fresh air-fuel charge enters the crankcase, the oil droplets cover the moving parts. The amount of oil that is mixed with the

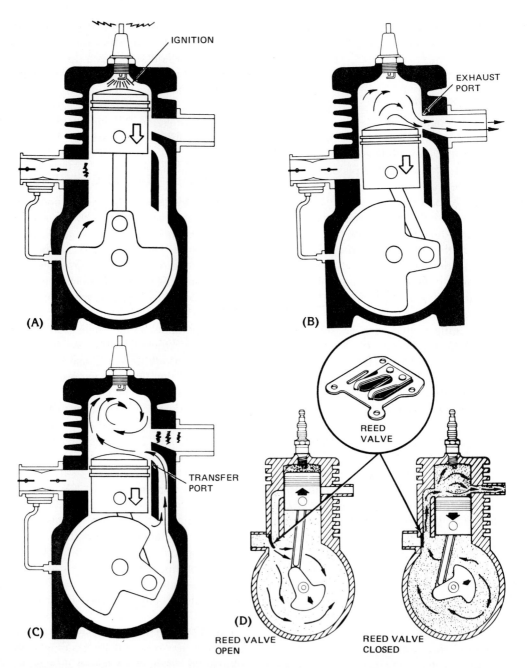

Figure 16-5. A diagram of the 2-cycle engine. (A) As the piston nears TDC ignition occurs. The resulting high combustion pressure forces the piston down. (B) As the piston moves down the cylinder, it opens the exhaust port, allowing the burnt exhaust gases to flow from the cylinder. (C) As the piston nears BDC, it opens the transfer port, allowing a fresh air-fuel mixture to flow from the crankcase into the cylinder. (D) A reed valve is used in the intake port of many 2-cycle engines.

gasoline is *very critical*. Depending on the type and design of the engine, the mixture ranges from 1 part oil to 16 parts of gasoline (½ pt of oil per gal of gasoline) to 1 part of oil to 40 parts of gasoline (1 qt of oil per 10 gal of gasoline). It is important to follow the manufacturer's recommendations as to both the amount and the type of oil to use. Many manufacturers will void the engine warranty if their oil is not used.

Small Engine Maintenance

Every operator's manual is designed to encourage good preventive maintenance practices. The manufacturers know that these practices extend engine life and result in better service. A good experience with a particular engine will result in the customer's buying the same brand again. Good maintenance is not a complicated job that requires many tools. Often it is checking the machine over before starting the engine. A loose nut is easy to tighten, but if the operator waits until it is lost it takes more time and money to repair. The handle on a push-mower has a great deal of vibration, which causes nuts to loosen. If the landscape worker does not check the mower, the handle may vibrate loose and cause a delay in the job.

Always inspect the belts and chains on a daily basis. Both should have the correct tension and be in good running order. Usually a belt shows signs of wear long before it breaks, and can be ordered and replaced without a loss of operating time. If the machine has a cutting blade, such as a lawn mower, the blade condition should be checked. A sharp blade does an easier, faster, and better job.

Servicing the Air Cleaner

The small gasoline engine uses over 9,000 gal of air for each gallon of fuel. It becomes obvious that both the fuel and the air must be clean. Three different types of air cleaners are common

on small engines, and each type is very effective in filtering the dirt from the air if serviced properly. Many operators do not realize the importance of keeping these filters serviced on a regular basis.

The Oil Foam Element. The oil foam air cleaner is commonly used on late-model engines. Each manufacturer specifies the frequency for service, but the majority recommend cleaning and reoiling the element once for every 25 hours of normal operation. If the engine is operated under extremely dusty conditions, the element should be cleaned more frequently—even on a daily basis.

To service the oil foam element, remove the complete air cleaner assembly. Take the assembly apart and inspect the condition of the foam element. If it is extremely dirty, it should be serviced more frequently. Wash the foam element in a liquid detergent and water until it is clean. At the same time, wash the cover and body. *Do not* use gasoline as a washing solvent, because it may catch fire and it does not remove the fine abrasive grit from the element. Rinse the element in clean warm water, then squeeze in a dry towel. Be sure all the water is removed, then saturate the element with SAE 30 engine oil. Using another shop towel or cloth, squeeze the element to remove all excess oil. Reassemble the foam element, the air cleaner body, and the cover, and fasten to the carburetor. Make sure the gasket forms an airtight seal.

The Oil Bath Cleaner. The second most common type of air cleaner is the oil bath type. Remove the complete cleaner from the carburetor and pour the old oil from the bowl. Using a scraper, remove the caked dirt from the bottom, then wash the complete cleaner in a cleaning solvent or in kerosene. *Do not* use gasoline to wash the air cleaner! Wipe the air cleaner with a shop towel and refill with the same oil that is used in the crankcase. Do not overfill (above the line marked "oil level,")

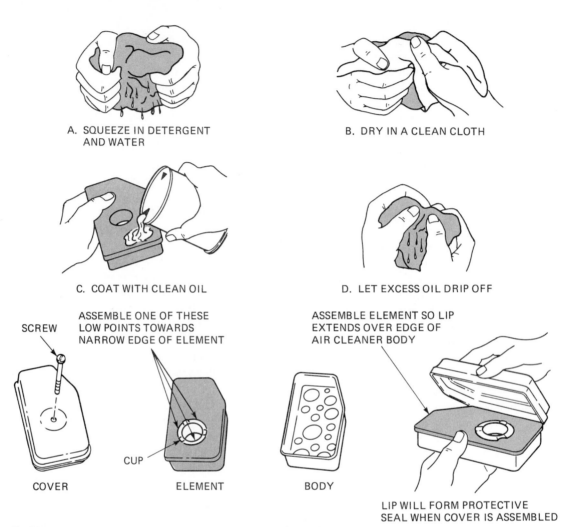

A. SQUEEZE IN DETERGENT AND WATER

B. DRY IN A CLEAN CLOTH

C. COAT WITH CLEAN OIL

D. LET EXCESS OIL DRIP OFF

SCREW

ASSEMBLE ONE OF THESE LOW POINTS TOWARDS NARROW EDGE OF ELEMENT

ASSEMBLE ELEMENT SO LIP EXTENDS OVER EDGE OF AIR CLEANER BODY

CUP

COVER

ELEMENT

BODY

LIP WILL FORM PROTECTIVE SEAL WHEN COVER IS ASSEMBLED

E. REASSEMBLE THE OIL-FOAM ELEMENT AND THE BODY CORRECTLY

Figure 16-6. Cleaning the oil-foam cleaner.

since this will affect the engine performance. In commercial applications, the oil bath air cleaner should be serviced on a daily basis.

The Dry-Type Element. The dry-type air cleaner is becoming more popular, but is still not common on most small engines because of the extra costs. It does not require as much

service, and does a superior job of filtering small particles. The element is more expensive as a replacement cost. To clean, remove the cover and paper element. Tap the element against the heel of the hand until the element is clean. Many manufacturers specify that the element be washed in a nonsudsing detergent similar to those used in automatic dishwashers. The element should be air-dried for 48 hours prior to use. Do not use oil or more than 35 lb per square inch of compressed air to clean this type of air cleaner element. You should use a drop-light to inspect the element for holes. Any hole will allow dirt to pass through the filter and damage the engine. Hold the light inside and rotate the element to determine whether there are any small holes. If one is found, discard the element.

Servicing the Fuel System

The engine uses air and gasoline to produce a combustible fuel mixture. In the 2-cycle engine, oil is also mixed with the fuel. It is important that each of these is clean. If the engine appears to lose power or to operate at irregular speeds, the fuel strainer should be inspected and serviced. Most small engines use one of three types of fuel strainers: (1) an in-line screen attached to the fuel pick-up hose, (2) a screen in the fuel tank, or (3) a combination glass sediment bowl and screen. The operator's manual will describe the system used for a particular engine.

In-Line Screen. To service an in-line screen, simply shut off the fuel supply and remove the fuel line and the cup-shaped screen. Flush the screen with clean solvent and replace it. Many manufacturers recommend that the screen be replaced with a new element annually.

Fuel Tank Screen. The screen in the fuel tank is also serviced by flushing with a clean solvent. However, it is usually more difficult to remove. For this reason, a new element is generally used.

Glass Sediment Bowl and Screen. The glass sediment bowl and screen is superior to either the in-line or the fuel tank filter. However, it is also more expensive, so its use is limited to larger engines. The glass bowl provides a water trap and should be inspected daily. If either water or dirt appears in the bowl, the fuel supply

should be turned off and the bowl removed and washed. Care should be taken not to damage the gasket between the bowl and the body of the strainer. Also, most have a flat screen in the top of the strainer body. Using a small screwdriver, remove and wash the screen in clean solvent. Replace both the bowl and the screen, and check for leaks before starting the engine.

Servicing the Crankcase Breather

The crankcase breather is similar to a one-way valve that allows gases to pass out of the engine but not into the engine. The 4-cycle engine incorporates a crankcase breather to (1) prevent excessive pressure in the crankcase, (2) maintain a partial vacuum during the compression and exhaust strokes, (3) recirculate combustible vapors from the crankcase to the carburetor, and (4) prevent dirt from entering the engine. If proper service is not performed, there is a tendency to blow oil out of the engine around the seals or the dipstick.

Many horticulture companies are small and do not have specialized workers. If you work in a small firm, you will probably be asked to help with the regular maintenance of equipment. Servicing the crankcase breather is a task that is done on the job.

Small engines use one of three types of crankcase breathers: (1) the reed-valve, (2) the ball-check, or (3) the floating-disc type. The floating-disc type is the most common today

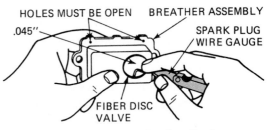

Figure 16-7. Checking the floating-disc type breather to assure proper crankcase vacuum.

and is easy to service. This fiber disc limits the direction of air flow caused by the strokes of the piston. The air can escape from the crankcase, but the one-way valve prevents a return flow. This maintains a partial vacuum in the crankcase. To check, a .045 in wire gauge should *not* enter between the fiber disc and the body of the breather. If it does, replace the complete breather. Also, a new gasket should be installed, making sure that the holes in the body match those in the gasket. The dirty breather should be flushed with a cleaning solvent.

The reed-type and the ball-check breather should also be removed and flushed in a clean solvent. The reed-type breather should have a clearance between the reed and the valve body of .015 to .030 in. If the clearance is not within this specification, the breather should be replaced. The ball-check breather should work freely by alternately blowing and sucking on the crankcase side of the breather. If the ball does not seat properly, it should be replaced.

Changing the Oil

Every horticulture worker who uses small engines should be able to change the oil. Most manufacturers recommend changing the oil every 25 hours of operations under normal conditions. As with other service recommendations, this should be more frequent when operating under dirty conditions. Also, the operator's manual should be followed for specific recommendations. The recommended viscosity, or weight of the oil, varies with the temperature and the engine manufacturer. Table 16-1 gives the recommendation of five common engine manufacturers. Oil should also be selected using the American Petroleum Institute (API) classifications. Most manufacturers recommend that the oil be at least an SC or SD class. This is stamped on top of the oil can and indicates that the oil is suited for engines operating under severe conditions. Oil is a low-cost maintenance item

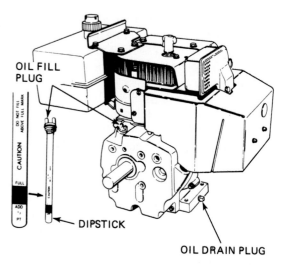

Figure 16-8. To change the oil in a 4-cycle engine, remove the oil drain plug and drain into a container.

when compared to the cost of a new engine. Always check the oil level when fuel is added!

Procedures. To change the oil, locate a container that will hold at least the capacity of the

TABLE 16-1
FUEL MIXING CHART
2-Cycle Engines

Mix Ratio	Oil (Qts)	U.S. Gal	Volume of Gasoline Liters
12:1	1	3	11
16:1	1	4	15
20:1	1	5	19
24:1	1	6	23
40:1	1	10	38

DIRECTIONS

1. Add 1 gallon of gasoline in mixing can.
2. Add correct amount of oil and close can. Mix thoroughly.
3. Add remainder of gasoline for recommended mix. Mix completely by shaking the mixing can.

TABLE 16-2
RECOMMENDED OIL VISCOSITY VARIES WITH TEMPERATURE. ALWAYS CHECK THE OPERATOR'S MANUAL FOR SPECIFIC MODEL RECOMMENDATIONS.

Manufacturer	Above 40° F [4.4° C]	Above 32° F [0° C]	Below 5° F [−15.5° C]	Below 0° F [−17.8° C]	Below −10° F [−23.3° C]
Briggs and Stratton	SAE 30 or 10W-30	5W-20 or 5W-30		SAE 10W Diluted with 10 percent Kerosene	
Clinton	SAE 30		10W		5W
Kohler	SAE 30		SAE 10W	5W or 5W-20	
Tecumseh	SAE 30 10W-30 10W-40		5W-20 5W-30	10W Diluted with 10 percent kerosene	
Wisconsin	SAE 30	SAE 20 or SAE 20W	10W		

crankcase. Run the engine until it reaches operating temperature, then stop the engine and remove the spark plug wire. Remove the drain plug which is usually located in the bottom of the oil sump. Allow the oil to drain completely. It may be necessary to tilt the engine slightly to remove all the oil. Wipe any excess oil off the engine or the machine and replace the drain plug. With the engine on a level surface, fill the engine to the recommended level with the correct viscosity and type of oil. Reconnect the spark plug wire and start the engine, while checking for oil leaks. After stopping the engine, recheck the oil level. Do not overfill the engine.

Since the 2-cycle engine uses a fuel-oil mixture to provide lubrication, the oil does not have to be changed. The oil must be mixed with the gasoline in the correct proportions. Otherwise, engine damage is sure to occur. A special fuel can is available with a premeasured cup that screws into the neck of the can. These red metal safety cans are a must for the commercial operator or the serious owner who wants the maximum service life from the investment.

Servicing the Ignition System

The spark plug conducts a 20,000-volt spark into a combustion chamber that may reach over 1,000° F [537° C]. This occurs about 1800 times per minute during normal operating speeds. Ignition problems and carburetor problems are the two most frequent causes for poor starting and running.

If the engine will not start, disconnect the spark plug wire from the spark plug with a special deep-socket wrench by turning counterclockwise. Engine manufacturers recommend a special tester to determine whether the spark will jump a .166-in test gap. If it does, you can assume the problem is not in the ignition system. Color of the spark is *not* an indicator of quality, since atmospheric conditions determine whether the spark is blue, red, or even yellow. The .166-in gap simulates a normal spark plug gap under compression. If a spark tester is not available, the spark plug may be removed and reconnected to the spark plug wire. While you are holding the spark plug against the cylinder head, observe the spark characteristics. The spark should have a sharp clear snap when the engine is rotated. Check that gasoline, fumes, or other combustibles are not present when testing the ignition.

The Spark Plug. To operate correctly, the spark plug must have the correct gap between the electrode and the ground. Most engines use a .030-in plug gap; however, the operator's manual should be used.

The spark plug must also be the correct "heat type" for the application. Spark plugs are available for specific operating conditions and are designated in "cold," "normal," and "hot" ranges. This refers to the operating temperature of the spark plug and not to the intensity of the spark itself. An engine used for short periods and at low operating temperatures should use a "hot" spark plug, while engines that operate under load for long periods of time should use a "cold" plug. Figure 16-9 illustrates the differences between the hot and cold types.

Spark plugs may also be selected as to the "reach." *Reach* is the term to describe the distance from the lower end of the shell to the bottom of the ground electrode. The common 14-mm spark plug is available with 4 different reaches to fit different thicknesses of cylinder

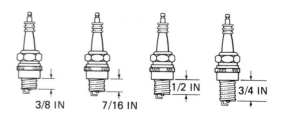

Figure 16-10. Spark plugs are designed for different thicknesses of cylinder head. The small engine usually takes plugs with a shorter reach. Always check the owner's manual for specific recommendations.

heads: ⅜ in, ⁷⁄₁₆ in, ½ in, and ¾ in. Figure 16-10 illustrates the parts of the spark plug.

To recondition the spark plug, inspect it for cracks in the insulator. Using a wire gauge, set the gap between the center electrode and the ground electrode to specifications (usually .030 in). Replace the copper gasket and tighten it with your fingers. Table 16-3 gives general recommendations for tightening the spark plug. If the engine still does not have a spark, the problem is in the points or the condenser. If you do not have the correct tools and repair manuals, this job should be performed by a small engine mechanic.

Servicing the Carburetor

The carburetor is designed to:

1. atomize the gasoline into very small droplets
2. mix droplets with the correct amount of air
3. regulate the volume of the air-fuel mixture going into the cylinder.

If the fuel is clean and the air cleaner is well serviced, there are few carburetion problems.

When you are starting a cold engine, it may be necessary to "choke" the engine. The choke is a butterfly-type valve near the top of the carburetor that restricts the volume of entering air. This results in a richer fuel mixture and improves the vaporization of the fuel. After the engine is at operating speeds, or high idle, the choke should be opened. Many engines use an automatic choke.

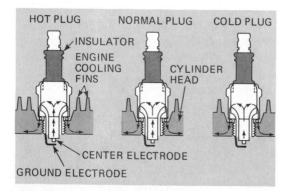

Figure 16-9. The heat range refers to the operating temperature of the spark plug. This is determined by the distance the heat must travel along the insulator nose before escaping through the cylinder head or block.

TABLE 16-3
TORQUE RECOMMENDED FOR SMALL ENGINE SPARK PLUGS

Spark Plug Thread Sizes	TORQUE WRENCH		Without Torque Wrench Engine Cool*
	Cast Iron Head	Aluminum Head	
(Size, mm)	(ft–lb)	(ft–lb)	Turns with wrench (Beyond finger tight)*
10 mm [⅜ in]	12–15	10–11	¾–1
14 mm [⁹/₁₆ in]	25–30	22–27	½
18 mm [¹¹/₁₆ in]	30–40	25–35	½–¾

* Only with new spark plug and gasket. This provides clearance for the gasket to crush and form a seal.

Two needle valves control the air-fuel mixture during operation. The idle-adjustment valve regulates the air-fuel mixture at low operating speeds, usually 1750 revolutions per minute (RPM's). The high-speed needle valve regulates the air-fuel mixture at full speed, usually 3600

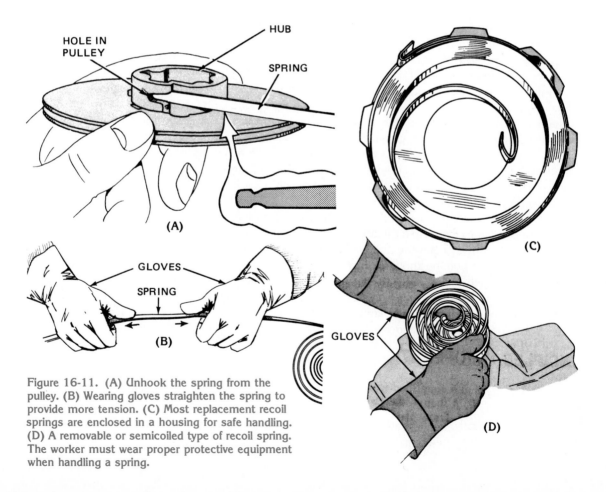

Figure 16-11. (A) Unhook the spring from the pulley. (B) Wearing gloves straighten the spring to provide more tension. (C) Most replacement recoil springs are enclosed in a housing for safe handling. (D) A removable or semicoiled type of recoil spring. The worker must wear proper protective equipment when handling a spring.

RPM's. Since there are four basic types of carburetors and several models of each type, carburetor repair is best performed by the specialized small engine mechanic. Many new carburetors are preset and cannot be adjusted by the operator.

Servicing the Starter

Today, most small engines incorporate a rope rewind starter or a battery-operated-starter motor. Used correctly, both types will provide trouble-free service. The rope rewind starter may be either a horizontal pull or a vertical pull model. If you observe the condition of the rope, simple maintenance may replace expensive repair. If the rope appears frayed, it can be replaced with one of the correct length and diameter. These are available from most small engine suppliers. Always use care when working with spring-wound starters. They can fly apart and injure the operator. Eye protection should always be worn.

To replace the frayed rope, pull it as far as possible. Using lock grip pliers, clamp the pulley to the starter housing. Most models require the starter to be removed from the engine. Cut the knot near the starter pulley and remove the frayed rope. Be sure that the spring has the correct tension. Select the new rope and singe the end to prevent fraying. Then thread the rope through the housing eye and the pulley. Tie the end of the rope and pull the knob into the recess of the pulley. Slip the rubber handle on the rope and tie the opposite end, using a figure 8 knot with the pin inserted through the knot. Slowly release the lock grip pliers and feed the new rope into the pulley. If necessary, bolt the starter on the engine and test the starter action. Each manufacturer can provide you with a repair manual that gives step-by-step instructions for each model.

The battery-operated-starter problems usually originate from two maintenance problems: (1) improper belt adjustment or (2) lack of battery service. Check the belts every 25 hours of operation. Belts should generally have about ¼-in [6.5 mm] deflection per foot when 30 to 35 lb [13.6 to 15.9 kg] of pressure is applied at the midpoint. Also, inspect the face of the belt for checks, cracks, or glazing. A timely replacement will prevent costly downtime. Do not run belts too tight, since they will stretch and shorten the service life. It is also very hard on the pulley bearings.

The battery should also be checked every 25 hours. The battery uses a liquid that is a mixture of water and sulfuric acid, called an *electrolyte*. If the battery is serviceable, remove the battery caps and check the electrolyte level. If necessary, use clean distilled water to bring the electrolyte up to the correct level. Do not expose the battery to a spark or flame, because the hydrogen gas produced by the battery may explode. Clean the terminals, cables, and support brackets with a solution of baking soda and water. *Do not* allow the solution to get into the cells of the battery! This will "kill" the chemical action of the battery. Most suppliers have a silicon spray that protects the terminals from corrosion. If it is necessary to remove the cables, use the correct tools to prevent damage to the post.

Many new engines are equipped with a "no-maintenance" or "low-maintenance" battery. These do not require the addition of electrolyte; however, they will need to have the terminals and cables cleaned every 100 hours.

When cranking the engine that is equipped with a battery-operated starter motor, do not allow the starter to operate for longer than 15-second intervals. If the engine does not start, wait 30 seconds before running the starter motor again. During cranking, a large volume of heat is created by the starter motor. Also, the battery is quickly discharged. By operating in 15-second intervals, both the starter motor and the battery will have a much longer life and you can avoid unnecessary breakdowns.

Troubleshooting the Small Engine

By combining the knowledge of basic engine principles with a logical system of engine analysis, many operators can develop diagnostic skills to determine whether the engine can be repaired in the field or whether it should be sent to the shop. Often the problem can be solved and the engine back on the job with a minimum of lost time. Most engine problems can be classed into several groups:

1. will not start
2. difficult to start
3. kicks back when starting
4. lack of power
5. excessive vibration
6. erratic speed
7. overheating
8. high oil consumption.

When these conditions occur, a simple systematic check should be made of the compression, ignition, and carburetion systems.

Checking Compression

Using the starter, spin the flywheel against the compression stroke of the engine. The flywheel provides weight to maintain even engine speed. It should have a sharp rebound against the compression. Some engines are equipped with various compression release systems. If the engine has the "easy-spin" or other types of compression releases, spin the flywheel counterclockwise to determine the actual compression. If there is very little compression, check for a loose spark plug or loose cylinder head bolts. Several other problems may result in poor compression, but each requires a specialized small engine mechanic for satisfactory repair.

Checking Ignition

Using a special spark tester, determine whether the spark will jump a .166-in gap. This represents the correct spark plug gap under compression. If the tester is not available, hold the spark plug wire ⅛ in [3.1 mm] from the cylinder head and try to start the engine. The spark should have a sharp clean snap as it jumps the ⅛-in [3.1 mm] gap. If the spark does not occur, check for a shorted kill switch or a shorted ground wire. Other ignition problems should be handled in the shop.

Checking Carburetion

Be sure the engine has fresh, clean gasoline in the fuel tank. Also, check that fuel is available to the carburetor. The strainer may prevent the flow of fuel. Be sure that the throttle and the choke are operating properly and that the speed control cable is free to move. If the engine will not start after five pulls, remove and inspect the spark plug. If the spark plug is dry, add 1 tablespoon of gasoline directly into the engine. If the engine fires a few times but does not run, the engine should be sent to the shop. If the spark plug is wet:

1. check the choke to see whether it is stuck in the closed position
2. look for water in the fuel tank
3. check to see whether the idle-adjusting valve is too open, or
4. check whether the float is stuck open.

Other problems causing wet or dry plug conditions will require the services of a small engine mechanic.

Checking Excessive Vibration

If the engine or the machine develops excessive vibration, the machine should be stopped and the blade and crankshaft should be examined. Many engines use the blade as a part of the flywheel weight. If the blade is loose, excessive vibration and kickback will occur. Also, check all mounting bolts, belts, pulley alignment, or other misalignment. Many of these problems can be eliminated by routine maintenance.

Care and Maintenance of Small Engines: A Review

The small internal combustion engine has replaced much of the "human power" in horticulture. These engines may be either the 4-stroke cycle or the 2-stroke cycle. Both types use gasoline or diesel, depending on the model. The 4-cycle engine runs in a sequence of intake, compression, power, and exhaust strokes to complete one cycle. The 2-cycle combines the four sequences during two strokes, thus completing one cycle. These engines can be distinguished by several means.

The 4-cycle engine has both the carburetor intake and the muffler attached near the top of the cylinder. It also has an oil sump, a dipstick, or a visual oil-level check. The 2-cycle engine has a muffler that is mounted near the center of the cylinder. The carburetor may be attached on the opposite end to the spark plug. Its fuel tank will also have instructions concerning the ratio of oil to gasoline. Since this is the only method of engine lubrication, it is important to use the correct fuel mixture!

Small engine maintenance is a combination of knowledge of the manufacturer's recommendations and the common sense to check the engine each time it is operated. The small engine uses a series of subsystems to change chemical energy to mechanical energy. The air cleaner, fuel system, crankcase breather, lubrication system, and ignition system require periodic routine service to obtain maximum performance. These are step-by-step maintenance activities that any operator can perform.

Often, the operator is called on to combine the technical knowledge of basic engine principles and a systematic process of engine analysis to diagnose engine problems. These decisions determine whether the repairs can be made in the field or whether the machine should be sent to the shop for work by a specialized small engine mechanic. These problems may be grouped into four activities:

1. checking compresssion
2. checking ignition
3. checking carburetion
4. checking excessive vibration

Many of these problems can be solved in the field with common tools. The design life of the engine and the machine is of little importance if the operator does not practice good preventive maintenance!

THINKING IT THROUGH

1. Describe the sequence of events of the 4-stroke cycle engine.
2. Describe four major differences between the 2-stroke engine and the 4-stroke engine.
3. Give two reasons why all chainsaws use the 2-stroke engine.
4. Explain the difference between the fuel used in the 2-cycle and the fuel used in the 4-cycle engine.
5. List the service required for the three common air cleaners.
6. Describe the two criteria used to select engine oil: (a) viscosity and (b) API classifications.
7. Using a small 4-cycle engine, demonstrate the procedures used to troubleshoot the following systems:
   (a) Compression
   (b) Ignition
   (c) Carburetion

17

Horticultural Plant Tools and Equipment

As you visit a horticulture business, try to imagine the operation without any tools and equipment. Impossible? Then imagine the operation using hand tools but without the benefits of power tools and equipment. It was only a few years ago that most of the work was done with either hand tools or horses as a power source. The large draft horse was used in the nursery and in the orchard for tillage, planting, cultivation, and harvesting. The hand lawnmower was used in landscape maintenance. Wood stoves were tended around-the-clock in the greenhouse or the hothouses to maintain even temperatures for young seedlings. And all trees and shrubs were balled and burlapped by hand, often taking several workers several hours to dig one 4–6 inch (in) [10.16–15.24 cm (centimeters)] tree. It is easy to see how machines have improved jobs in the horticulture industry and increased the productivity of the worker.

Recently, a horticulture research firm conducted a series of interviews with nursery managers. This survey reported that manual labor accounted for slightly over one-half of the production costs, a figure that is several times greater than those for other segments of the agricultural industry. The managers agreed that increased mechanization was necessary, but that machines must triple the productivity of the worker to be economically justified. Machines such as the hydraulic digger are being developed to reduce the amount of manual labor in ball and burlap operations. This machine can dig a 4–6 in [10.16–15.24 cm] diameter tree in a matter of minutes. However, the horticulture worker must be better trained to operate and maintain this expensive equipment. The nursery worker of tomorrow must be able to operate mechanical harvesters and potting equipment.

Often the little jobs are the ones left undone. Yet many times it is those small details that determine the final outcome. The Olympic athlete works on details to improve performance. The difference between a gold medal and an "also ran" is the daily attention to details and practice. Preventive maintenance includes those jobs with tools and equipment that are done on a daily or periodic basis. Yet these small service jobs affect the economic life of the machine, the timeliness of the operation, and the safety of the operator.

Check with the local equipment dealer and price the used equipment. It becomes obvious that good preventive maintenance pays dividends. The equipment that has been maintained will always sell faster and bring more money because it looks better and will have a longer life. Proper storage may double the resale value.

Timeliness is also affected by preventive maintenance. Because of improper belt tension, the bearings of a motor on the soil shredder fail. This results in seedbed and planting delays. By moving the sale date back, the business loses the "early crop premiums." Several hundred dollars may be lost because a very simple maintenance job was postponed. Because there was not time to check the bolts in the PTO-shaft, a rototiller was returned to service. After a few hours of operation, a bolt came out, allowing the PTO-shaft to sling loose. The tiller could not be used until a new shaft was ordered—which took several weeks. And if it had struck people, they would have been hurt—or possibly killed! Good preventive maintenance defies the adage "There is never time to do the job right, but there is always time to do it over."

CHAPTER GOALS

Your goals in this chapter are:

- To identify why preventive maintenance is important
- To use electric energy safely and to maintain overload protection devices
- To perform preventive maintenance on AC and DC motors and storage batteries
- To describe and demonstrate preventive maintenance procedures for the engine
- To describe and demonstrate the preventive maintenance of the power train

- To service the hydraulic system and describe the safety rules of hydraulics
- To service bearings and seals
- To service cutting and mowing equipment
- To maintain the hand and power sprayer
- To prepare tools and equipment for storage

Electric Motors and Controls

Electricity is a very good source of power for many horticulture operations, because it has the advantage of being economical. For most ap-

Figure 17-1. Regular servicing of machines will help maintain the machine's longevity and safety.

plications, the electric motor has a lower initial cost than the internal combustion engine. Then too, the electric motor usually has less than one-half the operating costs of a gasoline engine. A second major advantage is the ease of auto-mation of electric power. Many types of auto-matic controllers are available. Using a ther-mostat, a motor-driven fan can cool a greenhouse. Or a humidistat can automatically regulate the amount of moisture in the air. A timer may be used to control the lighting, watering, misting, or a variety of other jobs. The third advantage of electric power is the quiet operation and absence of smoke and fumes. The power source can operate in the same area as plant materials and not affect the air quality. The major disadvantage of electric energy is the problem of using it in mobile operations such as tractors, tillers, or mowers. Although it can be used in mobile applications, costs and main-tenance are greatly increased. Electricity, like other forms of energy, requires safe use. Horti-culture workers should understand the basics of electricity for their own safety and for the safety of others. Three basic terms should be under-stood to help explain how electricity is used.

Electrical Pressure

The electrical term used to describe the pressure of electricity is *voltage*. The common voltage available for most applications is 120 volts, 240 volts, or 460 volts. The 120-volt circuit is used for most small hand tools and equipment, such as edgers, trimmers, shears, and small mowers. This is the same voltage that is used in residential and office buildings. A 240-volt source is usually used to drive one-half horsepower and larger motors, electric heaters, and other electric tools that require a lot of power. The 460-volt circuit is reserved for big jobs. The electric motor that drives the irrigation pump may be 460 volts.

Figure 17-2. Always use a test lamp to determine if the power is on. Connect the black lead of the test lamp to the black or red wire in the box and the white lead of the lamp to the white wire in the box to determine if the circuit is hot.

the greenhouse or classroom may be 100 watts. Thus, *watt* is an electrical term that describes the rate at which electricity performs work. This unit can be compared to horsepower. In fact, new tractors may be rated both in horsepower and watts or kilowatts. Seven hundred forty-six watts are equivalent to one horsepower. As you can see, the watt is a small unit. Because of this, most references are to kilowatt (Kw) or 1,000 watts. Many electrical loads are referred to in kilowatts. A kilowatthour (KwH) is a reference to a load that consumes 1,000 watts in one hour. The final determination of the electricity is in KwH or the equivalent number of watts used over a time period.

A water system has water pressure rated in pounds per square inch, a flow rated in gallons, and an output rated in gallons per hour. The basic electrical system has equivalent terms. The electrical pressure is rated in voltage, the flow in amperage, and the output in kilowatt-hours.

Electrical Flow

Amperes or "amps" is the term used to describe the amount of electrical flow through a circuit. The nameplates of all electric motors and hand tools will report the amperage required to operate the tool. This value may range from less than 2 amps in the case of hand shears to more than 30 amps in an electric hot water heater. It is important to match the amperage of the motor to the capacity of the electrical wire and the fuse. The fuse is a safety check to prevent overcurrent to the motor. If a one-quarter horsepower motor requires 6¼ amperes and the fuse will allow 15 amperes to pass through the circuit, the motor will be damaged.

Electrical Power

The hand-held hair dryer that many people use is rated at 1,000 watts. The bulbs used to light

Electrical Safety

Each year, for every 100 accidental deaths, 4 are the result of electrical shock. Also, electrical fires destroy over $150 million of property each year. Certainly, electrical safety is important in the horticulture industry. A few simple rules will help make its use safe for you and other workers.

Always be sure that the power is off before you make adjustments or service. Each load is controlled by a switch and is also controlled through a main distribution panel. If there are questions, use a test lamp to determine whether the power is off. Familiarize yourself with the location of the switches and the main distribution panel.

Always have a dry area to control electricity or to make adjustments. Water is a good conductor, so exercise extreme caution when electrical power tools or equipment are used in damp areas.

Use only approved wiring practices. Color of the wires is very important. Only black or red wires are used as current-carrying or "hot" wires. Both the black and the red wire may be switched or fused. The white wire is a grounding wire from the load back to the ground. It should not be switched, fused, or otherwise interrupted. The green or bare wire is an equipment ground wire. This wire is to prevent electrical shock to the worker. If the hand tool has a three-wire plug, always use a three-wire receptacle or an approved adapter.

Overload Protection

Just as electrical safety is important to protect the horticultural worker, it is necessary to protect the electrical motor from damage. Good tool care will help prevent problems, but the motor must be protected from too much amperage, short circuits, and overheating. The most common protection is from fuses or circuit breakers.

A fuse is a device designed to fail if exposed to too much amperage. Fuses are available in a variety of forms. The most common is the plug-type fuse. This may be either single element or a dual element. The single-element or "ribbon" fuse is designed for constant load circuits such as the lighting circuit of a greenhouse. The dual-element or "time-delay" fuse should be used on electric motors. These fuses allow larger amperage flow for short periods of time as the motor starts, but protect the motor from short circuits or overheating. Fuses are also available as cartridges or blades in either element type.

| 125 VOLTS
15 AMPERES | 125 VOLTS
20 AMPERES | 250 VOLTS
30 AMPERES |

Figure 17-4. Several types of plug caps are used to reduce the chance that equipment will be operated with the correct amperage and voltage. Plug caps should not be altered or changed.

Today, most electric circuits are protected by a circuit breaker. This switch-type mechanism is located in the distribution panel and "trips" when the circuit is overloaded. After the breaker cools and the reason for the interruption is fixed, the breaker can be reset without replacement of a fuse. These breakers are available either as single-element or dual-element types.

Many motors are protected by a bimetal thermal protector. This safety device is built into the motor and automatically interrupts the circuit if a short occurs or if the motor overheats. Two types that are available are the *automatic reset* and the *manual reset*. As the motor cools to normal operating temperatures, the automatic reset will restart the motor. The manual reset has a small button that must be firmly pushed to restart the motor. The nameplate will identify the type of protection for the motor.

The plug cap on the electrical cord also serves as an overload protector. Motors that are designed to operate from a 20-amp 120-volt circuit will have a three-wire plug with two flat blades parallel and one small round blade. Motors designed for a 30-amp 120-volt circuit have two flat blades at a 90° angle to each other and one round blade. Motors designed for 30-amp 240-volt circuits have two flat blades at a 30° angle and one round blade. These plug caps were designed to prevent possible motor damage from incorrect amperage or voltage and should not be altered.

PLUG FUSE   CARTRIDGE   BLADE OR
             FUSE   KNIFE FUSE

Figure 17-3. The three different kinds of fuses. The plug fuse is the most common in the horticulture industry.

Preventive Maintenance on Electric Motors

Electrical tools and equipment, like other tools, require care if they are to give extended service life. Unlike engines, motors do not require daily maintenance. They do not require checking the oil level, fuel level, and air cleaner. But they do require periodic service. The belt drive should be checked for tension and alignment every 50 hours of operation. If too much tension is applied, the bearings will have abnormal wear and short life. Belt misalignment will result in a short life of the belt, pulley, and motor. The motor should be cleaned and oiled seasonally. Dust, water, and excessive oil contribute to motor failure. Because of this, many motors are available in a totally enclosed frame.

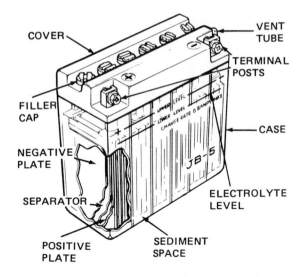

Figure 17-5. A sectional view of the battery shows the lead plates within each cell. The cell is filled with electrolyte which is composed of 36 percent sulfuric acid. The level of which must be maintained.

Battery Care

Motors that are used on mobile equipment such as electric carts or mowers require a battery for the power source. The small race car uses two "AA" dry-cell batteries to drive the motor. When they no longer have enough power, they are discarded and new batteries are installed. Horticultural equipment uses a rechargeable battery that is filled with an electrolyte solution. This solution is about 36 percent sulfuric acid and 64 percent water. A chemical reaction produces electrical current that is used to drive the motor. When insufficient current is produced by the battery, it can be recharged from a 120-volt source.

Charging the Battery.   Always check the electrolyte level to be sure it covers the lead plates and that the top of the battery is clean. Batteries may be cleaned by using a commercial cleaner or with common baking soda and water. Be sure that the cleaning solution does not get into the battery, since it will stop the chemical reaction. Also remove the negative (−) cable

before charging. Many times it is best to remove the battery from the machine before charging it.

Determine the battery voltage by counting the number of cells or filler caps. A 12-volt battery has six cells, while a 6-volt battery has three cells. Remove the filler caps before charging. Using a battery charger, select the charging rate. Generally, the slower the charging rate, the more effective the charge will be. Do not charge at a rate exceeding 70 amps for a 6-volt battery or 40 amps for a 12-volt battery.

Connect the charger to the battery. The positive (+) lead cable should be attached to the positive terminal and the negative (−) cable to the negative terminal. Each battery has a (+) stamped on the case near the terminal and the cable is usually red. The negative cable is usually black. Since the gases from the battery are very explosive, keep open flames and electrical sparks away from the area. Do not smoke near charging batteries. Start the charger and adjust the charging rate and the timer. A fast charge should be set between 15 and 45 minutes. A slow charge should be set from 14 to 24 hours. Most commercial chargers will automatically stop when the battery is fully charged. Always check the electrolyte level before and after the charging cycle. A safe level ranges from the top of the plates inside the cell to the filler neck of the battery case. If necessary, add distilled water to maintain the proper level.

After the battery is fully charged, turn the charger off and remove the lead cables. Replace the negative battery cable. A commercial battery protector or petroleum jelly may be applied to the cables to reduce corrosion. After the battery cover or shields are replaced, the battery is ready for service.

Servicing Horticultural Equipment

There are many types of machines used in the industry, and they are constantly changing to fit the industry's needs. For many horticultural applications, the internal combustion engine is the primary power unit. Engines are found on rotary lawn mowers, power scythes, tractors, and many other machines that require mobility. The preceding chapter explained basic engine principles and the maintenance of the small single-cylinder air-cooled engine. Most of those principles also apply to the multicylinder engines used on larger machines.

Preventive Maintenance of Multicylinder Engines

As with the smaller engine, daily and periodic care is very important. Each manufacturer publishes an operator's and service manual, which specifies the nature and the frequency of re-

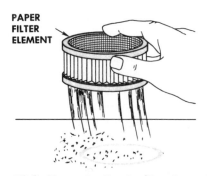

PAPER FILTER ELEMENT

Figure 17-6. To service the dry-filter element air cleaner, bump it firmly against the heel of your hand until all the dirt is removed. Do not strike it against hard objects since it is easy to damage the seal.

quired service. Most recommend that the engine be serviced before each use. Checks of the oil, fuel, coolant, and electrolyte levels should be made. The grease fittings are identified and should be lubricated according to manufacturer's recommendations. The engine should be examined for signs of leaks. Also, the belts, radiator, fan, and air cleaner should be checked.

Air Cleaner Service

The operator's manual will specify the frequency of service that is required for air cleaners. As a general rule, you should check the element, hoses, bowl, and precleaner weekly. Any loss of power, excessive exhaust smoke, or excessive oil consumption is an indication that the air cleaner needs cleaning. The operator's manual will provide a detailed procedure to follow that includes either washing the element or using compressed air. Compressed air with less than 35 psi pressure [2.45 kg/cm²] is recommended when the element must be used immediately after service. A washed element must be allowed to completely dry before it is used. Many operators keep two elements on hand and use one while the other is washed and dried. Do not use high air or water pressure or high temperatures (160° F [71° C]) when servicing the element.

Fuel System Service

Safety and cleanliness are two keys to remember. Do not refuel the machine when it is running or extremely hot. Also, do not expose fuel vapor to sparks or open flame. Either of these conditions will result in a fire. Many manufacturers recommend that the machine be refueled at the end of the work day to prevent condensation in the tank. This should be done after the machine has had time to cool off to reduce the chance of a fire.

As you refuel the machine, keep all foreign matter such as grass clippings or dirt away from the tank. Always use clean fuel and containers. Many times, a 5-gallon can is used to refuel horticultural machines. The can should always be clean and should be used only as a fuel container. A red metal can with a safety cap is preferred and will reduce both contamination and fire hazards.

Most machines use a fuel filter to prevent small particles of dirt or water from entering the carburetor. The filter should be replaced periodically as specified in the operator's manual. If fuel flow is restricted to the carburetor, you should check the fuel filter as a possible cause. Notice that most filters have an arrow on the body to ensure that it is installed correctly. If the filter requires service more frequently than

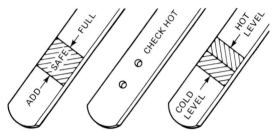

Figure 17-7. The blade of the dipstick will have markings showing the correct level of oil for safe engine operation. Always replace the dipstick correctly.

is recommended, you should check the fuel storage system and the fuel supplier for possible problems.

Lubrication of the Machine

Each machine has special requirements that are outlined in the operator's manual. The engine is designed to use a specific oil. Most manufacturers recommend an SC, SD, or SE classification oil. This classification is stamped on the top of each can. The hydraulic system has different requirements and thus may require a different type of lubricant. Some machines are designed to use the same lubricant for the hydraulic system, transmission, and the final drive. Some even use the same lubricant for the engine,

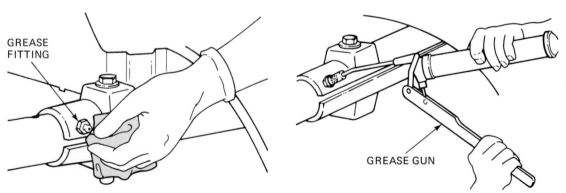

Figure 17-8. Using the operator's manual determine the type and frequency of service. Always clean the grease fitting before coupling the grease gun.

transmission, final drive, and hydraulic systems. The *only* way to know is to check the operator's manual.

You should follow the same general procedures when checking any of these components. Always place the machine on a level surface. Locate the dipstick and remove any foreign material from around the opening. Remove the dipstick and wipe with a clean shop towel. Reinsert the dipstick and remove it for the final reading. The blade of the dipstick will have markings indicating the level range. If lubricant is required, it is important that it be added slowly so that the machine is not overfilled. Always be sure that the dipstick and filler cap are firmly in place after checking the level.

Most bearings and bushings require periodic lubrication. The operator's manual will contain a guide that identifies the location and frequency of lubrication. Most pivot points, joints, and bearings require a SAE No. 2 multipurpose lithium base grease. Some bearings that operate at high speeds and temperatures specify a SAE EP grease or a "wheel bearing" grease. Like oils, specific greases are designed for specific applications. It is wise to use the correct grease for the job.

After the type of grease and the location of the fittings are determined, wipe the dirt from around the grease fitting with a shop towel. Be sure that the grease gun is also clean. Place the coupling of the gun on the fitting and hold it in a straight line with the fitting. Slowly pump the handle two or three strokes as the bearing accepts the grease. This amount of grease is usually adequate and does not destroy the grease seal. Specific recommendations should be obtained from the operator's manual.

Preventive Maintenance on the Power Train

The power train may be as simple as a single belt running from the engine to the reel of a tiller, or as complex as a hydrostatic transmission with a limited-slip, two-speed rear axle on a sophisticated tractor. Regardless of the complexity, the basic preventative maintenance principles will increase the life of the machine.

Maintaining Belts

The service life of the belt depends on three factors: proper installation, tension, and alignment. Before replacing a belt, loosen the tension device and remove the old belt. If the belt is broken, determine how tension is applied. The drive pulley may move back and forth or an idler-pulley may be used. Be sure the replacement is both the correct size and length for the job. As belts are used, they become longer. Since a wedging action between the belt and the pulley is required, the proper tension must be maintained. Too little tension will result in slippage and overheating. Too much tension places too much stress on the belt, the pulleys, and the bearings. Most operator's manuals specify the amount of tension required. As a general rule, the belt should have ¼ in [6.5 mm] of deflection per foot when 20 lb [9 kg] of force is applied to the belt. You should check the tension of a new belt frequently, since initial seating and stretching occurs during the first 10 hours of operation.

Misalignment of a pulley forces the belt to wear and results in short service life. To check alignment, place a straightedge between the two pulleys. The straightedge must touch both edges of both pulleys. If the pulleys are out of alignment, the two shafts are not parallel or the pulley is bent. The operator's manual will give the necessary adjustment procedures.

Maintaining Chain Drives

The chain and sprocket drive reduce the slippage between the two shafts. Also, this system is used when timing of the parts is important. As with the belt, proper installation, tension, and alignment are very important. Chain tension is re-

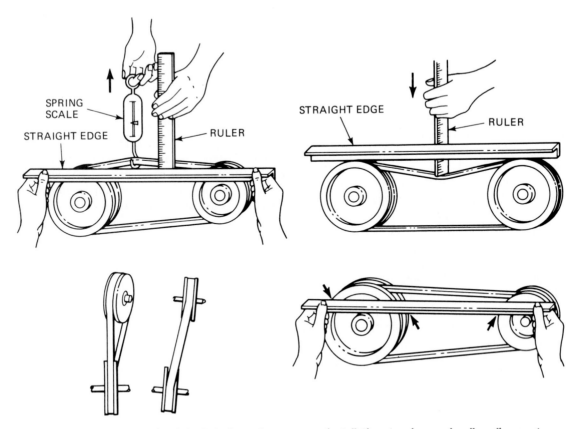

Figure 17-9. The service life of the belt depends on proper installation, tension, and pulley alignment.

ferred to as *slack*, since the chain should not have initial tension. A chain drive has a drive sprocket that pulls the chain from the driven sprocket. This side of the drive loop is called the *tension side*. The part of the loop from the driven sprocket back to the drive sprocket is called the *slack side*. When determining the amount of slack, it is important that the tension side of the loop be tight.

To determine the amount of sag, place a straightedge across the span of the two sprockets while pulling the chain down. Generally, the sag should be approximately ¼ in [6.5 mm] per 1 ft [26 mm] of distance. If proper slack can not be established, it is possible to add or remove a "half-link" or a link of chain.

Alignment of the chain drive is determined by using a straightedge between the sprockets. The straightedge should touch both edges of both sprockets. As with the vee-belt, the operator's manual will give the necessary adjustment procedures.

Safety is important. Be sure that the original shields are replaced after you have serviced either the chain or belt drive. Never attempt to adjust, service, or lubricate the machine while it is in operation. To reduce the chance of injury, always shut off the engine and remove the key before you attempt any maintenance operation.

Maintaining the Clutch

The *clutch* is a device that disconnects the power source from the load. In the case of the rototiller, it may be an idler-pulley that applies

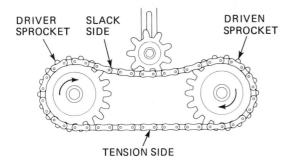

DRIVER SPROCKET SLACK SIDE DRIVEN SPROCKET

TENSION SIDE

Figure 17-10. A chain drive must have the correct alignment and tension. Usually a third sprocket or idler is used to remove the sag from the chain. Always stop the engine before adjusting a chain drive.

spring tension on the belt. The tractor may incorporate a centrifugal, electric, or a disk-type clutch to allow the operator to start and stop the machine. Some machines, such as a shredder, uses a slip or jump clutch to disconnect the power when an obstruction such as a large chunk enters the machine. Each clutch has specific advantages and requires particular maintenance for efficient operation.

The tension on the idler-pulley-type clutch is usually regulated by a threaded bolt attached to a tension spring. Most manufacturers give length specifications; however, the operator must periodically examine the belt and observe for any slippage during operations. If the sides of the belt appear worn or glazed, the tension should be checked and adjusted.

The centrifugal clutch uses a set of internal weights to expand and enlarge the diameter of the pulley. Some centrifugal clutches require periodic lubrication. The electric clutch uses an electromagnet to engage the power train. As increased electric current flows through an iron bar, the strength of the magnetic field increases. This magnetic strength holds the clutch plate to a plate attached to the input shaft of the transmission. By interrupting the current flow, the clutch is disengaged. Because this clutch is electrical rather than mechanical, it does not usually require lubrication. If slippage occurs in

either the centrifugal or electric clutch, replacement is usually necessary. The specific operator's manual should be used to determine the correct procedures.

The disk-type clutch uses a spring-loaded pressure plate and a clutch plate to connect the power source to the load. The power is disconnected when pressure is applied to a clutch pedal, which releases the spring pressure of the pressure plate. Each time the clutch is engaged, wear occurs in the clutch plate and allows the tension to be reduced. After sufficient wear has occurred, the clutch will begin to slip. Adjustments must be made periodically to prevent excessive wear.

The primary adjustment is in the free travel length of the clutch pedal. Each operator's manual will provide free travel specifications and the procedure to follow in making the adjustment. Usually, no lubrication is required. Good operator techniques can extend the life of the clutch. As the load is applied, it should be smooth with a minimum of slippage. A smooth release of the foot pedal is essential. The operator should not use the pedal as a foot rest. If the foot rides on the pedal, the throw-out bearing and the clutch plate will be damaged. Also, the operator should be sensitive to overloading of the machine. Overloads result in short machine life and increased repair costs.

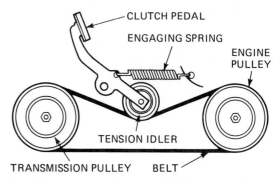

CLUTCH PEDAL
ENGAGING SPRING
ENGINE PULLEY
TENSION IDLER
TRANSMISSION PULLEY BELT

Figure 17-11. Clutch adjustment is very important. The operator's manual will give specific procedures for correct adjustment.

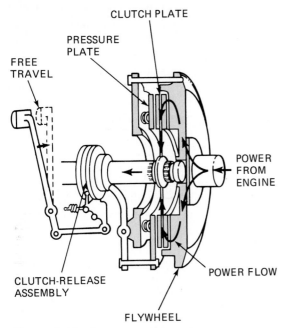

FREE TRAVEL

CLUTCH PLATE

PRESSURE PLATE

POWER FROM ENGINE

POWER FLOW

CLUTCH-RELEASE ASSEMBLY

FLYWHEEL

Figure 17-12. The disk-type clutch is more common on large equipment. The operator should be able to release the clutch with a smooth, even motion, reducing the amount of wear to the clutch plate.

The only clutches designed to slip are the "slip" or "jump" clutches. Using spring-loaded tension, these clutches allow the load to be disconnected only when an obstruction is encountered. The amount of slippage is regulated by the tension of the spring. At the beginning of each season, these clutches should be checked to be sure they will separate when excessive power is applied. You should reduce the tension and burnish the linings until the clutch begins to smoke. The operator's manual will give the procedure and the necessary torque settings for correct operation.

Servicing Tires

Although often neglected, tires play an important part in machine operation. The quality of the job, safety of the operator, and the amount of repair costs depend in part on proper tire service. Tire inflation is important. An underinflated tire

lowers the height of the machine. This may cause a mounted mower to scalp the turf. Or if the uninflated tire is on a precision planter, the planting rate will be increased and the seed may be planted too deep for proper emergence. The low-pressure tire will have a shorter service life. Because there is too much flex in the sidewall, the tire will overheat, wear on the outside edges of the tread, and eventually have premature sidewall failure. On the other hand, a tire with too much pressure will tend to slip under heavy loads, wear the center of the tread, and leave a deeper track. Overinflated tires will damage a smooth surface of a golf green and will cause mounted mowers to cut too high.

Visual inspection is a poor way to determine tire inflation. A 16 horsepower (hp) lawn and garden tractor has a 23 × 10.50-12 rear tire with a recommended pressure of 8 psi [.56 kg/cm²]. A loss of 2 psi [.14 kg/cm²] is one-fourth of the recommended pressure. A special low-pressure tire gauge is essential for maintaining quality and tire life. Always check the tire when it is cold. As the tire is used, the air is heated and the pressure increases. Do not bleed air pressure from a warm tire. This will result in too-low pressure during normal conditions. Each operator's manual will give the proper inflation for both front and rear tires.

Servicing the Hydraulic System

Many horticultural machines use a hydraulic system. The reel-mowers used in turf maintenance use hydraulics to power the reels, drive the machine, and control the steering and the brakes. Lawn and garden tractors use hydraulics for power steering and for driving accessory equipment such as a rototiller. These systems operate on the basic principles that a fluid cannot be compressed and that when pressure is applied at any point, it acts with equal force in all directions. The system is relatively trouble-free—if the proper preventive maintenance is performed.

Checking the Fluid Level

Most manufacturers recommend that the fluid levels be checked daily. This should be done before the machine is hot, since fluids expand with higher temperatures. Each machine is designed to use a specific fluid. One manufacturer uses SAE 10W-30 engine oil in the system. Another requires automatic transmission fluid, while a third requires its own brand of hydraulic oil, which is available only from the company dealer. Because of different designs, each of these fluids is required in the three different machines. The fluids should not be mixed or exchanged.

To check the fluid level, most systems use either a dipstick or a sight glass. Just as in the engine, the fluid level should be within the "SAFE" operating range. If not, you should check the operator's manual for the recommended type to add. While checking the fluid level, you should look for bubbles, foam, or a milky color. Bubbles or foam indicate that air has entered the system, probably from a leak. The milky color can result from water in the system. This may be caused by improper storage or a loose breather cap.

Checking for Leaks

Hydraulic fluid operates under very high pressures, often in the 2000–3000 psi range (140–210 kg/cm²). Because of these pressures, both eye protection and protective clothing should be worn. Hydraulic fluids can pierce the skin and cause a severe injury. When checking the system, you should observe for both pressure leaks and air leaks on the suction side of the system. Also, on a regular basis you should inspect the hydraulic lines and hoses that form the system. Any wear point should be adjusted before the line fails.

Because of necessary movement, most systems use several hydraulic hoses. These hoses are subject to damage from sunlight, chemicals, weather, and storage as well as the usual mechanical damage. It is much better to follow a good preventive maintenance program and to replace marginal hoses than to wait until the hose fails in use. This reduces the downtime as well as the expense in replacing the fluid lost in the rupture.

Keeping the System Clean

Cleanliness is very important in maintaining the hydraulic system. The fluid is pumped through valves, controls, and orifices, which have close-fitting parts. Any dirt, water, or other foreign material will result in wear and eventual failure. As you check the fluid level, be sure to clean around the dipstick opening and use only a clean shop towel. Any fluid that is added should be clean. Hydraulic filters are safety elements that help to reduce contamination. They should be serviced according to the manufacturer's recommendations. Care must be taken to prevent dirt from entering the system during filter changes.

Practicing Safety

As with all tools and equipment, good safety practices will help both the operator and the machine. Because hydraulic systems operate at high pressures, eye protection and proper clothing are essential. The machine should not be running while you are servicing the system. The working components such as reels, mowers, and blades should be lowered to the ground before servicing. Also, park the machine correctly where children will not climb on or under it. When transporting the machine between jobs, lock the cylinders with the stops or chains that are provided. Do not rely on the hydraulic system to carry and hold the load.

Servicing Cutting and Mowing Equipment

Cutting and mowing equipment play a major role in the horticulture industry. Properly serv-

Figure 17-13. Good safety practices are important for the machine, the operator, and other workers in the area.

iced equipment can do a better job with less power and time required. Workers who use hand shears may find that it is necessary to re-dress the blade several times each day. This maintenance job can be performed easily and quickly if the correct procedures are followed. The landscape maintenance worker will do a better job if the rotary mower blade is sharp and balanced. Many times this preventative maintenance job is a daily operation.

Redressing Hand Shears

The two basic types of pruning shears are the *anvil-type* and the *bypass* or *shear-type*. The anvil-type has one cutting blade, which works against a soft metal plate. The anvil-type shear is available both as a hand-held shear and as a lopper. The anvil-type shear has only one cutting blade and is sharpened like a pocket knife. The blade has a bevel on both sides. If possible, it should be disassembled before redressing.

Redressing the Anvil-Type Shear. If the blade is blunt or badly nicked, a fine-grit grinding wheel or a flat mill file should be used. Using the original angle as a guide, remove only enough metal to restore the edge. Excessive

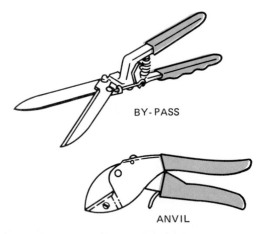

Figure 17-14. The two basic types of hand shears are the anvil type and the bypass or shear type.

grinding shortens the service life of the shear. Also, the cutting edge may be hardened and tempered. These blades should be redressed with only a hand file. A grinding wheel may remove the temper from the blade and shorten the service life of the shear.

Using eye and hand protection, hold one side of the blade against the grinding wheel so that the wheel is turning toward the blade. Make sure the original angle is maintained. Turn the grinder on and begin at the point of the blade. Move the blade across the face of the wheel with moderate pressure. Inspect the edge after every stroke to maintain the correct angle. Repeat the process with the other side of the blade. After all nicks are removed, the blade is ready to be whetted.

Whetting the Blade. To restore a sharp edge to the blade, you should use a whetstone. Whetstones may be referred to as oilstones, sandstones, Arkansas stones, or by other commercial names. The purpose is to add a microscopic sawtooth edge for improved cutting action. Larger blades should be placed on a flat whetstone, with the back of the blade slightly elevated. Using a sweeping motion, apply moderate to firm pressure, with the cutting edge of the blade leading. Make sure that the entire blade surface contacts the stone. When one stroke is completed, turn the blade over and repeat the process with the opposite side of the blade. Repeat this whetting process until there is a fine cutting edge. For those shears with curved or small blades, use either a round stone or a gouge slip. Both allow a curved surface to be whetted without nicking the blade.

Redressing the Bypass Shear. The bypass or shear-type blade is more difficult to sharpen because of the close tolerance between the two blades. This type uses a scissorlike cutting action. If possible, the blades should be disassembled to make the job easier. Each blade has a bevel only on one side. Always try to maintain the original angle, which is usually about 70

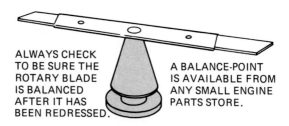

ALWAYS CHECK TO BE SURE THE ROTARY BLADE IS BALANCED AFTER IT HAS BEEN REDRESSED.

A BALANCE-POINT IS AVAILABLE FROM ANY SMALL ENGINE PARTS STORE.

Figure 17-15. Balancing a rotary blade. A balance point is available from most small-engine parts store.

degrees. If the blade is nicked, it should be ground with a fine-grit grinding wheel. Move the blade across the wheel, beginning at the point. Inspect the edge after every stroke, checking both the correct angle and the freedom of nicks. Do not remove more metal than necessary. Also, if the blades are not separated, *do not* close the shear during the grinding process. A small wire edge forms on the inside of the blade and must be removed before the shear is closed.

Whetting the Blade. When all nicks are removed, it is time to whet the blade. Holding the blade open, first place the angle against the whetstone. With the edge leading, make sure the entire edge of the blade comes into contact with the stone. Continue this process until a fine edge is produced. Then, carefully lay the inside face of the blade on the whetstone. Using light pressure and a circular motion, remove the wire edge.

Repeat the same procedure for the other blade. When both blades have been redressed, check the tolerance between the blades. Some shears may be adjusted so that the blades maintain the proper tolerance.

Redressing Rotary Mower Blades

Rotary mower blades operate at very high speeds—up to 19,000 ft/min [5,795 m/min] or 215 mph [346 km/hr]. Because of the effects of centrifugal force, it is very important that the blade remain balanced.

Remove the blade from the mower, noting which side is on the top. The bevel of the blade should be on the top side. Using a medium-grit grinding stone, position the blade on the tool rest so that one end of the blade contacts the wheel and the grinding action is toward the blade. Check the original angle (30° to 40°), and try to maintain that angle during the grinding.

To grind, move the blade across the wheel. It is important to count each stroke to make sure both ends of the blade will be in proper balance later. Inspect the blade after each stroke. When the nicks are removed, reverse the blade and grind the same number of strokes. After the grinding is done, it is advisable to check the blade on a balance point.

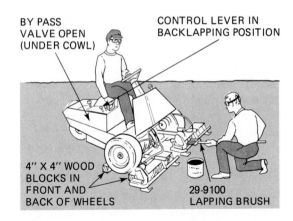

Figure 17-16. Reel lapping is necessary to maintain a sharp cut. Always use the correct safety precautions.

Redressing Reel-Mowers

When they are properly maintained, reel-mowers do a superior job with most lawns and turfgrasses. The redressing operation may be divided into two maintenance activities, lapping and grinding. Grinding is a special job that requires special equipment and is usually done by the horticultural mechanic or by a specialty shop, so it will not be discussed here. Lapping, however, is a regular maintenance job that can be performed on the machine. Lapping is often done on the job by the machine operator and a helper.

Reel Lapping. Usually when the cut edge of the blade of grass appears ragged or torn, it is time to lap the reel. This process adds a small "land" to the blade of the reel. It establishes a match between the reel and the bedknife cutting edges. Lapping will not correct nicked or rounded blades or uneven wear. If these conditions exist, you should begin by replacing or regrinding the reel and bedknife. Proper cutting action requires a reel clearance of 0.002–0.003 in [0.00508–0.00762 cm], approximately the thickness of a page of newspaper.

Lapping Compounds. In many areas, a commercial paste-type lapping compound is available. These compounds may either be premixed or require the addition of a liquid carrier. If neither of these is available, a satisfactory compound can be made from a mixture of medium-grit (80) lapping material and liquid dishwashing detergent. Add the dry grit until the mixture is of a free-flowing consistency. This solution should be applied with a 2½–3 in [6.35–7.62 cm] wide paint brush and an extension handle.

Lapping Procedure. Since the lapping compound is applied while the reel is turning, two people should work together to do this job. Extreme caution should be taken to prevent serious injury. Before starting the machine, make a precise bedknife-to-reel adjustment to be sure that the two components are parallel and that *light* contact is made. Some machines have a separate power lapping attachment. Others have a "backlap" setting, which reverses the reel direction. In both situations, the reel direction is reversed during the lapping operation.

With the reel direction reversed, have the second person start the reel slowly. Using the extension brush, apply the lapping solution evenly over the full length of the reel. If the mower has a gang, all the reels should be lapped at the same time. Continue to apply the solution, while listening to the contact noise between the reel and bedknife. When there is little noise, stop the reel. Examine the land area and determine whether it meets the manufacturer's specification. As a general rule, a minimum of $1/32$ in [.079 cm] should be visible on the blade of the reel. If this margin is not visible, reduce the clearance between the reel and bedknife and repeat the lapping procedure.

When a sufficient land area has been attained, stop the reels and the power source. Use a small, flat mill file to remove any burr that may be on the bedknife. Use low water pressure to completely wash the lapping solution from the unit. By hand, rotate the reel to check the sharpness. Use a sheet of newspaper; the sheet should be sheared cleanly across the width of the reel. If it is not, repeat the lapping procedure.

Servicing Chemical Application Equipment

The horticulture industry uses large quantities of agricultural chemicals to increase plant growth, remove weeds, and control insects. Many states require applicators to be certificated. You should always check before applying any chemical.

Whether dry or a liquid, the chemical should be applied with proper equipment. This may be as simple as a hand-operated knapsack sprayer

Figure 17-17. Chemicals are applied to horticultural crops using devices as simple as a tank sprayer and as complex as a helicopter.

or spinner, or as complex as a self-propelled high-clearance sprayer or even a helicopter. Preventative maintenance affects both the accuracy of the application and the life of the equipment.

The Air Sprayer

The hand-operated compressed-air sprayer is the most common piece of application equipment. It is simple yet often abused. The tank may be made of polyethylene, fiberglass, galvanized steel, or stainless steel. Most have a screen in the tank and an in-line strainer prior to the nozzle. The sprayer should be cleaned after each day's use. Many chemicals tend to settle out of the solution. Care should always be taken to ensure that the tank and screens are clean.

Procedure for Cleaning. When cleaning, be sure you know what material is being applied and the compatibility of the chemical with other materials. The best source of information is the original chemical container, since manufacturers usually recommend the best cleaning procedures. Generally, the tank should be rinsed with clear water. This should be followed by washing the tank with a mild dishwashing detergent and water. The tank and components should be completely rinsed with plenty of water. Some chemicals require an acetic acid rinse following the washing. The nozzle, screen, and strainer should be removed from the sprayer and washed, using the same procedure as the tank.

The Power Sprayer

The power sprayer uses either a PTO-driven or engine-driven pump to apply pressure to the solution. Because of the additional parts, the power sprayer is more difficult to clean than the hand sprayer. As with the hand sprayer, refer to the original chemical container for specific cleaning procedures.

At the end of the season, the sprayer should be completely cleaned. Follow this cleaning with the circulation of diesel fuel through the system. You should then remove all screens, strainers, and nozzles and store them in a jar of diesel fuel until the next season.

The Dry Chemical Applicator

Regardless of whether the material is fertilizer, pesticides, or herbicides, the chemical tends to react with the applicator parts. This results in peeling, rusting, or scaling of the applicator and reduces its life. Much of the time required for cleaning can be reduced by loading only enough material to do the job, otherwise the excess must be cleaned from the machine. After the job has been completed, the applicator should be washed with clean water. Be sure all the caked material is removed from the corners and the agitators. After the season has been completed, the applicator should be washed with diesel fuel to protect the metal parts from corrosion.

Chemical Safety

The same precautions are necessary when servicing the chemical applicator as when you are operating the equipment. You should use the recommended protective clothing, including eye protection. Do not eat, drink, or smoke until you have washed all the chemical residue from your hands.

Horticultural Plant Tools and Equipment: A Review

Tools and equipment have become a very important part of the horticulture industry. As machines are substituted for human labor, the need for trained workers with mechanical skills increases. The use of tractors, mowers, and hydraulic machines that ball and burlap large

Figure 17-18. Careful cleaning of the dry-chemical applicator is necessary. Make sure any worker performing this task is protected with personal safety equipment.

trees are examples of the types of equipment that tomorrow's workers will operate and maintain.

Preventive maintenance, which includes those jobs that are done on a daily or periodic basis, is important when working with tools and equipment. It may include checking the oil, coolant, fuel, and air cleaner element. It may also include servicing belts, chains, or tires. These small service jobs affect the economic life of the machine, the timeliness of the operation, and the safety of the operator.

Electric energy is widely used in today's modern horticulture business. Electricity is used to light, heat, and mechanize the industry. To understand basic electricity, you may contrast the electric circuit with the water system. The water system has pressure, flow, and quantity. The electric circuit has a pressure that is measured in volts. Common voltages are 120 volts, 240 volts, and 460 volts. The circuit also has a flow, described in amperes or just "amps." This is the flow of electrons over a conductor and may vary from less than 1 to more than 50 amps. The quantity of electricity is calculated in watts. A watt is 1 amp flowing at 1 volt pressure and 1 ohm of resistance. This very small quantity is seldom used. More common is the kilowatt, or 1,000 watts. An electrical device that uses 1,000 watts in a 1-hour period uses an electrical quantity called a kilowatthour (kWh). Safety is very important when working with electricity. Always be sure that the power is off before you make adjustments or give service. Use caution when working in damp areas. Be sure that only approved wiring practices are used.

Overload protection is necessary to protect both the electric motor and the worker. A fuse is a device designed to fail if exposed to too much amperage. Fuses are available in a variety of sizes and types for specific applications. The plug cap on the electrical cord, in a limited sense, protects the motor from too much amperage. Plug caps should not be changed or altered.

Many electric motors operate from a storage battery, which is filled with electrolyte and uses a chemical reaction to produce electrical current. These batteries require periodic maintenance and recharging. Care must be taken to eliminate sparks or flames from the area when you are recharging the battery. The vapors produced by the charging are very explosive.

Every machine is designed to do a specific job. Although every machine is different, the basic principles of preventive maintenance are the same. Every internal combustion engine has check points that require service. The power train also has common service requirements. Little jobs like checking belts or hydraulic fluid

levels may make the difference between effective maintenance and machine failure.

Cutting and mowing equipment require preventive maintenance. The periodic redressing of hand shears, mower blades, sickles, and reel-mowers is required to do a good job. Also, chemical application equipment performs better if it is serviced properly. This preventive maintenance not only improves the accuracy of the application but also increases the service life of the equipment.

THINKING IT THROUGH

1. What percentage of total production costs results from the charges of manual labor? What is the trend of this cost?
2. What three factors are affected by good preventive maintenance?
3. Name three electrical terms that describe the circuit. How do these compare to the water system?
4. Which type of fuse should be used to protect the electric motor? Why?
5. What preventive maintenance should be performed on electric motors?
6. What safety precautions should be observed when you are recharging a battery?
7. What preventive maintenance should be performed on multicylinder engines?
8. What preventive maintenance should be performed on the power train?

9. What should be checked on the hydraulic system?
10. What special precaution should be observed when you are redressing the hand shear?
11. How can you be sure that the rotary mower blade is balanced after it has been redressed?
12. Why must an extension handle be used when you are hand-lapping the reel-mower?
13. What is the general recommended procedure for cleaning the chemical sprayer?
14. How should the dry applicator be cleaned at the end of the season?

SIX

Horticultural Plant Products and Services

A horticulturist involved in a wedding? Not only do horticulturists marry like other people, but many horticulturists are directly involved in the process of *conducting* weddings. Starting with the design of bridal bouquets, and the lapel carnations and orchid corsages for the attendants, and extending to the decorations of the wedding place and reception, the horticulturist's work is ever present.

Of course, the flowers are grown in the production phase of horticulture. Yet the delivery of the flowers and floral arrangements is the business of a floral shop. These shops become directly involved in the preparation of weddings. Many floral shops have employee(s) who assist couples in planning the wedding floral displays, bouquets, and corsages for the wedding. These employees must have a thorough knowledge of floral plants, especially flower colors and sizes. Likewise, they must have some artistic ability to assist in the overall floral planning for the wedding. Many floral shops build much of their reputation on providing total floral service and advice in planning weddings.

Funeral services and many other events require the use of floral arrangements and plants. The employees of these establishments must be able to meet the customer and provide quality floral services.

Just as floral shops serve special services, such as weddings, lawn maintenance firms service homes and businesses with lawn-care help. A lawn service business might help a new homeowner with a lawn for the home, establish a lawn around a new commercial building, or provide maintenance services for existing lawns.

These types of businesses require personnel with expertise in lawn grasses, grading, seeding, watering, fertilization, pest control, mowing, and reseeding. Garden centers also provide advice to customers about the care and maintenance of lawns; in many cases they even provide the actual care and maintenance service to the customer, in addition to the supplies and advice. These workers also need specific knowledge and skills in lawn care and maintenance.

Unit Six is designed to provide you, the potential employees in horticultural product and service businesses, with the skills and knowledge you need. Floral and floriculture businesses and lawn care and maintenance are the two service occupational areas highlighted in Chapters 18 and 19. With careful study and practice of the skills outlined in these chapters, you will possess the ability to secure and keep jobs in the horticultural product and service areas.

18

The Care and Maintenance of Lawns

A group of neighbors had gathered in John's back yard and were admiring his lawn. The grass was thick, green, lush, and weed free. John had just finished mowing the lawn and it had the appearance of a deep, thick, green carpet. "How does he do it?" one neighbor asked.

"He must just have a green thumb!" someone responded.

Does John have a green thumb or is his secret to a green lawn something else? In most cases the key to a healthy green lawn is simply proper care and good maintenance practices.

A lawn care service <u>must</u> have many green thumbs, since it wants to develop many envy-producing lawns similar to the one John has developed. Employees of lawn care service firms will need skills in all aspects of quality lawn control and maintenance.

The management of the lawns begins when the lawn

is first installed, and continues with each application of fertilizer, each watering, each spraying for weeds, and each mowing. A good lawn such as John's does not come about as a result of the mythical green thumb. You can be sure that he spent many hours maintaining his lawn.

CHAPTER GOALS

Your goals in this chapter are:

- To describe the process of establishing a new lawn
- To determine the type of lawn grass seed that best suits different lawns in varying locations
- To identify the methods of sodding
- To analyze the nutrition needs of newly established and existing lawns
- To outline the conditions under which watering lawns should be undertaken
- To identify the pests that invade lawns

A good lawn can add many dollars to the value of a piece of property. Homes and other buildings are much more attractive if their grassed areas are in excellent condition and have the characteristics that John's neighbors saw in his lawn.

Homeowners and businesses with lawns invest a great deal of money in lawn maintenance. Have you ever noticed that commercial buildings often have well-established green lawns? This is because there are businesses that maintain lawns. If you look in your local telephone directory yellow pages, you will find many lawn care services. These firms will care for and maintain your lawn and will contract with commercial businesses to care for their lawns.

Establishing a Lawn

A good lawn starts at the time when the lawn is first being installed. Proper installation can save many hours of needless work later for the person who is trying to grow a good lawn. The first part of good lawn installation is the preparation of the site.

Preparing the Site

The site must be cleared of debris and trash, including rock and other potential mowing hazards. After the debris has been cleared, the site is ready to be graded. The objective of grading is to loosen the top portion of the soil and to smooth the soil. Depending upon the amount of investment you wish to make, you might remove the topsoil and replace it with a fertile soil more conducive to the growth of good grass. Normally, 4 to 6 in (inches) [10 to 15 cm (centimeters)] of topsoil would need to be added, which could result in a sizable investment. If your budget is limited, it is best to grade the site as smoothly as possible and use the existing topsoil.

As you grade the site, remember that it is very important that the slope of the site allow the water to flow *away* from houses and other buildings. A slope of 3 to 5 percent, sloping away from the house, will ensure that the water flows away properly. As a worker for a lawn maintenance firm you will develop the skills necessary to grade a lawn properly.

Care should be taken to protect existing trees on the site. Equipment used for grading can harm trees if workers are not careful. Also, when grading around existing trees, you should not add extra soil over the roots of these trees because a few extra inches of soil can actually kill a tree.

Nutrition

After the soil has been graded and is being readied for planting, you should take a soil test.

In Chapter 6 you were provided with the proper technique for conducting soil tests. As you learned, the soil test will provide you information about pH and about the basic soil nutrient levels.

For most lawns, a pH level between 6.0 and 7.0 is best. Lime is needed on lawns whose pH is near 6.0 or less. If the pH is close to 7.0 or higher, various forms of sulfur can be added. It is important to spread lime and sulfur materials evenly, because concentrated amounts of these materials may kill the lawn.

The soil test will provide you with information about the amount of organic matter in the soil. Soils with low organic matter content (usually less than 5 percent) should have organic matter added to the soil. Materials such as peat moss, manure, compost, and leaf mold are examples of organic matter sources. These materials should be worked into the soil, not just added to the surface. Organic matter in the soil increases the soil's ability to hold moisture and to hold nutrients. Thus, soils high in organic matter content are better able to grow a good lawn.

In addition to pH and organic matter information, the soil test will also provide you with information about the nutrient level of the soil. Most likely the site for a new lawn will need ample additions of the three primary nutrients: nitrogen (N), phosphorus (P), and potassium (K). Refer again to Chapter 6 for additional details and a review of soil nutrients.

Grasses such as Kentucky bluegrass require large amounts of nitrogen. In fact, a new lawn needs a fertilizer that provides twice as much nitrogen as it does the other primary nutrients. A fertilizer that has a 2:1:1 ratio is often used for newly installed lawns. This ratio gives 2 parts nitrogen to 1 part phosphorus to 1 part potassium.

Applications of fertilizer for lawns are usually determined by the amount of fertilizer needed per 1,000 square feet [90 square meters] of lawn. How much fertilizer would you apply to a 1,000-square-foot lawn if your soil test indicated that the soil needed 2 lb (pounds) [0.90 kg (kilograms)] nitrogen, 1 lb [0.45 kg] phosphorus, and 1 lb [0.45 kg] potassium per 1,000 square feet [90 square meters]? Assume that you plan to use a 20-10-10 dry fertilizer. If your answer as 10 lb [4.5 kg], then your are correct! Remember that a 20-10-10 analysis fertilizer has 20 percent nitrogen, 10 percent phosphorus, and 10 percent potassium. Thus, 10 lb [4.5 kg] of fertilizer spread over 1,000 square feet [90 square meters] will give the needed nutrients.

As you apply the fertilizer, it is best to work it into the soil so that it is available to the roots of the grass seedlings. The application of fertilizer is a common task of many lawn or landscape maintenance firms. If you seek employment with these types of businesses, you will need to know how to apply fertilizer correctly.

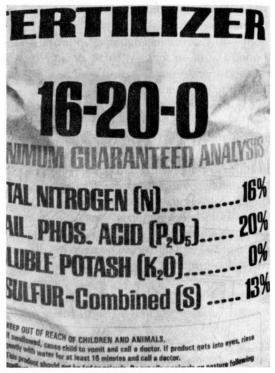

Figure 18-1. Make sure you check the fertilizer analysis to ensure it contains the proper ratio of nutrients for your needs.

After you have provided the proper nutrition to the soil, you are now ready for the final grading. Hand raking is usually necessary to prevent changing the grade established in the initial grading. After the final grade has been smoothed, an application of water is given to the soil to help it settle. This watering will aid in identifying low areas in the grade. The low areas can be filled prior to seeding so that a smooth lawn results.

Selecting Lawn Grasses

Turfgrasses are divided into two groups: warm-season and cool-season. Cool-season grasses such as Kentucky bluegrasses are widely used in the Midwest and Northern regions. A cool-season grass will turn green with the warmer temperatures of spring. Once the hot days of summer arrive, the cool-season grass will become semidormant, but its greenness will return when fall approaches.

A cool-season grass would not do well in Southern regions. The temperatures in the South are too hot and the grass would disappear. If, on the other hand, Bermuda grass (a warm-season grass) were grown in the Southern regions, it would give a nice lawn, because Bermuda grass makes a nice lawn in warmer-temperature areas. However, Bermuda grass would not be able to survive the harsh winters in Northern areas.

Figure 18-2 shows the adapted zones for various turfgrasses. Kentucky bluegrass, bentgrass, and fine fescue are all adapted to zone A and give a permanent lawn. For a permanent lawn in the warmer areas of zone B, Bermuda grass and zoysia grass are recommended. However, for the cooler regions of zone B, use cool-season grasses such as bentgrass, fine fescue, or Kentucky bluegrass. Warm-season grasses such as St. Augustine grass, Bermuda grass, zoysia grass, and centipede grass will provide a permanent lawn for regions in zone C.

In selecting the proper lawn grass, several factors should be considered:

1. Whether lawn maintenance will be considered enjoyable, or whether the homeowner would rather not have to bother with it
2. The use of the lawn (that is, will it receive heavy use from children and pets?)
3. The quality of lawn desired
4. The amount of maintenance required to keep the lawn looking nice
5. How quickly the lawn is desired
6. Other types of lawns found throughout the neighborhood
7. Whether the lawn will be started and growing under sun or shade

Each species of grass has its own unique characteristics that must be considered before it is selected for a special purpose. For example, when an extremely fine-texture grass is needed, such as for putting greens, creeping bentgrass is commonly used. Other characteristics to consider include the facts that:

1. The amount of time required to obtain good coverage will vary, depending on species.
2. Each species varies in its resistance to diseases and insects.
3. Species vary in their requirements for sun and shade.

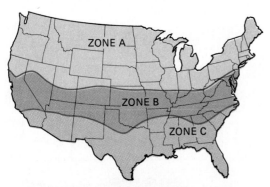

ZONE A—COOL SEASON GRASSES
ZONE B—WARM OR COOL SEASON GRASSES
ZONE C—WARM SEASON GRASSES

Figure 18-2. The three zones adapted for various turf grasses.

4. Some species may be started from seed; others must be started vegetatively by small clumps of grass called plugs (that is, zoysia grass).
5. Some species requires little maintenance, while others require a great deal of maintenance to retain a desirable appearance.
6. The texture of the blade as well as the color of green will vary with the species of important turfgrass.

It is a good idea to hand-rake the soil just prior to planting. Raking up the soil helps provide better contact between the seeds and the soil, which improves germination. Slightly raking over the seeds with about ¼ in [0.625 cm] of soil also helps to ensure germination. Grass seeds are small, so be careful not to cover them with too much soil. You may choose to save some of the seed for touch-up the following season. Fall and spring are the best times to seed lawns. There is usually more moisture and the temperatures are neither too hot or too cold to prevent germination. A push-type or a "cyclone-type" hand seeder can be used to sow the seeds.

Seeds must be kept moist, and yet must not be drowned in water. When a seed dries between rains, weeks could pass before it would germinate. If they are allowed to dry out after sprouting, the entire lawn could fail.

In order to germinate, seedlings must absorb water and swell. This whole process may take from a few days to 2 weeks in good weather, or longer when it is cold. Start irrigating the seedlings immediately, moving the sprinklers often, and water all areas on dry, windy days.

Seed Mixtures.   The importance of using high-quality certified seed cannot be overemphasized. If you use poor-quality, low-cost seed your lawn will show it. The amount of money you pay for seed is a small part of the total cost of establishing a lawn, so why ruin the entire project by purchasing poor seed? Only use seed that is free from weed seed and has a high germination percentage.

The exact seed mixture will depend on the particular use of the turf. Mixes with annual grasses such as annual rye grass are helpful in establishing quick cover. However, this seeding should be reseeded during the next season with a proper permanent grass mixture. Various differences in characteristics already mentioned will determine the ratio of grasses in the mixture.

The amount of seed to be sown will be determined in part by the purity of the seed and the germination percentage. Seeding rates should provide a satisfactory stand of seedings.

To prevent the loss of moisture, to reduce erosion, and to control the drying out of the surface, mulches are often applied. Straw at 1 to 2 bales for each 1,000 square feet [90 square meters] is often used as a mulch. The mulch should be removed as soon as seedlings emerge (see Figure 18-3).

Sodding

Many lawns are established by means of a method called *sodding*. When you are estab-

Figure 18-3. Straw is placed on a freshly seeded lawn to provide a mulch.

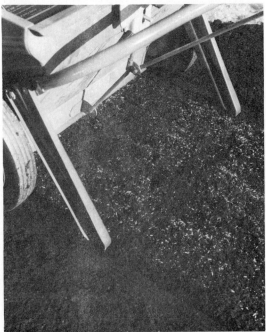

Figure 18-4. (Top) A worker applies a strip of seed to lawn. (Bottom) Plug planting machines are used to seed large areas.

lishing a lawn by sodding, grass sod is applied to the site. Sodding usually is accomplished by one of three techniques: (1) strip sodding, (2) plugs, and (3) sprigging.

Landscape firms quite often sod lawns for commercial as well as residential customers. As an employee of these firms you will be given the task of laying the sod and properly caring for it once it is in place. You might also be involved in cutting the sod at the sod farm.

Strip Sodding. *Strip sodding* refers to covering the entire lawn site with strips of sod. Established grass sod is cut into strips along with the root system. These strips can be placed directly on the prepared seedbed. The sod is usually cut with ½-to-¾ in [0.625-to-0.3125 cm] of soil and roots left under the grass itself. The strips are usually 1 to 2 ft [30 to 60 cm] wide and 4 to 6 ft [1.2 to 1.8 m] long (see Figure 18-4). Although sodding is the least troublesome way to provide an excellent lawn in a short period of time, it is also the most expensive method.

Strip sodding is used when you want a grassed area immediately. Also, sodding is desirable for steep slopes where erosion is a problem. A major disadvantage of sodding is its cost, which is often seven times greater than the cost of seeding.

Plugs. *Plugs* are blocks of sod cut into squares of approximately 2 by 2 in [5 by 5 cm]. Plugs are usually placed one per square foot [0.09 square meter] and firmed into the soil. Figure 18-4 shows a plug-planting machine used for large areas such as athletic fields and fairways.

Sprigs. Starting grass from sprigs of the grass is called *sprigging*. This technique is used mainly with warm-season grasses. A sprig must have a piece of the rhizome with roots attached.

Sprigs are planted 6 to 12 in [15 to 30 cm] apart (see Figure 18-4), in such a way that the roots are covered with soil and the stem section is above the soil surface. The restriction on time of planting is not quite as severe as with seeding, but sprigging should be done at a time that will allow newly developing grass plants to become established before cold weather arrives.

Watering Lawns

Proper watering is definitely the key to the successful establishment of a lawn. The requirements of newly seeded areas are demanding. Until the seeds have germinated and have become established, it is important that the soil surface not be allowed to dry out. It is often necessary to water two or three times a day on dry, windy days. After the lawn has been established, water applications should be heavy enough to ensure that the soil is rewetted to a depth of 6 in [15 cm]. A weekly inch [2.5 cm] of water should be sufficient. However, during a hot, sunny, windy week it may be necessary to apply as much as 2 in [5 cm] of water per week to prevent moisture stress.

A lawn care business often contracts with a firm to provide total care for the firm's lawn, including watering. If you took a job with a lawn care firm, you could be given the task of watering the lawns of a large commercial business. This task involves a great deal of responsibility and requires that you have the "know-how" to accomplish the job.

When you are watering, remember that sodded areas should not be allowed to dry out completely. Once again, enough water to penetrate to a depth of 6 in [15 cm] should be applied.

Maintaining Existing Lawns

In the chapter introduction, we saw that John's lawn was better maintained than most of the neighbors'. Good management practices were cited as the reason. Like John's lawn, any lawn can be maintained with a well-developed system of care and maintenance.

Lawn care businesses and maintenance firms also spend a great deal of time maintaining existing lawns. Good management practices are just as crucial for these businesses as they are for John's lawn. If John makes a serious error and harms his lawn, only one is hurt. However, if a commercial business makes an error and loses a lawn, it also loses its reputation, which is important for future business.

Fertilizer

The green thumb that John was supposed to have was in all likelihood directly related to his use of fertilizer. A good healthy lawn is continuously renewing itself. New roots and rhizomes are developing and constantly replacing old ones. The one ingredient necessary to keep this renewal process going is a regular program of fertilization. All three of the primary nutrients, nitrogen (N), phosphorus (P), and potassium (K), are needed for a healthy lawn.

Nitrogen is needed in larger quantities for growth than phosphorus and potassium. If all other elements are available, nitrogen alone can determine the amount of growth. *Phosphorus* is essential to a young seedling's rapid growth. A supply of phosphorus is usually adequate and is not needed in as large amounts as is nitrogen and potassium. *Potassium* is important throughout the life of the turf, for a healthy root system and general plant vigor. Potassium is needed in larger amounts than phosphorus, but lesser amounts than nitrogen.

To fertilize your existing lawn, you should choose a complete fertilizer, one that contains the three primary nutrients. The best ratio of a complete fertilizer for established lawns is one that is high in nitrogen, low in phosphorus, and medium in potassium. A 4:1:2 ratio (4 parts nitrogen, 1 part phosphorus, and 2 parts potassium) will meet the basic requirements.

To provide ample or adequate nitrogen, you should make the first application early in the growing season and continue throughout the season. Your cooperative extension service can provide guidelines for your area.

Reseeding

One practice that helps many people to maintain good lawns is to reseed occasionally, which means applying lawn grass seed to established lawns. This practice should be handled in the early part of the growing season.

Reseeding will assist in patching any bare spots in the turf that may have resulted from weed, disease, or insect damage. Also, animals and children can wear a thin spot in areas of the lawn where the traffic is greater. Reseeding can be a help in repairing these areas.

Reseed with a good-quality seed. Refer to the previous section for information about establishing lawns, and review the seed selection data.

Mowing

Many good lawns are made unattractive and are actually damaged by poor or improper mowing techniques. As a landscape maintenance worker you will be mowing lawns. There are several general rules to remember when mowing lawns.

Equipment used to mow lawns should be in good repair and adjusted properly. Chapter 17 discusses the adjustment of horticultural equipment, including mowers. Mowing blades should be sharp. A dull blade will actually tear the grass blades, so that after the grass dries it will be brown. A sharp blade cuts the grass evenly and will not result in a brown, torn appearance.

The mower should be set to the right height. As a general rule, the time to mow grass is when the grass is 2 in [5 cm] taller than its recommended height. For example, Kentucky bluegrass, which normally grows at a height of 2 in [5 cm], should be mowed when it reaches a height of 4 in [10 cm].

Before you mow, you should walk over the area to check for debris that might be caught and hurled by the mower blades, injuring by-

Figure 18-5. Grass should be mowed when it is about 2 in [5.08 cm] above its normal height.

standers. Mowing can be dangerous and should not be done when other people are present. A small rock can become a deadly missile when picked up and hurled by the mower blade.

Watering

Lawns require ample amounts of water to grow properly. Highly fertilized lawns need an ample supply of water, especially during the active growing season, if they are to develop into full, thick turf. Many beliefs exist as to the need for watering lawns, but there are some widely accepted general rules to follow. However, the supply and expense of watering a lawn in many areas of the country may force you to change your watering habits to suit local conditions.

Determining when a lawn is in need of water is difficult. Anyone can see that a deeply brown lawn is in need of water, especially during the advanced states of drought. But how can you tell whether a green lawn is approaching a water deficiency? One approach is to walk across the lawn and look at your footprints. Footprints in a water-deficient lawn will remain wilted.

Watering in the spring for a lawn that has not received ample rainfall should be avoided. Little rain will help the grass become deep rooted, which will help it to withstand drought later on.

When you water, it is best to do it in early morning. Mornings are usually characterized by lower humidity and less wind. There conditions will not evaporate the water as quickly, so more of the water can be used by the grass.

How much water should you apply? Many people do not monitor the amount of water. A general rule of thumb is to apply 1–2 in [2.5–5 cm] of water per week. Hot, windy summer days might demand 2 in [5 cm] or more, whereas cooler, less windy days will mean applying less water. Your lawn sprinkler will state the amount of water it applies per hour, so that you can monitor the amount applied. Generally, a sprinkler applies about ½ in [1.25 cm] per hour.

When water supplies are short and expensive, it is best to water the lawn only later in the fall (after September 15) so that the lawn will have ample time to develop sturdy growth to make it through the winter without damage.

Lawn Pest Control

A good lawn needs to always be free of weeds, diseases, and insects. This does not happen by accident. A good, thick turf actually helps to prevent these pests from developing. For example, a thick turf helps to crowd out weeds, healthy grass is less resistant to diseases, and insects do not cause damage to a good turf. These items relate to the development of a good turf.

Homeowners often turn to professionals when it comes time for diagnosing and controlling lawn pests. Garden centers, lawn care firms, and landscape maintenance firms provide expert advice on pest control. These businesses will also provide the pest control service themselves. As an employee you will be involved in the application of pesticides. This means that you will need to know the techniques for determining the proper pesticides and applying them.

The person who is managing the turf must always be on the alert for the pests in the lawn.

Figure 18-6. Keeping a lawn weed-free is part of good management.

Even with the best management practices, these pests can cause damage. You must be ready to take immediate corrective action to prevent the spread of weeds, diseases, and insects.

Weed Control. Remembering the old saying that a weed is any plant that is growing where it is not supposed to, you can see that the lawns are prime places for "weeds" to grow. Some weeds such as the dandelion seem to possess the ability to grow anywhere and under the most difficult conditions.

Crabgrass, dandelion, curly dock, and white clover are a few of the common lawn weed problems. Depending upon your location, the common lawn weed problems will vary. You should seek help in identifying the common weeds and their chemical control.

In Chapter 14 you learned about the careful use of pesticides. Local, state, and federal laws govern the use of certain pesticides, and new, more effective chemicals are being developed that will change the recommendations for pesticide use. Thus, prior to applying any *herbicide* (a pesticide that kills plants) you should check to determine the availability of the chemical and the conditions under which it can be used. You should also review the safety precautions involved with pesticide usage.

Turfgrass Diseases. Leaf spot, rust, blights, and molds are some of the diseases that can invade lawns. Specific diseases will be more predominant in certain geographic areas. Diagnosing lawn diseases requires skill and experience. Until you become knowledgeable in disease recognition you should consult a professional such as your extension service.

Most diseases can be prevented and/or controlled by chemicals called *fungicides,* that prevent or control most lawn diseases. Newer and even better chemicals are also being developed.

You should carefully follow the label recommendations for applying fungicides, and keep proper chemical safety precautions in mind.

Lawn Insects. Sod webworms, ants, chinch bugs, grubs, and mites are common insects that can cause damage to lawns. *Insecticides* are the chemicals that are used to control insect problems. As with herbicides and fungicides, you should consult a professional about the up-to-date recommendations for insecticide usage. Following recommendations and chemical safety precautions are equally important steps in the usage of insecticides.

The Care and Maintenance of Lawns: A Review

A green thumb is not the only ingredient in establishing and maintaining a good lawn. Good management practices are the major ingredient. Preparation of the site is the first step, and smoothness and proper grade should result. Applying the right amount of fertilizer, adjusting the pH, and (if needed) increasing the percentage of organic matter in the soil are important steps and are to be taken only after the results of the soil test recommend such measures. Selecting the right seed and seeding the lawn properly, including adding a mulch such as straw, enhance the lawn's growth. Finally, careful watering is crucial.

Many lawns are established by sodding. Strip sodding, plugging, and sprigging are used.

Once the lawn has been established, good maintenance is required to keep the lawn healthy. Fertilization, watering, reseeding, and pest control contribute to proper maintenance.

As an employee in lawn or landscape maintenance firms, you will be asked to perform many, if not all, of these tasks. Your employer is contracting with homeowners, commercial businesses, and recreational areas to provide quality lawn service. The firm's reputation is based on how well the job is done. You, the horticulture employee, are crucial to that job.

THINKING IT THROUGH

1. What is the process used to grade soil to its proper slope?
2. What materials are applied to a soil to alter pH?
3. List two warm-season and two cool-season grasses that are suitable for lawns.
4. How many pounds of a 10-5-5 dry fertilizer would you apply to a 2,000-square-foot [180-square-meter] lawn to obtain 2 [0.90 kg] nitrogen, 1 lb [0.45 kg] phosphorus, and 1 lb [0.45 kg] potassium per 1,000 square feet [90 square meters]?
5. Distinguish among strip sodding, plugging, and sprigging.
6. How much water would a lawn need in a hot, dry spell, in inches per week?
7. What does a mulch do when it is applied to a freshly seeded lawn?
8. Name some of the major pests that invade lawns, causing damage.

19

Floriculture Crops and Services

The flower show judge paused again in front of Jason's floral arrangement. After carefully examining the arrangement, the judge laid the blue ribbon on the arrangement. Jason's floral arrangement had been selected as the best arrangement in the county fair.

"Congratulations, Jason!" said Barbara as she came up and shook hands with Jason. Barbara had just moved to Oakdale and lived three houses down the street from Jason. She continued, "That flower arrangement is really beautiful. I wish I could do something like that."

Jason replied, "Designing floral arrangements like this is simple if you know a few guidelines. I'll be glad to show you how to design them. Why don't you drop by the house tomorrow afternoon?"

Jason had learned how to make floral arrangements in the horticulture class at his high school. His floral-arranging skills were sharpened by working after school for a local florist. Jason wants to work as a retail florist when he graduates from school.

A retail florist sells floral arrangements and houseplants to the general public. Some florists may even operate their own greenhouse. In this chapter you will learn the basic principles of floral design as they are used by florists.

CHAPTER GOALS

Your goals in this chapter are:

* To identify the materials needed to construct a floral arrangement
* To wire a flower
* To condition a flower to be used in a floral arrangement
* To identify flowers according to the four shapes used in floral arrangements
* To describe five principles of floral design
* To design a floral arrangement

Barbara met Jason the next afternoon to learn the basics of floral design. Jason started out by saying, ''In order to make a floral arrangement you need certain supplies and materials. The materials used are called *mechanics*. After you have the mechanics, you only need floral plants and a knowledge of what to do with them.''

Floral-Arranging Mechanics

A variety of materials and supplies is used in making floral arrangements. In order to make a floral arrangement, you need to know what mechanics are available and what each is used for.

Figure 19-1. The materials used in making floral arrangements.

Cutting Tools

Floral designers may use as few as two tools or as many as eight in designing a floral arrangement. A *knife* is needed to cut thin-stemmed flowers and foliage. Ordinary *scissors* or shears are used to cut medium-size stemmed flowers. *Tin snips* may be needed to cut the wire netting used in floral arrangements. Ribbon and lightweight fabrics are cut with *ribbon shears*. *Hedge clippers* are used to cut heavy stems. At times, *pinking shears* may be used to trim plants with serrated edges, such as hydrangeas. *Wire cutters* are used on silk and plastic flowers and wires. People who construct floral designs every day will probably have all these tools, while amateur designers can get by with fewer tools.

Wires

Wire is often wrapped around the stem of a flower by the florist. This is done to strengthen the stem so that it will remain in place in the floral arrangement. Wire is also used to straighten crooked stems. In some arrangements, a crooked stem may be desirable for a certain effect, so a straight stem could be made crooked through the use of wire.

The wire used by florists is normally sold in 12-lb [9.9-kg] boxes. Floral wire comes in a variety of thicknesses. The thickness of the wire is represented by a number that is called the *gauge number*. The gauges of the wires used by most florists are 18, 19, 20, 21, 22, 24, 26, 28, and 30. The smallest is 30-gauge wire, while 18-gauge wire is the thickest. Eighteen-gauge wire is about as thick as a pencil lead.

The cheapest wire is plain, and will rust in 2 or 3 months. The more expensive wire is covered with a green enamel paint and will not rust.

Containers

Most floral arrangements are placed in some type of container. The container can add to or detract from the arrangement.

Tall vases are often used with long-stemmed flowers such as roses. *Low bowls* or *oblong trays* are used with many arrangements, since they are very versatile. Center pieces are often made in the low bowls or oblong trays. *Goblets, urns,* and *pedestals* are used for designing medium-sized arrangements. *Baskets* and *deep bowls* are used for large arrangements, because they can hold more flowers.

The size of the container should match the size of the arrangement. A small arrangement does not look good in a large container. Likewise, a large arrangement in a small container will look out of balance and could fall over.

Plastic is used in many containers today. It is relatively inexpensive and comes in a variety of shapes and colors. Metals such as pewter and copper-tin alloys are used in some containers. These containers are moderately priced and add a look of richness to floral arrangements. Glass is a material that is used in many containers. It comes in a variety of colors and finishes but is breakable. Wood containers are used to create rustic and natural arrangements, but they must be lined with plastic in order to hold water. Marble is a very expensive container material but it does hold water. The cheapest type of container is made of papier maché. It is not attractive but can be spray-painted or covered with foil. It also needs to be lined in order to hold water.

The price of the container should be considered in designing a floral arrangement. Unless the customer requests otherwise, the container should be the least expensive part of the design.

Green, white, gray, black, and brown are good colors for containers, because they tend to blend into the arrangement. They do not detract from the color of the flowers.

Picks

Thin sticks with wires attached to them are called *picks*. The sticks may be made of wood or metal. Normal length of the picks are 2½, 3,

Figure 19-2. A pick.

4, and 6 in [6, 7.5, 10, and 15 cm]. A thin piece of wire about 6 in [15 cm] long is attached near the end of the stick. Picks are often painted green. The wire is wrapped around both the base or stem of the flower and the stick. It is very rigid and is stuck into the foundation of a floral arrangement. Figure 19-2 shows a pick.

Foundations

Flowers and plant materials are generally not placed loose into a container. They are inserted into a substance that holds the plant materials in place, which is know as the *foundation*. Various materials are used as foundations in floral arrangements.

Foam blocks are widely used in floral designs. They are very light when dry, but become heavy when wet. They are generally sold in bricklike blocks but are cut to fit the container. The stems of the flowers and other plant materials are inserted into the foam blocks after the block has been soaked in water and a flower preservative.

Not only does the foam hold the flowers in place but it also supplies water. Most foam blocks are known by their brand names. The major brands are Oasis™, Quickee™, and Niagra™.

Wire netting (also called chicken wire) is sometimes used as a foundation. The wire netting has a 1-in [2.54-cm] mesh and comes in rolls in widths of 12, 24, and 36 in [30.5, 61, 91.5 cm]. Rolls 12 in [30.5 cm] wide are easiest to use. Tin snips are used to cut the wire. The wire is placed in the container and the floral materials are placed in the various meshes. Wire netting is often used, along with other foundation materials.

Styrofoam is used as a foundation at times, because it does not hold water. Small tubes called *water tubes* may be used with styrofoam. These tubes are filled with water and a preservative, and are stuck into the styrofoam. The flower stems are then inserted into the tubes.

Several other items may be used for foundations. One made in California from bundles of pine needles is called a Needle-Pak. Flowers are inserted directly into the Needle-Pak; picks are not needed. Large, fleshy stems may be cut to the same length and bound together. To insert flowers into this foundation, picks are used.

Other Mechanics

Ribbons are used in many floral arrangements; they may be used to add color to the arrangement, or as a substitute for plant materials in some arrangements, primarily because they are cheaper. A bow made of ribbon will unify and tie the arrangement together.

Floral tape is another material needed in floral designing. Floral tape normally comes in green or white and should be waterproof.

Preservatives are materials used to extend the life of flowers and other plant materials. The cheapest preservatives come in powder form and are mixed with water, but they can also be bought in liquid form. Some come as sprays and are used to seal the stomates. Preservatives are normally placed in water with the flowers.

Preparing Floral Materials

Would you consider going on a long trip without first checking out your car and getting it ready for the trip? You might add water to the radiator, put air into the tires, and clean off the windshield. Before you make a floral arrangement, several things must be done to get the flowers and other plant materials ready to be used in an arrangement. Proper preparation of the plant materials will result in a better-looking and longer-lasting floral arrangement.

Conditioning Flowers

Cut flowers should undergo a process known as *conditioning* before they are used in floral arrangements. The conditioning process occurs soon after cut flowers arrive at the florist's. Cut flowers are often shipped dry for considerable periods of time over long distances. To help them regain their vigor, conditioning is needed.

The conditioning process is started by cutting the end of the stem off about ½ in [1.25 cm] from the bottom. A slanted cut is desirable, since this allows the plant to take in water more rapidly. This procedure is used with most flowers in floral designs, such as mums, carnations, asters, pompoms, marigolds, and gladiolas.

A few flowers with milky sap, such as poinsettias and poppies, are generally not cut because the sap clogs the cells that absorb water. Often these plants have the cut ends treated with a special substance before they are shipped. If you do cut a poinsettia or poppy, you should hold the end of the stem over a burning candle for a second. This prevents the cells from becoming clogged.

The flowers should next be cleaned by removing the bottom one-third of the foliage on long-stemmed flowers and the bottom one-half of foliage on medium-stemmed flowers. The reason why this is done is because otherwise

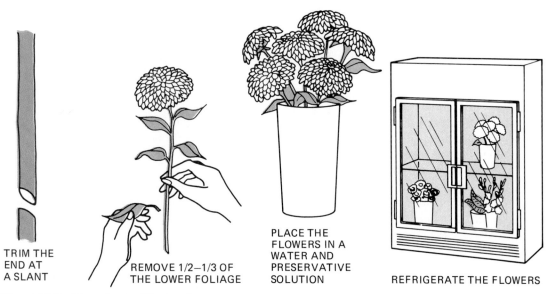

TRIM THE END AT A SLANT

REMOVE 1/2–1/3 OF THE LOWER FOLIAGE

PLACE THE FLOWERS IN A WATER AND PRESERVATIVE SOLUTION

REFRIGERATE THE FLOWERS

Figure 19-3. Steps in conditioning flowers.

when the flowers are placed in water for storage, these lower leaves will be underwater and will decay rapidly. Decayed and wilted leaves are unattractive and could detract from saleability.

The flowers are then placed into a bucket or container that contains warm water and a floral preservative. You should not use a metal container, because the preservative will react with the metal and will be useless. The container should be about one-third full of solution that is at a temperature of 100° F [38° C]. Roses and spring flowers are placed in cool water instead of warm water. The flowers should sit in this solution for at least 4 hours. Afterwards refrigerate the flowers for at least 8 hours before working with them.

Some cut foliages are conditioned also. Eucalyptus and palms are placed in warm water for a period of time before they are refrigerated. Salal, huckleberry, leatherleaf, and ivy are sprinkled lightly with water then placed into refrigeration.

After the flowers and foliages are conditioned, the container used should be thoroughly cleaned with a disinfectant before it is used again.

Wiring Flowers

Flowers are often wired before they are used in a floral arrangement. The stem is either strengthened with wire or replaced by wire. The wire stem makes the flower easier to position and holds the head upright. A basic principle to remember in wiring flowers is that large flowers, such as standard mums, need large wire, while smaller flowers need smaller wire. Flowers that will be placed in water should not be wired unless they have a weak or crooked stem, with the only exception being roses. Roses are nearly always wired, because of their heavy head. Several different techniques are used to wire flowers.

Insertion. With insertion, the end of the floral wire is inserted into the *calyx,* the enlarged

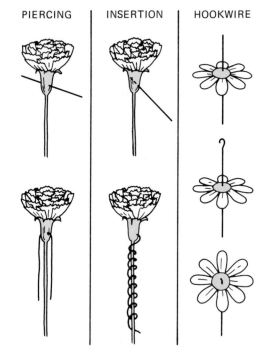

Figure 19-4. Different methods of wiring flowers.

portion of the stem right below the flower. The wire is then held with the right hand and the flower is twisted with the left hand. This results in the wire's twisting around the stem.

Piercing. With piercing, the piece of floral wire is inserted crosswise through the calyxes. The two protruding ends are then bent down next to the stem, and the wire and stem are taped. Both piercing and insertions are used with flowers with large clayxes, such as carnations and roses.

Hook Wire. For flowers with buttonlike centers, such as daisies and mums, the hook-wire method is often used. The floral wire is inserted into the center of the flower head and stem. The end of the wire above the flower is bent into a small hook resembling a fishhook, which is then pulled down into the flower. The hook will nearly be invisible in the head of the flower, while the other end of the wire will protrude from the end of the stem.

Floral Arrangements and Designs

After assembling the mechanics needed for a floral design and preparing the floral materials, you will be nearly ready to start making a floral arrangement. However, before work on the arrangement begins, you should be aware of the standard designs used in floral arrangements. Before placing the first flower in an arrangement, you should have some idea of what the finished arrangement will look like.

Standard Floral Arrangements

There are three basic types of floral arrangements: line arrangements, mass arrangements,

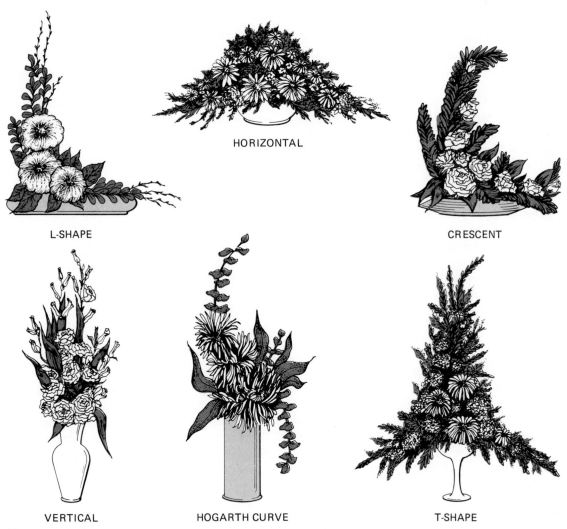

L-SHAPE

HORIZONTAL

CRESCENT

VERTICAL

HOGARTH CURVE

T-SHAPE

Figure 19-5. Examples of line arrangements.

and line-mass arrangements, with several variations for each type of arrangement.

Line Arrangements. Arrangements emphasizing curving or straight lines are called *line arrangements*. They are made with slender branches and flowers that tend to be long, such as snapdragons, delphiniums, and gladiolas.

There are several styles of line arrangements. The *vertical* style is primarily an arrangement that runs straight up and down and is generally fairly slender. A modification of the vertical style is the *Hogarth curve* or *lazy S*. This style resembles a long S that has been straightened out a little. The *crescent* is another curved style. It looks like a C that has been opened up. The *horizontal* style is achieved by arranging the flowers in a line that is parallel with the horizon or ground. The center of the horizontal arrangement is about as tall as the length of one-half the arrangement. If an arrangement is 20 in long [50 cm], it will be about 10 in [25 cm] in the center. Another line arrangement is the *L-shape*. As the name implies, the arrangement is in the shape of an L. The *T-shape* is another type of line arrangement. It looks like an upside down T, and is sometimes called an open triangle. See Figure 19-5.

Line arrangements are fairly inexpensive to make, because they do not require large amounts of plant material. Only a few different types of flowers and foliages are used.

Mass Arrangements. Floral arrangements using a large number of flowers and foliage are called *mass arrangements*. Mass arrangements come in two basic shapes, *circular* and *triangular*. Circular arrangements are sometimes called oval arrangements because the overall outline of the arrangement will resemble an egg or a circle.

The triangular arrangement can be balanced triangle (symmetrical) or an unbalanced triangle (asymmetrical). The difference between the mass arrangement triangle and the line arrangement

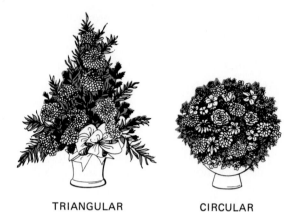

TRIANGULAR   CIRCULAR

Figure 19-6. Examples of mass arrangements.

T-shape is that the triangle is filled in with plants, whereas the T-shape is more open.

The mass arrangements may be more expensive than the line arrangements, because more flowers are used.

Line-Mass Arrangements. A combination of parts of a line arrangement and parts of a mass arrangement results in a line-mass arrangement. This arrangement requires more flowers than the line arrangement, but fewer than the mass arrangement. An example of a line-mass arrangement is the modified L-shape, where the basic L-shape of the line arrangement is used but a mass of flowers is placed at the intersection of the two lines that form the L.

Flower Shapes

When you are designing floral arrangements, it is important to realize that flowers come in different shapes. The shape of the flower will help to determine where it should be used in a floral arrangement. Flowers come in four basic shapes, and a good floral arrangement will often have all four types of flowers.

Line or Spike Flowers. One group of flowers, called *line flowers*, tends to be long and slender. They are often used to form the framework or skeleton of a floral arrangement. Examples of line flowers are gladiolas, snapdragons, delphiniums, larkspurs, bells-of-Ireland, stock, celosia, and lily-of-the-valley. Line flowers are in both line and mass arrangements.

Mass Flowers.   Flowers that are round and compact are called *mass flowers*. They are placed inside the framework formed by line flowers. If mass flowers are used by themselves in an arrangement, the larger flowers are placed in the center of the arrangement. Some typical mass flowers are carnations, asters, mums, tulips, and daisies.

Form Flowers.   Form flowers have unusual shapes and are noted for their beauty. They are used in floral arrangements to enhance the beauty of a design and to focus attention on the design. Form flowers are used sparingly and should not be grouped together. Flowers used

as form flowers include roses, orchids, callas, anthuriums, birds-of-paradise, and irises.

Filler Flowers.   Small, delicate flowers are used to fill in and complete the arrangement. They are often fuzzy or feathery. Commonly used filler flowers are baby's breath, statice, and acacia.

Foliage is often used as filler and occasionally for lines. Emerald palm, huckleberry, asparagus fern, eucalpytus, leatherleaf, and evergreens are commonly used foliages.

Basic Principles of Design

Casual observers often look at a floral design and think, "That is certainly a pretty arrangement." Unknown to the observer, there are several reasons why the arrangement is "pretty"; it did not happen by accident. The reason why floral arrangements are pretty is because the floral designer used one or more of five basic guidelines or principles of floral design.

Balance.   The floral arrangement should appear to be steady. It should not look as if it could tip over in any direction. When this is achieved the design has *balance*. Small, pale flowers placed near the edges of the arrangement, with the brighter, larger flowers placed in the center of the arrangement, help to give balance to an arrangement.

Proportion.   The leaves, flowers, foliages, and container used in a floral arrangement should have a good-size relationship. This is known as *proportion*. If large flowers are used in an arrangement, then large foliage and a large container are needed. Small flowers require small containers.

Rhythm.   Just as music has rhythm, so should a floral arrangement. An arrangement has rhythm when you look at it and find your eyes moving smoothly over the arrangement and returning to the starting spot. *Rhythm* is created

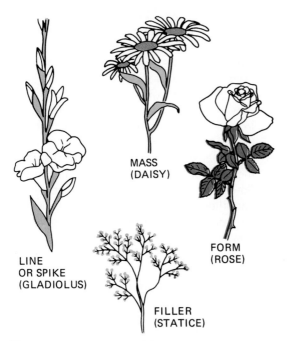

MASS
(DAISY)

FORM
(ROSE)

LINE
OR SPIKE
(GLADIOLUS)

FILLER
(STATICE)

Figure 19-7. Examples of flowers shapes.

by carefully spacing flowers of different colors, sizes, and forms. You do not want all the same color or size flowers in one area. They should be spaced out in the arrangement at various levels and various distances from the viewer.

Focal Point.   Most floral designs will have one area to which your eyes are drawn as soon as you view the arrangement. This is known as the *focal point*. The focal point is often low and in the center of the design. It is often created by placing a single, large, brightly colored flower or a grouping of brightly colored flowers in the center of the arrangement. Sometimes a bow made of ribbon is used as a focal point. All the lines and curves of the arrangement should meet at the focal point.

Harmony.   When all parts of the floral design look as if they belong together, *harmony* is achieved. The colors of the flowers and ribbons should go together. The arrangement should look appropriate for the occasion and location. Shape, balance, rhythm, proportion, and color are all aspects that go into giving an arrangement harmony.

Building the Arrangement

There are several basic steps to follow in building a floral arrangement. If these are followed, it will not be difficult to design a nice floral arrangement.

1. Develop a picture in your mind of what the finished arrangement should look like.
2. Choose the container and plant materials. Make sure that the container is appropriate for the occasion. They types of flowers and plant materials selected will depend upon the occasion, price, and availability of materials.
3. Assemble the equipment needed to build the arrangement.

4. Soak the foam block in water, or prepare the mechanics if you are using something other than a foam block.
5. Place the Oasis or other mechanics into the container, secure the mechanics in the container, and fill the container with water.
6. Cover the mechanics with greenery.
7. Establish the main lines of the arrangement. For example, if you were constructing a horizontal line arrangement to be used as a centerpiece, you would insert one or two of the longest flowers in the sides of the foam block parallel to the ground to establish the width or length of the arrangement. Then you would place one or two flowers in the center of the foam block to establish the height of the arrangement (see Figure 19-8). Normally, either line flowers or foliages are used for establishing the main lines. What you have done here is to establish the skeleton of the arrangement.
8. Add mass or form flowers to fill in the main skeleton. Mass flowers of different sizes and lengths are added to the arrangement. They are within the imaginary line established by the line flowers, and they are placed at different depths. Form flowers should be used sparingly. If the arrangement is to be seen primarily from one side, several form flowers or large mass flowers may be placed low and in the center of the arrangement to create a focal point.
9. Add filler flowers and foliage. The arrangement is completed by adding delicate filler flowers or foliage to fill it out. The foliage and filler flowers should not extend beyond the imaginary line created by the main line flowers.

After building the arrangement, you should carefully analyze it to make sure the basic principles of balance, proportion, harmony, and rhythm have been met. Do not expect your first arrangement to be perfect; it takes practice to become a top-notch floral designer.

1. PICTURE WHAT THE FINISHED
   ARRANGEMENT SHOULD LOOK LIKE

2-3. ASSEMBLE THE MATERIALS

4-5. SOAK THE FOAM BLOCK IN WATER,
     THEN SECURE TO THE CONTAINER

6. COVER THE MECHANICS WITH GREENERY

7. ESTABLISH THE
   MAIN LINES

8. ADD MASS FLOWERS

9. USE FILLER MATERIAL TO
   COMPLETE THE ARRANGEMENT

Figure 19-8. Steps in constructing a floral arrangement.

Designing Floral Arrangements: A Review

A large portion of the business done by retail florists is making floral arrangements. In order to make a floral arrangement, cutting tools, wire, florist tape, containers, and foundations are needed. Before flowers are used in an arrangement, they should be conditioned so that they will last longer. Flower stems are often strengthened through wiring or through the use of picks.

The basic floral designs used by florists are line arrangements, mass arrangements, or line-mass arrangements. The flowers used by florists are classified as line, mass, form, or filler flowers. The finished floral design will have balance, proportion, harmony, rhythm, and a focal point.

By following simple guidelines, most people can design good floral arrangements.

THINKING IT THROUGH

1. What type of cutting tools may be used in designing a floral arrangement?
2. Why is wire used in floral arrangements?
3. What shapes of containers are most commonly used in floral designs?
4. Why are foundations needed in floral arrangements?
5. What are the commonly used foundations for floral arrangements?
6. How should flowers be conditioned if they are to be used in a floral arrangement?
7. How does the insertion method of wiring a flower differ from the hook-wire method of wiring a flower?

8. How does a line arrangement differ from a mass arrangement?
9. List the four types of flowers used in floral arrangements and name three flowers found in each group.
10. How would you define balance, proportion, rhythm, harmony, and focal point as they relate to floral arrangements?
11. Assume that you had to explain to a friend how to make a floral arrangement. What would you say?

SEVEN

Horticultural Experience and Leadership

Would you like to spend several years working at a job and never receive a promotion or pay raise? Probably not! Do you know what you can do to increase your chances of promotion?

One thing that will help you find a job and may help you with future promotions is to have work experience that is varied. Employers have a tendency to hire experienced people. As a student you should make plans for gaining as much work experience in horticulture as possible.

A person possessing leadership skills may also be promoted rapidly. Speaking effectively and working well with others are two leadership skills that are beneficial to horticulturists.

In this unit you will learn how you can obtain work experience in horticulture. You will also learn what types of leadership skills are needed in the horticulture industry and how these skills can be developed to help you find and keep a job.

20

Horticultural Supervised Occupational Experience Programs

Have you ever read the help wanted ads in the newspaper? If you have, you may have noticed that employers often look for people with experience when it comes time to hire employees. The following advertisements are typical of those involving horticulture jobs:

Help Wanted. Harmony Hill Country Club needs an experienced greenskeeper to mow greens and fairways, apply fertilizers and herbicides, and maintain the equipment. Apply in person to Mr. Mitchell in the clubhouse.

Modern floral shop needs a dependable floral designer. Forty hours per week. Good pay and benefits. Previous experience needed. Call Ms. Stidham at 743-4654 for an appointment.

What do these two advertisements have in common? Both ads mention that the applicants should have experience. Most employers in horticulture want to hire workers with previous experience, so you will be able to find a horticulture job more easily if you have some practical experience in that field.

How do you go about getting this practical work experience? Your horticulture teacher will be able to help you gain this experience through a Supervised Occupational Experience (SOE) Program. The SOE program is an essential part of your horticulture class. In this chapter you will learn what SOE programs are, why they are needed, and how to plan your own SOE program.

CHAPTER GOALS

Your goals in this chapter are:

- To determine why Supervised Occupational Experience (SOE) programs are important
- To determine which type of SOE program is best for you
- To plan an SOE program for yourself
- To keep records of your SOE program

What is a Supervised Occupational Experience Program?

A Supervised Occupational Experience (SOE) program is a planned sequence of practical activities that enables you to put into practice what you learned in the horticulture classroom. An SOE program provides for an opportunity for hands-on work.

For example, Mark was interested in turf. During his senior year his vocational agriculture teacher arranged for him to gain actual work experience at a golf course, through a program in which Mark was released from school classes in the afternoon. Mark, his parents, his teacher, and the golf course superintendent signed an agreement about the hours, wages, and conditions under which Mark would work. The vocational agriculture teacher visited Mark regularly on the job, and Mark kept a detailed record of his experiences.

Brenda, a sophomore, lived on a dairy farm but was interested in horticulture, so she decided to grow strawberries for her SOE program. She obtained permission from her father to use a piece of ground 50 ft (feet) [15.25 m (meters)] long and 50 ft [15.25 m] wide.

During the first year she treated the soil with methyl bromide, bought certified plants, and carefully set the crowns at the right depth. The plants grew well. Brenda picked the flowers off

Figure 20-1. Supervised occupational experience (SOE) programs help prepare students for employment.

the plants the first year, fertilized them in August, and covered them with straw during the winter. The following spring the strawberry plants started producing. Brenda took a sample box of strawberries to the manager of the local grocery store. The manager liked the strawberries and bought Brenda's entire crop. Brenda kept accurate records of her income and expenses. She made nearly $600 from the 50 × 50 ft [15.25 × 15.25 m] patch of strawberries.

Cathy's interest in horticulture, on the other hand, was in the area of nursery plant production. She started 100 juniper plants in the school greenhouse. They were potted up in mid-March and placed on black plastic film between the school greenhouses. She used a slow-release fertilizer in the potting mix. In early November the junipers were moved to a cold frame. While polyethylene plastic was placed over the cold frame in December so that it could serve as a wintering house. During mild spells Cathy checked for watering. In the spring Cathy sold the junipers to a local nursery.

As these three examples show, SOE programs provide students with a chance to learn skills by actually performing them. Mark learned more about turf by working on the golf course, Brenda learned about growing strawberries by raising them at home, and Cathy learned how to produce container-grown junipers by using the school greenhouse and cold frame.

We may gain a better understanding of what an SOE program is if we examine the four words that are used in the term *Supervised Occupational Experience program*.

The first word is *supervised*. This indicates that someone will oversee the occupational experience program. That someone is the vocational horticulture teacher, who will give advice and help you set up the program. From time to time, the vocational horticulture teacher will provide individualized instruction at your home, at your place of employment, or at school. This will depend upon where the occupational experience program is conducted.

Figure 20-2. Your vocational horticulture teacher will supervise the SOE program by demonstrating basic skills.

Mark was visited at the golf course by his vocational horticulture teacher. The teacher went to Brenda's farm several times to see how well the strawberries were growing and to offer technical assistance on strawberry production. Cathy was helped at school.

The second word is *occupational*. This means that the experience you gain is related to a job. The activities are not general in nature but relate specifically to types of work performed in the horticulture industry. Cathy learned how to produce container plants just as she would if she worked in a nursery. Mark performed the tasks that most other employees at the golf course performed. The techniques used by Brenda in growing strawberries are the same as those used by commercial strawberry producers.

Experience is the next word. This means that you actually participate in activities. Only by actually performing a job can you gain experience. You apply in an actual situation what is learned in the horticulture class. At the golf course, Mark did not sit around and watch people work. You can learn some things by watching others, but the best learning comes by actually doing the job. Mark got to run the mowing machines, sprayers, and sod cutters. Both Cathy and Brenda did the actual work in their projects.

The final word is *program*, which indicates that this is a planned series of activities. It is not a one-shot or hit-and-miss type of activity, but rather one that is carefully thought out and has several parts to it. The vocational agriculture teacher helped Cathy and Brenda to plan what they were going to do. They prepared a budget, developed a plan, and then produced the crop. A training agreement involving the vocational agriculture teacher, Mark, his parents, and the golf course superintendent was developed to ensure that Mark learned while working at the golf course.

In essence, an SOE program is a planned series of events in which actual work experience is gained, which helps to prepare students for success in the occupation of their choice. The program is under the supervision of the vocational agriculture teacher.

Why Are SOE Programs Needed?

The primary reason for having an SOE program is for students to learn. Actually performing a task is the best way to learn. Would you want a mechanic to work on your car if that mechanic had only watched others and never worked on a car before? Would you entrust the task of pulling your teeth to a dentist whose only experience was reading a manual on how to pull teeth? What would you think of your school football team if they never practiced? They probably would not win very many games, because practice is essential. Occupational ex-

perience programs help you learn by providing a chance for practice, which is one of the best ways to learn.

Participating in an SOE program may help you find a job. Though there are no guarantees, usually an employer would rather hire a person who has had some prior experience in horticulture.

Another reason for having SOE programs is that you will probably earn money for the work you do. If you have a home project such as a vegetable garden or raising plants for sale, you should be able to make a profit. If you are placed on a job, you will probably be paid for your work.

In addition, a number of FFA awards are based on the SOE program. If you have an outstanding program you could be a district, state, or even national winner (see Chapter 21).

As you can see, there are at least four reasons for participating in an SOE program: (1) It is an excellent way to learn, (2) It may help you find

Figure 20-3. FFA awards can be won based on a student's supervised occupational experience (SOE) program.

Figure 20-4. Supervised occupational experience can be gained at school, at home, or in the community.

employment, (3) You may earn money, and (4) Many FFA awards are based on such programs.

Conducting SOE Programs

There are a number of ways in which an SOE program can be conducted. The three most common ways are through a home project, work in the school laboratory, or placement in a job.

Home Project

Many students gain occupational experience by conducting some type of horticulture operation in their home or in their home community. The student accomplishes this by applying at home what was learned at school. Brenda's strawberry patch is an example of a home project. A student could plan and maintain a lawn, landscape the home, grow a vegetable garden, build a small greenhouse and grow plants in it, operate a custom lawn care service, grow a crop of hanging plants in trees during the summer for sale as indoor plants during the fall, start pansies in a cold frame for sale in early spring, or even operate a business as a plant sitter for people who go on vacations. These are just a few examples of how students can participate in SOE programs in their home community.

There are several advantages to having an SOE program at home:

1. You gain experience in making management decisions. The success or failure of a project may depend upon your management decisions.
2. You realize all the profits from a successful project.
3. A home project can be tailored to fit your

individual interests, so there is a possibility for a variety of experiences ranging from turf care to vegetable production to nursery management to houseplant care.

School Laboratory

Another method of gaining occupational experience is in the school laboratory. The school laboratory may include a greenhouse and nursery in addition to the school grounds. In such cases students usually spend part of the time in class and part of the time working in the laboratory. Practical experience can be gained in the school laboratory in such tasks as budding, pinching, watering, pruning, and transplanting. Experience may be gained in operating specialized horticultural equipment that would not be available at home. In programs emphasizing landscaping, students may plan and actually install a landscape. In floriculture programs, students will design floral arrangements. Cathy's container-grown junipers were the product of a school-based SOE program.

In a number of schools, the students will operate a retail horticulture outlet. People in the community come to the school to purchase bedding plants, floral arrangements, and potted plants. Students gain practical experience in making sales and working with customers. Activities in the school laboratory can be part of a student's SOE program.

One advantage of an SOE program at school is that the vocational horticulture teacher would be readily available for consultation if there were a problem. Also, specialized facilities and equipment are available that might not be available at home.

Placement in a Job

One of the best ways to gain experience in horticulture is to work for a horticulture business. Some students obtain part-time work during school in a horticulture business. In many schools, a program is operated cooperatively between the school and local businesses, so that students can spend part of the school day working in a horticulture business. This program is often limited to juniors or seniors.

Mark's work at the golf course is one example of this type of program. A plan is developed that outlines what the student is to do on the job. The vocational horticulture teacher then supervises the student at work. Students who work in horticulture businesses gain firsthand knowledge. This is an excellent way to learn, and this type of SOE program may lead to a full-time job.

In planning an SOE program you may want to participate in a variety of projects. It is possible to have several different types of projects. Each year that you are in the horticulture class, you should expand your SOE program to include new experiences. When you graduate from the program, you will have the experience that is needed to get a good job.

Essentials of a Good SOE Program

A quality SOE program has five characteristics. You should be able to guess several of the characteristics if you reexamine the information already presented. Characteristics of quality SOE programs are:

1. They are planned.
2. They offer a variety of experiences.
3. They are challenging.
4. Records are kept.
5. They are supervised.

Planning

If you were to make a trip between Miami, Florida, and Los Angeles, California, by car, what would you do first? Would you just get

Figure 20-5. Your teacher will help you plan the SOE program.

into your car and start driving? Probably not! You would first look at a road map to determine which route you should follow to reach your destination. Prior planning is needed.

Prior planning is also needed for an SOE program. You should decide what kind of experiences you would like to gain and then develop a plan that will help you reach this goal. The plan should contain a starting date, location, agreement between parties involved, and tasks to be performed or skills to be learned.

If the SOE program is to be conducted at ·home, you should develop a budget. In this budget, possible expenses and income will be listed. You could list the supplies that will be needed. A calendar listing of the times at which various horticultural operations should be performed should be in your plan.

Brenda developed a plan for growing her strawberries. On the first page of Brenda's plan she stated that she planned on growing strawberries on a 50 × 50 ft [15.25m × 15.25m] plot of ground to be provided by her father. She

listed all the expenses she planned on having. Some of the expenses were for purchase of strawberry plants, methyl bromide, fertilizer, straw, and baskets. Brenda also estimated how much the plants would produce and what she thought the plants could be sold for.

On the second page of Brenda's plan was a list of all the work that had to be done in order to produce a crop of strawberries. The dates when the work should be done were also listed. She listed when the plants were to be bought, set out, fertilized, and covered for winter. She also listed how she planned on preventing diseases, and pest problems.

The last part of the plan was a section where all the people involved in the home project signed their names, indicating that they agreed with Brenda's plan. The vocational agriculture teacher, Brenda's parents, and Brenda all signed the plan.

The plan developed by Cathy for raising junipers at school was similar to Brenda's. Cathy estimated her expenses and income and listed all the work that had to be done. In Cathy's plan she agreed to pay rent to the school for use of a section of the greenhouse and cold frame. Cathy's parents, the teacher, and Cathy all signed the plan.

A different type of plan is needed if your SOE program involves work in a horticulture business. A training plan is developed with the employer and vocational agriculture teacher. The list of skills you plan on learning is included in the training plan.

Mark had a training plan at the golf course. The training plan stated that Mark would learn to operate the green and fairway mowers, to apply fertilizers and pesticides, to operate the irrigation system, to perform general maintenance on the equipment, to operate a sod cutter and aerator, and to overhaul small gasoline engines. On the training plan a blank column was placed next to each skill to be learned. When Mark became proficient at the skill, a check would be placed in the column. The training plan was signed by Mark, his parents, the vocational horticulture teacher, and the employers.

A Variety of Experiences

The second characteristic of a quality SOE program is that it contains a variety of experiences. It should be planned so that many new skills are learned. The more different skills you learn, the more employable you will be.

If your SOE program is at home, you will want more than just one type of activity. If you decide to grow roses, you will certainly learn how to do that; but then all you will know is how to grow roses. If possible, you should get experience in landscaping, gardening, lawn and plant maintenance, and plant care in the home. The more different types of home projects you have, the more you will learn. This will be to your advantage in the future. Next year Brenda plans on growing cantaloupes in addition to the strawberries. Cathy plans on growing poinsettias and Easter lilies for her SOE program next year. They both will be learning new skills.

If you work in a horticulture business, you will want to perform a variety of tasks. Doing the same job over and over gets stale, and you are not learning anything new. A training plan helps to assure you a variety of experiences.

Challenging

Your SOE program should be challenging, meaning that it should be neither too difficult nor too easy. Having two tomato plants as a project may be a good start, but it needs to expand, since this would not take long to learn. The project should be expanded to include more tomato plants and additional crops.

Likewise, you do not want an SOE program so large that you cannot keep up with it and perform the work properly. Holding down a part-time job in a horticulture business, doing your schoolwork, and having a lawn service plus a greenhouse may be too much. You should plan your SOE program so that it will be challenging but not overwhelming.

Each year that you are in vocational agriculture or horticulture, your SOE program should increase in size. In your first year, you might have a garden. The next year you may want to expand the garden or add different types of experiences, as Brenda and Cathy plan on doing. Having the same project year after year is not desirable.

Records

A fourth characteristic of a quality SOE program is that records are kept. There are several reasons why records should be kept.

Keeping records is part of an overall management plan in horticulture. Records let you know whether you are making or losing money. Every time Brenda spent money on supplies for her

OCCUPATIONAL EXPERIENCES AND FINANCIAL RECORDS

Month __April__ 19 ____

SUMMARY OF THE WORK I DID (List the *different kinds* of jobs you did this month. This should reflect jobs planned on pages 4 and 5.)	Date*	Hours Worked		Wages Received*
		No. Hours		
		Unpaid	Paid	
Transplanted geraniums	4/3		3	$ 7.50
Mixed potting soil & sterilized	4/4		2	5.00
Washed greenhouse windows	4/6		3	7.50
Balled and Burlaped shrubs	4/8		8	20.00
Watered geraniums	4/10		1	2.50
Moved plants to the coldframe	4/11		3	7.50
Sprayed Malathion for insect control	4/12		3	7.50
Waited on customers and sold plants	4/15		8	20.00
Pinched geraniums	4/17		2	5.00
Cleaned out storeroom	4/18		3	7.50
Repaired seed display rack	4/20		2	5.00
TOTAL FOR MONTH Forward to page 21				

*Record work performed by the day, several days, week, or month as practical and meaningful. Record wages by pay period on the date received.

Figure 20-6. Tasks performed should be recorded in the record book.

strawberry project she wrote the amount down in her record book. She also recorded all the money she received for the sale of strawberries. The amount of strawberries produced and the selling price went into the record book.

Whenever Brenda performed any work on the strawberry crop, such as weeding or water-ing, she recorded the time that was spent. She also made observations about such matters as weather or plant problems in the record book. The type of records kept by Cathy on the junipers she grew at school were similar to Brenda's.

The records kept by Mark were different. Each day he recorded the number of hours he worked and what type of work he performed. If he had any expenses, such as buying work clothes, he recorded them in his record book.

In planning for future horticulture enterprises, remember that records can provide helpful information. If you own a horticulture business but need to borrow money to expand or operate it, your records can help in obtaining a loan. Records are also used for selecting FFA award winners. If plants are unhealthy, the records may provide clues as to what the problem is. The type of records you should keep will be discussed later in this chapter.

Supervised

The final characteristic of a good SOE program is that it is supervised. Your vocational horticulture teacher will make visits to your home or place of employment on a regular basis. The purpose of these visits is to assist you in conducting your SOE program.

If you have a home project, the vocational horticulture teacher can help you with disease or insect problems. The teacher can show you how to prune or trim shrubs at your home. The teacher will be able to give you advice, based on experience, about ways to improve or manage your SOE program.

If your SOE program is at a horticulture job, then the vocational horticulture teacher will visit you at work. The teacher will help to solve any problems that might exist between you and the employer, and can also provide tips on how to be a better employee.

Many vocational horticulture teachers are employed by their schools during the summer, when as part of their duties they often supervise SOE programs.

SOE Program Records

Several types of information should be contained in your record book.

Work Performed

One type of information that should be recorded on a daily basis is the type of work that was performed. Did you water? Prune trees? Install a lawn? Design a floral arrangement? Sell plants? Apply herbicides? Sterilize soil? By keeping a record of the tasks you perform, you will have a complete list of the skills you have practiced. When you seek employment, you may want to list on the job application form all the skills you can perform.

Hours Worked

Along with the list of tasks performed, a record of the hours worked must be kept. When you apply for a job, the number of hours of experience you have may be as important as the type of experience you have. Whether it is a home project, a school project, or a job, the number of hours you work should be recorded. Mark, Brenda, and Cathy kept records of the amount of time they worked.

Expenses and Income

Whether you have a home project or a project at school; own your own business; or are working for a business, you will want to record expenses and income. How much money was spent on fertilizers, pesticides, and seed? What was the income on the produce? How much money was spent on transportation to and from the job? Did you have to buy your own uniforms or tools? In order to determine whether you are making a profit, it is important that an up-to-date list of your expenses and income be kept.

You will need to keep a record of the wages you have earned. This will be useful for income tax purposes and for budgeting.

Agreement

Many record books have an agreement section. This section is to be filled out and signed by

Figure 20-7. Your record book should be kept up-to-date.

you, and by your employer if your SOE program is in a business. If you are conducting a home project, your parents should sign the agreement. The agreement outlines what each person is to do. This prevents misunderstandings. No matter what type of project you have, there should be a completed agreement form. Your vocational horticulture teacher will help fill out the agreement section and will probably sign the agreement.

Training Plan

The importance of a training plan was mentioned earlier in this chapter. A training plan is simply a list of the different skills you will acquire while you are employed in a business. This helps you gain a variety of experience. From time to time, the training plan should be checked to see that you are gaining all the experience you are supposed to. A training plan is used when you are placed on a job to gain experience.

Extracurricular Activities

In addition to looking at the total hours of work and the types of experience that an applicant has, many employers look at extracurricular (after-school) activities when they are hiring.

Therefore, you should record your extracurricular activities in your record book. Examples of extracurricular activities are playing sports, the band, Student Council, Junior Achievement, Boy or Girl Scouts, church youth groups, and participation in other community activities. Most record books will have a section where FFA activities and other types of activities can be recorded. The activities section of the record book is especially important if you are applying for an FFA award.

Notes

Most record books contain a section for jotting down notes. The note could be anything from "there was a killing frost this morning" to "the seeds had an 80 percent germination rate." This section of a notebook can be extremely helpful. Some of the items that should be recorded are planting dates, spray dates, fertilizer applications, watering schedules, temperatures, and specific cultural practices that should be performed. If you have any problems with your plants, you may be able to determine why by looking at your notes.

Horticultural Supervised Occupational Experience Programs: A Review

An SOE program will provide you with the experience that employers look for when they are hiring employees. All students enrolled in horticulture or agriculture courses should have an SOE program. The SOE program allows people to gain practical experience; what is learned in the classroom is put into action in an actual working situation.

SOE programs may be conducted at home, in the school laboratory, or at a job. Students benefit from an SOE program by learning practical skills, improving their chance to get a job, earning money, and becoming eligible for FFA awards.

A good SOE program will be planned, contain a variety of experiences, be challenging, have records kept, and be supervised.

THINKING IT THROUGH

1. What is a Supervised Occupational Experience (SOE) program?
2. Explain in detail why you should have an SOE program.
3. Give 10 examples of how occupational experience may be gained.
4. What are five essentials of a quality SOE program?
5. What types of records should be kept for an SOE program?
6. How are records useful?
7. Outline an SOE program that would be good for you, considering your career goals.

21

Horticulture and Leadership Skills

The horticulture industry needs people who possess both horticultural skills and leadership skills. The importance of both types of skills is illustrated in the following example.

Ms. Brown, owner of Mid-City Flowers, asked Sue to come to her office. "I have been asked to speak to the Bud and Blossom Club next week on flower arranging," Ms. Brown said, "but I have to make a business trip out of town. I have been very pleased with your work. You are a skillful floral designer. Would you be willing to speak to the club for me?"

"I . . . I . . . I couldn't do that," stammered Sue. "Whenever I get in front of a crowd or have to speak, I get real nervous. I wouldn't know how to put on a . . . a demonstration for the club."

Sue was good at arranging flowers but she lacked skills in public speaking. If Sue possessed the leadership skill of public speaking, she would be even more valuable as an employee.

As a horticulture worker you will often work with customers. You will need to be able to communicate with customers. At times you may be expected to supervise other workers. Certain skills are needed when you do this. These personal relations skills can be classified as leadership skills.

In this chapter you will discover what type of leadership skills will be useful in horticulture and how they can be developed. One way in which leadership skills can be developed is through the FFA. You will learn in this chapter how both leadership skills and horticultural skills can be improved through the FFA.

CHAPTER GOALS

In this chapter your goals are:

- To define leadership and determine the importance of leadership in the horticulture industry.
- To identify FFA activities that will improve your leadership skills and participate in these activities.

- To identify opportunities that are available in your community for developing leadership skills and make use of them.

What is Leadership?

Leadership means different things to different people. Some people think of leadership as the

Figure 21-1. Calling plays in a huddle and introducing a new student are both examples of being a leader.

quality possessed by a general who leads an army in charging up a hill. Other people may think of a quarterback on a football team. The president of a large corporation may be what some think of when they hear the term "leader." These are all types of leadership, but there are other types of leadership. A person who introduces a new student in school to other students is a leader. A leader could be the person who suggests that the group go to the malt shop after school.

A simple definition of *leadership* is the ability to take action at the appropriate time. A leader is a person who sees to it that action is taken when something needs to be done.

Types of Leadership

There are two types of leadership, formal and informal.

Formal Leadership. *Formal leadership* is exercised when a person is elected or appointed to a position. The president of the FFA, the captain of the basketball team, and a lieutenant in the army would all be examples of formal leadership. These people are expected to lead, because they have been placed in leadership positions by other people. In the horticulture industry, the manger or owner of a business is a formal leader. Such people are expected to make decisions and provide leadership because of their positions.

Informal Leadership. *Informal leadership* is involved when a person who is not the formal leader takes charge because the situation demands it. For example, Ron, Raoul, and Bob were employed by a landscaping and nursery firm. One morning Ms. Roberts, the supervisor, asked the three to load the tractor onto the trailer, which was hooked to the truck, and drive out to a house in the country that had just been built. They were to start grading and leveling the soil in preparation for seeding the lawn. Ms.

Roberts had some business errands to run, but would catch up with them later in the morning. Ron, Raoul, and Bob loaded the tractor (which had a front end loader and back blade) onto the trailer and were soon headed for the work site. Bob was driving the truck and soon noticed that it was hard to steer. He stopped the truck and got out to look at the tires. The right front tire was flat. "Well, it looks like we'll have to change the tire," said Bob. After searching all through the truck, they couldn't find a jack. They were out in the country with no houses in sight.

"Since we don't have a jack and there are no houses around, I guess we'll just have to wait until Ms. Roberts comes along before we can change the tire," said Bob.

After sitting down for a few minutes, Ron jumped up. "I know how we can change the tire. Let's unload the tractor, drive it around to the front of the truck, place the front end loader under the bumper, raise the truck up using the loader, and change the tire." Ron took charge. Soon the tire was changed, the tractor was reloaded, and they continued on their way.

Here Ron became the leader. This is the informal type of leadership, where a person who is not the appointed or elected leader becomes the leader when the situation demands it.

As a horticulturist, you may have the opportunity to be a formal leader, or you may be an informal leader. Both types of leadership are desirable and are needed.

Leadership Skills

Just as there are a number of technical skills that a person in horticulture should possess, such as pruning, budding, propagation, and caring for plants properly, there are a number of leadership skills that should be mastered. These leadership skills must be learned and practiced in the same way that any horticultural skill would be. People who demonstrate leadership skills in horticulture are often promoted more rapidly and receive good pay raises. Employers want to have leaders working for them. By developing your leadership skills you will be improving your chances of succeeding in horticulture. A leader should possess communication skills, social skills, and citizenship skills.

Communication Skills

The ability to communicate is a key ingredient in leadership. No matter whether you are a formal or informal leader, you will need to communicate your ideas to others. If you are an officer in an organization, you will need to communicate with your fellow officers and members. If you are a worker in a horticulture business, you will be talking with customers or with other workers.

With Other Workers. Managers, supervisors, and crew leaders must communicate effectively

Figure 21-2. Public speaking is an important leadership skill.

with the people who work under their supervision. The following example illustrates this point.

One morning Robin, the greenhouse supervisor, told Tom and Joan, "Your task for this morning is to propagate the mother-in-law plants. You'll probably want to make leaf cuttings." With those directions, Robin left.

"That doesn't sound too hard," thought Tom as he started gathering up the sansevieria.

"What are you doing?" asked Joan. "That's the wrong plant. That's the mother-in-law's tongue. The mother-in-law plant is the diffenbachia."

Tom replied, "I think you are right, but the sansevieria is propagated by leaf cuttings and I think this is what Robin meant." Do you know which plants Robin had in mind? Tom and Joan are confused because Robin did not communicate effectively.

With the Public. If Robin continues to communicate in this manner problems could occur in the greenhouse. It is possible that Robin could even lose his job because of poor communication skills. At times you may be called on to give presentations, as Sue was. The ability to speak in public would be to your advantage and might result in promotions or salary raises. Employees in horticulture jobs that deal directly with the public must be able to speak effectively. Horticulturists are often asked to make presentations to various groups. Skills in public speaking can be used throughout your lifetime.

The ability to use the telephone properly is another important communication skill. Horticulturists, especially florists, conduct a great deal of business over the telephone. Answering the telephone with the name of the business, such as "Mid-City Flowers," is the proper way. Good phone manners will help the business to gain new customers. Poor phone manners may lose customers.

"Hello," answered Fred as the phone rang. The voice on the other end of the line asked,

"Is this the Northside Florist's?"

"Yep, it sure is," replied Fred.

"Do you have any Christmas wreaths?" asked the potential customer.

"Yep, we got all kinds," answered Fred.

The customer asked, "Do you have any that cost around $20?"

"Yep, we sure do," was Fred's reply.

"What do they look like?" asked the customer.

"Well, they look like Christmas wreaths," said Fred. The potential customer thanked Fred for his help and hung up. Fred had just lost a sale because of his poor telephone manners. He should have answered the phone, "Northside Florist's, may I help you?" When the customer asked questions, Fred should have provided detailed answers. Good communication with customers is essential in horticulture. Poor communication skills could result in dismissal from the job.

Figure 21-3. Leaders need to possess parliamentary procedure skills.

Human Relations Skills

What do you think is the number one reason why people lose their jobs? Is it a lack of technical knowledge? No. The number one reason why people lose their jobs is because they cannot get along with others. The ability to get along with others is very important in the horticulture industry. You need to know how to work with your fellow workers and how to get along with the public. You have probably known people who were very disagreeable and hard to get along with. No one likes to be around people like that. It is possible to get along with others and not be a leader; however, one characteristic of a good leader is that the leader will strive to get along with others. A leader must develop human relations skills.

Good manners are another important social skill. A leader will be courteous and considerate of others. While eating out, a leader will use proper table manners.

A leader will also dress properly and practice good grooming habits. This includes bathing

often, brushing teeth, and having clean, combed hair. Deodorant should be used. Clothing should be appropriate for the occasion. Your appearance makes an impression on others. If you are a skillful horticulture worker but look sloppy, you may not be placed in leadership positions.

Citizenship Skills

A leader will participate in community activities. A knowledge of parliamentary procedure is needed for community activities. Professional societies conduct business by parliamentary procedure. In a democracy, parliamentary procedure is widely used to ensure that meetings are run smoothly and business is transacted in an orderly fashion. Organizations that do not use parliamentary procedure may have problems.

Randy did not know what to do. He had just been elected club president and was in charge of his first meeting. Marion was trying to give a committee report but Leon wanted to talk about having a hayride. Steve wanted to discuss ways to earn money for the club. Nothing was being accomplished. Randy and the members of the club needed a knowledge of parliamentary procedure. If they used parliamentary procedure, everyone would get a chance to speak and the business would be transacted.

FFA chapters use parliamentary procedure in their meetings. There are even parliamentary procedure contests in which FFA members must compete. A working knowledge of parliamentary procedure will be to your advantage, whether you are the leader or one of the members at the meeting.

Both you and the horticulture industry will benefit if you develop your leadership skills. You as an individual will be more secure in your work. You may be in line for promotions and pay raises if you demonstrate leadership skills as well as horticulture skills.

The horticulture business you work for will profit because you are able to work with other people and the public. Professional organizations in horticulture will also benefit from your leadership abilities.

Developing Leadership Skills

How does one go about developing leadership skills? Can they be developed by reading a book or doing a report? Partly, but not completely. To develop leadership skills effectively, a person must receive some basic instruction and then practice the skill.

The importance of getting occupational experience in the technical aspects of horticulture has already been discussed, but it is equally important to gain occupational experience in leadership development. How is this done? One way in which leadership is developed in high school horticulture students is through an organization called the FFA. The FFA is a labo-

ratory for learning leadership skills, just as the greenhouse may be a laboratory for learning horticultural skills.

The FFA

The FFA is an intracurricular (in school) organization composed of students who are enrolled in agriculture classes. When the FFA was originally founded, it was an organization for male farm students, but today it is much more than that. The FFA includes students who are enrolled in all phases of agriculture, such as agribusiness, forestry, agricultural mechanics, and horticulture. Special contests and activities have been developed specifically for members who are in horticulture and the other specialized phases of agriculture. Today the FFA is for both city students and country students.

The FFA was established in 1928 in Kansas City. A primary purpose of the FFA is to develop leadership skills. The leadership skills of students are developed through a variety of FFA activities.

FFA Chapter Activities. There are numerous FFA chapter activities designed to develop leadership skills in students. FFA chapters are led by a team of officers who practice such leadership skills as parliamentary procedure and public speaking. By becoming an officer, a person learns more about being a formal leader.

Each FFA chapter has a program of activities that is an outline of what the chapter hopes to accomplish during the year. The members of the chapter are divided into committees. Each committee has responsibility for planning and conducting part of the program of activities. The committees that are normally found in FFA chapters are:

1. Supervised Occupational Experience (SOE)
2. Cooperation
3. Community service
4. Leadership
5. Earnings, savings, and investments

Figure 21-4. Leadership experiences are gained through committee work.

6. Conduct of meetings
7. Scholarship
8. Recreation
9. Public relations
10. State and national activities
11. Alumni relations

Each committee has a chairperson. These committees meet regularly to plan and implement their activities. Leadership experiences are gained through the committee meetings. You learn how to work with other members on your committee and plan activities that will improve your human relations skills. This will be helpful to you on the job.

The FFA chapter holds monthly meetings that are conducted through parliamentary procedure. Items of business are transacted. By attending the chapter meetings, you can become more familiar with the proper use of parliamentary procedure and can develop parliamentary skills. If you are elected to an office you should become very proficient in the use of parliamen-

tary procedure. In many states there are contests in parliamentary procedure. Each school will have a team and the teams will compete in district, area, or state contests. This increases an individual's proficiency in using parliamentary procedure, and you will be surprised as you go through life that many people will look to you as an expert in parliamentary procedure because of your FFA experience.

FFA Degrees. There are four degrees of membership in the FFA: greenhand, chapter degree, state degree, and American degree. In order to advance from one to the next degree, you must demonstrate proficiency in leadership and also conduct quality SOE programs. By striving for each higher degree, you will develop additional leadership skills. The state and American degrees are very high honors. Each time you obtain a new degree, you will be given a medal that can be worn on your jacket or displayed at home. By earning each new degree, you are demonstrating initiative and a capacity for hard work. Those who are successful in earning FFA degrees are often successful in their work.

Achievement Award. In addition to earning degrees, FFA members may be involved in an achievement award program. The achievement award is based on the development of agricultural skills, leadership development, career understanding, and safety practices in a specific area of agriculture. Every FFA member can qualify for these awards by completing certain requirements on a simple checklist that can be obtained from the FFA advisor.

Proficiency Awards. FFA members may also compete for proficiency awards. A proficiency award is an honor given to a student who excels in a specific area of agriculture, such as floriculture. Students are recognized at the chapter, district, state, and national levels. A local winner receives a medal or plaque. State winners receive a certificate and a cash award. Regional and national winners receive a plaque and a cash award. There is a total of 22 proficiency awards. Of these 22 proficiency awards, 4 are specifically designed for horticulture students. If you are a proficiency award winner, this certainly will be to your advantage when you seek employment.

The *floriculture* proficiency award is for students who have developed knowledge and skills in the production, arrangement, and marketing of floral products. In order to qualify for this award, a student must have participated in an SOE program in floriculture. Accurate records must be kept on the income, expenses, and accomplishments in the SOE program. FFA activities and leadership activities in the school and community are also considered.

Barry won the floriculture proficiency award in his FFA chapter. Barry's SOE program in floriculture consisted of working for a local florist and growing African violets at home. Barry had spent 1,782 hours working for the local florist. Barry waited on customers and made corsages and floral arrangements. At home he grew African violets under grow lights. He sold between 50 and 60 African violets every year. Barry was chairperson of the FFA community service committee and participated in the FFA public speaking contest. All these activities were considered in selecting Barry as the floriculture proficiency award winner.

A proficiency award is also available in *fruit and/or vegetable production*. This award is for students who use modern management practices in the production and marketing of fruit and vegetable crops. An SOE program involving the production of fruits and/or vegetables is needed in addition to leadership activities.

Rhonda was excited when she learned that she had won the chapter proficiency award in fruit and vegetable production. She had grown a large vegetable garden during her first year of vocational agriculture. In her second year she rented ½ acre of land from a neighbor and raised sweet corn. She sold the corn to local merchants for $.60 per dozen ears. Rhonda added the production of pumpkins to her sweet corn business during the third year of agriculture. Rhonda was an officer in her Sunday School class and was a member of the student council.

A *nursery operations* proficiency award is available for those who are involved in the production and marketing of seedlings, shrubs, bedding plants, and other nursery stock. The SOE program should relate to this area, and leadership activities are needed.

Laura was the winner of the chapter proficiency award in nursery operations. She had spent over 1,500 hours working for a local nursery, and gained experience in setting out seedlings, pruning, fertilizing, transplanting, mowing, growing bedding plants, controlling plant pests, and waiting on customers. She was active in Junior Achievement and was on the FFA parliamentary procedure team.

The fourth horticulture proficiency award is in the area of *turf and landscape management*. This award is designed to encourage the devel-

Figure 21-5. A combination of leadership and technical skills are needed to win some awards in horticulture. A National FFA contest in horticulture is held annually.

opment of production, marketing, and management skills in the operation of turf and landscape occupations. FFA and leadership activities and an SOE program in turf or landscaping is needed to win this award. The winner of the chapter proficiency award for turf and landscape management was Albert. He had started his own lawn care service, and had contracts with a number of businesses and many private home owners to care for their lawns. He would mow, fertilize, aerate, and trim shrubs. During an average year Albert would make over $4,000 before his expenses were subtracted. Albert was a chapter FFA officer and had attended the state FFA convention.

Several other proficiency awards in which horticulture students might be interested are home and/or farmstead improvement, agricultural sales and/or service, outdoor recreation, and soil and water management.

Your vocational horticulture instructor will have the application forms for these awards. You may want to look at the application form

to see what is expected or what is required to qualify for the proficiency award. The earlier you start planning to win the award, the better your chances will be.

Contests. In the FFA there are a number of competitive contests involving both individuals and teams. One of these is a contest in public speaking. There are basically three different kinds of public-speaking contests. First-year students may enter the creed contest. In this contest the students recite the FFA creed. The student who does the best job is the winner.

Another public-speaking contest is called the prepared public-speaking contest. In this contest a student prepares a speech, writes it out, and practices it. Then the student gives the speech in front of a group of judges. The judges evaluate the speech on the basis of its content, its presentation, and the poise of the speaker.

The third public-speaking contest is called extemporaneous public speaking. In extemporaneous public speaking, a student is assigned a topic and is given a short period of time to prepare for the speech. In this type of speaking, students need a good background in agriculture and the ability to think fast on their feet.

Prepared and extemporaneous public-speaking contests are conducted at the chapter, district, regional, state, and national levels. There is a national public-speaking contest. The four regional winners are in the national public-speaking contest. Public-speaking skills will help you become a success in horticulture.

There are several FFA judging contests that horticulture students may enter. There are two national FFA horticulture contests that involve competition among the state winners. One is the floriculture contest, while the other is the nursery/landscape contest.

The floriculture contest is composed of four parts. In part one the contestants have to identify 50 common floral plants. During the second part of the contest, five classes of floral products are judged. Typically, four classes consisting of carnations, roses, gladioli, bedding plants, and foliage plants are judged. There is one class of floral arrangements to be judged. Fifty multiple-choice questions designed to test knowledge in floriculture are found in part three. The final part of the contest requires you to perform skills such as designing a floral arrangement, planting cuttings, or taking a telephone order.

The nursery/landscape contest is similar to the floriculture contest. Fifty common nursery landscaping plants have to be identified. There are four classes of landscaping materials to be judged. Models of landscaped homes are also judged. There is a 50-item multiple-choice test on nursery production and landscaping. In the final phase, students are given a landscape drawing and asked 10 questions about the drawing, such as how much sod or fence would be required in the landscape drawing.

Some states have an FFA entomology contest.

Figure 21-6. FFA members are encouraged to participate in horticulture exhibits and demonstrations.

Figure 21-7. The FFA sponsors a number of activities to develop leadership, such as the National FFA Convention held each year in Kansas City, Missouri.

Knowledge and skills needed for this contest include the ability to identify insects and insect damage and answer questions about insects and their life habits. Insects of both horticultural crops and field crops such as corn are included in the questions.

Fairs and Shows. Many county and state fairs have special divisions for horticultural exhibits, and FFA members are encouraged to participate in the fairs. The types of horticultural exhibits will vary from fair to fair. In many fairs, students may exhibit vegetables, fruits, and floral arrangements. In California there is a special section of the state fair for landscape plots where students design and install landscapes. In other states, landscape plans are judged at the state fair.

Giving demonstrations is another popular activity at many fairs. FFA members give demonstrations on such topics as floral arrangements, terrariums, gardening, care of houseplants, and lawn care.

Students benefit from participating in fairs. In addition to winning ribbons and possibly prize money, students develop leadership skills.

State and National Conventions. An excellent place to see leadership in action is at the state and national FFA conventions. Most states hold a state FFA convention every year. At the convention, state awards are given out, chapters are recognized for their outstanding achievements, and students may compete in state-level competition in such contests as public speaking and parliamentary procedure. By participating in the state conventions as observers or as contestants, students increase their leadership abilities.

The national FFA convention is held annually in Kansas City in the fall. The convention is very inspiring, because leadership is seen at its fullest there. Attending the national FFA convention is a once-in-a-lifetime opportunity.

Leadership Conferences and Camp. The FFA

sponsors several activities specially designed to improve leadership skills. One of these is the national leadership training conference in Washington, D.C., where several week-long sessions are held each summer. FFA members from all over the nation attend the conferences. Leadership and personal development sessions are conducted by national FFA officers and experienced staff counselors. FFA members who attend the conference also get to visit many historical sites.

Many states have a state FFA camp. These camps, which normally operate during the summer, provide an opportunity to gain and develop leadership skills. Students also enjoy camping and recreational activities.

Leadership Opportunities in the Community

There are many opportunities for developing leadership skills in most communities; all anyone has to do is look around. Many community organizations will help develop individuals' leadership abilities.

Most communities have 4-H clubs. These clubs have elected officers and use parliamentary procedures. Members of the 4-H club have various projects; some may have horticulture projects. For more information about 4-H clubs you can call your county agricultural extension office.

The National Junior Horticultural Association (NJHA) is an organization for young people interested in horticulture. Various contests are conducted that members may compete in.

Boy Scouts, Girl Scouts, and Explorers provide opportunities for developing leadership skills. Each troop has elected officers. Merit badges may be earned in such areas as gardening and fruit and nut production.

Junior Achievement is an organization in which groups of young people form a company and produce a product for sale. Members gain experience in organizing and operating a business. The product could be a horticultural product, such as silk or dried flower arrangements.

Most communities have many other organizations that help to develop leadership skills, such as church groups, the YMCA, and the YWCA. If you would like to improve your leadership skills you should become involved in some of these out-of-school organizations. Employers have a tendency to hire people who are members of groups where leadership abilities can be acquired.

Horticulture and Leadership Skills: A Review

The horticulture industry needs people who demonstrate leadership skills. Both formal and informal leaders are needed. Leaders should practice their communication and human relations skills. One excellent method of developing these leadership skills is through the FFA. Members of the FFA may work toward degrees and proficiency awards in horticulture. There are horticulture and leadership contests in which students may compete. By actively participating in FFA activities, you can improve your leadership skills.

THINKING IT THROUGH

1. How would you define leadership?
2. Name two types of leadership and tell how they differ.
3. What types of leadership skills are needed in horticulture? Why are they needed?
4. What is the FFA?
5. How can the FFA help develop leadership skills?
6. What opportunities are available at school, at home, at work, and in the community to develop leadership skills?

EIGHT
Employment in the Horticulture Industry

In a football game a team doesn't score a touchdown without crossing the goal line. Getting the ball to the 1-yard line doesn't put points on the scoreboard. A team has to be able to get the ball into the end zone. Likewise, having an in-depth knowledge of horticulture is not very helpful if you want a job in horticulture but don't know how to get one.

Finding a job in horticulture requires skills, just as designing a floral arrangement does. A number of steps are undertaken when you try to find a job. A person who knows the steps will have an advantage in finding a job.

At some time in the future you may decide to start your own business. A number of factors should be considered before you do this. The owner of a business has many responsibilities.

In this unit you will learn how to undertake the task of finding and keeping a job. You will also learn how to go about starting your own business and will learn to handle the responsibilities of a business owner.

22

Managing a Horticulture Business

Have you ever wanted to own your own business? Many young people have the dream of someday owning their own floral shop, nursery, or landscaping company. Owing a horticulture business is enjoyable, but it requires hard work and good decision making.

The owner of a business must make difficult decisions, keep records, keep up with state and federal regulations, maintain a safe business, and market the product. Financing has to be available to operate the business. Owning a horticulture business requires more than just a knowledge of horticulture. An example of this is the way Thomas operated his business.

Thomas had recently started a lawn and garden store. He knew a lot about horticulture, but he didn't know much about business. About 1 month after the store had opened, a large truck pulled up near closing time to unload plants that Thomas had ordered. Thomas was surprised. He had forgotten that he ordered them and that they were to be delivered today. The entire store was already full of plants and supplies, and there was no place to unload the new plants. Thomas ordered a couple of his employees to move several hundred sacks of fertilizer to make room for the plants. The workers were not happy with the way Thomas ordered them around and with the fact that they had to work when it was time to quit since they both had plans for the evening. After finally getting all the plants unloaded, Thomas wrote a check to pay for the plants. A few days later Thomas received a call from the bank. He did not have enough money in his account to cover the check. He had forgotten to record several checks he had written. Even though Thomas was a knowledgeable horticulturist, his business was on the brink of failure because he didn't keep good records and was a poor manager of people and money.

In this chapter you will learn how businesses are organized and what the responsibilities of an owner or manager of a horticulture business are.

CHAPTER GOALS

In this chapter your goals are:

- To outline the procedures for determining whether there is a need for a horticulture business
- To determine the legal requirements for starting a horticulture business
- To explain the need for developing a business financial plan
- To describe the different ways in which businesses may be organized, and point out the advantages and disadvantages of each
- To determine which business organization method would be best for you if you desired to start your own business
- To explain the roles and responsibilities of owners and managers of horticulture businesses
- To determine the types of records kept in horticulture businesses

Starting a Horticulture Business

How do you think someone would go about starting a horticulture business—just erect a sign that said "open for business"? Needless to say, the answer is no; there are several important steps that need to be followed if a horticulture business is to succeed. Let's see how one person went about starting a horticulture business.

Ms. Martinez had worked a number of years

Figure 22-1. Owning a horticulture business is rewarding but requires much hard work. Before starting a business a person should determine: if there is a need for the business, the necessary legal requirements, and the financial plan.

for a nursery that specialized in growing bedding plants, and she had learned a great deal about bedding plants. With her experience from this business and with the money she had saved, Ms. Martinez thought she could start a business of her own. She was thinking about starting a nursery that specialized in ground cover plants, which are low-growing, spreading perennials (such as myrtle, Japanese spurge, and ivies) that are used on slopes or wherever little maintenance is desired. She had located 5 acres of land near town that was for sale. She thought this would be a good location for the nursery, but one question bothered her: "Is there a need for a bedding plant nursery that grows ground cover?"

Determining the Need for the Business

One of the first things Ms. Martinez needed to do was to determine whether there was need for a ground cover nursery. Before starting any type of business, the first step is to determine if there is a need for the business. Are there other firms offering the product or service you plan to offer? If there are, how many are there? Would there be enough customers to support another business of the same type? Could your business operate more efficiently or provide better products or services than the other businesses of the same type? Before starting any business, you must first determine the type of competition you will have and how strong it is.

Along with this, you must determine whether there will be a demand for your product or service. For a business to be successful, there must be sales. If no one wants to buy your product or service, you will soon be out of business. A careful analysis of the community in which you plan to locate your business will help to determine whether there is a need for it. How do you think Ms. Martinez went about determining whether there was a need for a nursery specializing in ground cover plants?

The first step was to determine whether there were other bedding plant nurseries in the community. Ms. Martinez looked in the yellow pages of the telephone book to identify all the nurseries in the community. She then went to visit the firms to see what types of plants they grew. None of the firms she visited grew ground cover plants on a large scale. It appeared that there might be a need for a nursery specializing in ground cover plants.

Next, Ms. Martinez called several landscaping firms and made an appointment to see the manager of each. She wanted to know where they were obtaining their ground cover plants and whether they were pleased with the supply. She learned that most of the landscaping firms were shipping the plants. The managers were interested in Ms. Martinez's plans. Her plants would cost less because they would not have to be shipped. Also, delivery time would be cut, since the nursery would be located in the area. After talking with the managers of the landscaping firms, Ms. Martinez believed there would be enough demand for ground cover plants to start the nursery. Now, what was the next step?

Determining Legal Requirements

Before starting a business, a person needs to be aware of any legal regulations that may affect it. In most states a license to operate a business must be obtained before the business can be started. If the business handles food items, such as those in a fruit or vegetable stand, special health permits may have to be obtained. Zoning laws in many communities restrict the type of businesses that can operate in certain locations. If workers are to be hired to help in the business, then certain legal requirements must be met concerning minimum wages, hours of work, and safety conditions. How do you think a person goes about learning these regulations?

Ms. Martinez did two things to learn about the legal requirements of starting a business. First she went to the county courthouse. Here she checked with the county zoning commissioner to determine whether the land she intended to buy was zoned in such a way that a nursery could be located there. She also checked with the county clerk and treasurer to determine what types of licenses and fees would be required to start the business. After this, she went to see a lawyer who specialized in business law. Ms. Martinez asked the lawyer about the legal requirements for starting a business. It is wise to check with the the county courthouse and a lawyer before starting a business.

Developing a Financial Plan

After determining that there was a need for the business and learning of the legal requirements of the business, Ms. Martinez was ready to develop a financial plan. Before a business is started, the owner or owners must carefully analyze how much money will be needed to start the business and keep it running. Expenses for land, buildings, equipment, supplies, taxes, and salaries have to be figured. Projections of how much money will be coming into the business from sales of products or services also have to be made. For a business to operate, it must make a profit. A financial plan is used to help predict whether the business is going to be profitable. If the figures show that a business will not be profitable, it would not be wise to go ahead and start the business.

A financial plan often reveals that more money will be needed to get the business started than the prospective businessperson has. In this case, money may have to be borrowed. It is not uncommon to borrow money to start a business. Money is often borrowed from banks, savings and loan companies, and other firms that lend money. The Small Business Administration, part of the U.S. Government, is another source.

After determining that there is a need for the business, checking out the legal requirements, and developing a financial plan, the next step is to organize the business.

Figure 22-2. The owner of a business is called the proprietor or entreprenuer.

Organizing the Horticulture Business

There are three basic ways in which horticulture businesses can be organized: sole ownership, partnership, and corporation. Each type has certain advantages and disadvantages.

Sole Owner

A business that is owned and managed by one person is said to be a *sole proprietorship*. Sometimes the sole owner of a business is called an *entrepreneur* or *proprietor*. One person owns the business, makes most of the management decisions, and receives all the profits. Ms. Martinez's nursery is an example of a sole proprietorship. She has invested her own money in the business and makes all decisions concerning the operation of the nursery.

There are a number of advantages to being the single owner of a business. One is that the owner is the boss and is free to make any type of decision. If a decision needs to be made immediately, the owner does not have to consult with partners. Another advantage of a sole proprietorship is that the owner receives all of the profit. This may cause the owner to work harder and look for ways to serve the customer better. Many sole proprietorships are small; thus, the owner personally knows the customers and employees. This helps develop good working relationships among owners, employees, and customers.

Even though there are a number of advantages to being the sole owner of a horticulture business, there are also several disadvantages. One is that a single owner may not possess all the skills and abilities needed to operate a horticulture business. An owner needs both business and horticulture skills. If the owner lacks management skills, the business could go bankrupt.

One management skill needed by an owner is that of maintaining an inventory of stock. Whenever the stock on items for sale gets low, the owner must reorder the item. Customers are unhappy when they need to buy something and the store does not have the item.

The owner often has to give estimates on jobs. People want to know in advance how much it will cost to have a landscape installed, a lawn seeded, or a wedding planned. The owner will need to be able to figure out how much it will cost to do the job.

The ability to get along with people is a management skill needed by owners. An owner should be able to meet people and remain calm when customers make demands.

The owner needs skill in supervising employees. At times it is necessary to correct employees and give directions to them. This should be done in a tactful manner.

The Small Business Administration is willing to help people who own businesses to improve their management skills.

Another disadvantage of a sole proprietorship is that the owner may not have enough money to operate the business properly. Opportunities to expand the business may be hindered if the owner does not have enough money to take advantage of opportunities as they occur.

Just as a sole owner receives all the profits from the business, the sole owner also has the sole responsibility for making good the debts of the business. If the business has borrowed large amounts of money and goes broke, then the owner of the business has to repay the loans. If the business loses money, so does the owner.

Partnership

A *partnership* is a business owned by two or more people. In a partnership, the partners

Figure 22-3. In a partnership the owners jointly make decisions and share financial responsibility.

jointly make decisions and share financial responsibility for the business.

There are several advantages to a partnership. For one thing, different abilities and skills can be combined. A person with a strong background in horticulture could form a partnership with a person who is a good business manager. This combination of skills should result in a more efficient and prosperous business than one where either partner was in business alone.

Additionally, more money is available to operate the business. Each business partner provides money for operation of the business. By combining the assets of two or more partners, partnerships are able to increase the amount of credit available to the business.

Ray and David were close friends and often talked about starting their own horticulture business. Ray was the manager of a shoe store but grew roses and other flowers as a hobby. David worked as a floral designer for a florist. After several years of planning, they decided to form a partnership and start their own floral shop. By combining their money, they had enough to purchase a building that was easily converted into a retail flower shop. Neither Ray nor David would have been able to buy the building himself. Ray's experience as a shoe store manager was important in keeping records, hiring employees, and working with customers, while David's experience as a floral designer was needed to operate the business. Ray and David soon had a successful business. By forming a partnership, they combined the skills and money of each person into a profitable venture.

There are four main disadvantages to a partnership. One is unlimited financial liability. This means that if one partner cannot pay his or her share of business debts, the other partners must pay the full debt. If the business were to fail, one partner might have to pay all the debts if the other partners could not help pay them.

Whenever two or more people have to make joint decisions, there is always the possibility of a disagreement. Partners may disagree on how the business should be operated or how much work each should do. All partners must be in agreement before changes can occur in the business. One partner might want to expand the business while the other partner does not. A plan could be developed in advance for solving such problems, with procedures to be followed outlined in the plan. A third person whom both partners trust could be asked for advice.

A third disadvantage of a partnership is that each partner is bound by contracts made by the other partner or partners. If one partner enters into a contract that is unprofitable, the other partners must abide by the contract. This may cause hard feelings among the partners. An example of this problem is in the partnership formed by Ray and David. One day while David was gone a customer wanted an estimate on the cost of wedding floral arrangements. The customer described the arrangements and named the flowers that were to be used. Even though David always figured wedding arrangements, Ray gave an estimate based on what the last couple of weddings had cost. He didn't realize that the flowers the customer wanted were very expensive. David was upset when he returned, because they would lose money on the wedding. The best way to prevent this is to have each partner agree in advance that no contracts will be signed unless they first talk it over.

If one of the partners decides to leave the partnership, this could be a problem. The remaining partner would have to buy out the first partner's share of the business or find another partner. If a new partner could not be found and the remaining partner did not have the money to buy out the departing partner, then the old partner would not be able to get out of the business. This problem could be prevented by having a written agreement before the partnership is formed. In this agreement the procedures when one of the partners leaves the business could be outlined in detail. A partnership agreement will help prevent future problems.

Figure 22-4. Some businesses are organized as corporations.

Corporation

A third type of business organization is the corporation. A *corporation* is a business organization composed of individuals who have been granted a charter from the state to operate a business. The ownership of the business is shared by a number of people who have bought shares of stock, and the income from the sale of stocks is used to finance the business. The stockholders elect officers who are placed in charge of operating the business. In small corporations the number of stockholders may be limited to a few individuals who also serve as officers.

An example of a corporation is Home Lawn Care, Inc. This corporation specializes in caring for lawns in several large cities. It fertilizes lawns and applies herbicides on a regular basis to its customers. Each customer pays a set fee, and in return Home Lawn Care, Inc., keeps the lawn healthy and green. A fleet of large trucks is used to apply the fertilizers and herbicides.

This business was started by several people who saw the need for such a service. They studied the situation and determined that a lawn care service would make money, but it would have to operate on a large scale and would need people who knew what they were doing. In order to raise enough money to buy the trucks and start such a business, they determined that a corporation would have to be formed. Shares of stock were sold to the public to raise money. The organizers of the corporation also bought stock. A board of directors was elected by the stockholders to make decisions on behalf of the stockholders and hire a manager to oversee the daily operation. The profits from the business were divided up among the stockholders.

There are three main advantages to forming a corporation. Stockholders are not personally liable for debts incurred by the corporation. If it goes bankrupt, the stockholders may lose the money they have invested in the stock but they are not responsible for debts the corporation owes. If Home Lawn Care, Inc., went bankrupt, the stockholders would only lose the amount of money they had invested in stocks. They would not have to pay the company debts.

Corporations are able to borrow money more easily in many instances than individuals or partnerships. This enables corporations to operate on a large scale.

The third advantage of operating as a corporation is the ease of transferring ownership. Shareholders simply sell their stock to other individuals or groups of individuals. The shareholder does not have to have the approval of the other shareholders to sell the stock.

Just as there are advantages, there are also disadvantages for a horticulture business organized as a corporation. The tax structure for a corporation is different from the structure for an individual. Often corporations are taxed at a higher rate. However, this varies according to the type and size of corporation. Additionally, there are special state and federal rules and regulations governing corporations. In most states corporations must make special reports to the state on an annual basis.

Figure 22-5. In every type of horticulture business people have responsibilities.

Work Responsibilities in Horticulture Businesses

In every type of horticulture business, specific people have specific responsibilities. A horticulture business is somewhat like a football team. Each player has a specific job. However, for the team to win everyone must work together. The owner of a business must do certain things if the business is to be successful. Likewise, the workers or employees in the business have certain responsibilities. If a person is hired to manage a business, then that person has specific job duties. What do you think are the specific responsibilities of the owner and manager?

Owner Responsibilities

The responsibilities of the owner of a horticulture business will vary, depending upon the size and type of business. In large corporations the owners (stockholders) may hire a general manager to operate the business. In this situation the owners generally rely on the manager to assume many responsibilities. In smaller businesses, such as the ground cover nursery planned by Ms. Martinez, the owner has a number of responsibilities. Let's see what type of responsibilities Ms. Martinez might have.

Making Decisions. The owner of a horticulture business is constantly making decisions. These decisions may center on growing plants or operating a business. Some of the decisions that may need to be made by Ms. Martinez include:

1. What prices should be charged?
2. What type and how many plants should be grown?
3. How much should employees be paid?
4. What margin of profit is acceptable?

5. When should plants be started?
6. What type of equipment will be needed?
7. How should the business be advertised?
8. Should the plants be sold only to landscaping firms or should they be sold to the public?

An owner must be willing to make decisions. At times the decision may be wrong, and the owner should be willing to accept the outcome of the decision. The success or failure of a horticulture business often depends on how good the owner is at making decisions.

One way to gain experience in making decisions is through your SOE program. If you have a school or home project, you will be making management decisions similar to those being made by Ms. Martinez.

Keeping Records. Businesses must keep a variety of records. A complete record of expenses and income is needed to determine whether the business is making or losing money. These records will be useful in borrowing money and paying taxes. They also help in setting up cash flow. The cash flow of a business is a schedule showing when money is to be spent and when money will be coming in. This helps the owner or manager plan how to manage the business.

Many horticulture businesses will let customers charge their purchases. At the end of the month, the owner will send the customer a bill. Florists often take telephone orders for floral arrangements. A bill is then sent to the person who ordered the arrangement. Accurate records of the charges will need to be kept.

Ms. Martinez will have to pay bills. She will pay the companies that provide her with supplies. Water and electric bills have to be paid monthly. Often loans are repaid on a monthly basis. The owner of a horticulture business is constantly sending out and paying bills.

Have you ever wondered what happens to the federal income tax that is withheld from paychecks? The business owner must take the right amount of money out of the paycheck and send it to the government on a regular basis. The Internal Revenue Service should be contacted for forms and regulations. If the state you work in has a state income tax, then the owner must also withhold a certain amount for the state income tax. But this is not the only tax that comes out of paychecks. The owner of a horticulture business has to take a certain percentage of an employee's salary and send it to the government for social security tax. You may not be aware of the fact, but the owner has to contribute an equal amount to social security for each employee. For example, if Joe went to work for Ms. Martinez in her bedding plant nursery and had $50 taken out of his check for social security, then Ms. Martinez would have to add another $50 from her funds to send to the government. Joe would be credited with paying $100 to social security. The Social Security Administration should be contacted for forms and regulations.

The owner of a horticulture business has to keep records of the sales tax that is collected. The sales tax is sent to the state on a regular basis, along with a report of sales.

You can probably see by now that the owner of a horticulture business has a major responsibility in keeping records. A good mathematics background is helpful. Records have to be kept neatly and accurately. The practice you gain by keeping records on your SOE program will be helpful if you start your own business someday. It is not uncommon to find a bookkeeper or accountant employed part-time in a horticulture business.

Maintaining a Safe Business. The owner of a horticulture business is responsible for maintaining a safe working environment and a safe business. Regulations established by the Occupational Safety and Health Administration (OSHA) must be followed. Many states have health and safety regulations in addition to the OSHA requirements. Chemicals should be stored in a safe area and locked up. Customers should not be allowed in areas where they could encounter sharp or dangerous equipment. Employees should be encouraged to follow safe working practices. Operating a safe business is an important responsibility of the owner.

You may want to start learning about safety regulations now. One way this can be accomplished is by urging your FFA chapter to participate in the chapter safety contest. This FFA contest is available for participation by all FFA chapters. Members of the chapter put on safety demonstrations, conduct community safety campaigns, erect signs and posters, or promote safety in various other ways.

Being Aware of State and Federal Regulations. In addition to the OSHA requirements there are a number of state and federal regulations that owners of horticulture businesses must be aware of. The Fair Labor Standards Act is a federal law that contains regulations concerning the working hours and conditions for people under the age of 18. Ms. Martinez must abide by these regulations.

Most states have regulations concerning the transportation of horticulture plants across state lines. For example, the California Department of Food and Agriculture has extensive regulations that relate to horticulture. Nursery stock, plants, seeds, and bulbs that are shipped to California must be labeled with the name and address of the shipper plus the name of the state where the plants were grown. Ms. Martinez will need to be aware of this regulation because it may affect the plants she grows. All plant materials are subject to inspection for freedom from pests and diseases when they arrive. Certain plants are not allowed to be shipped into California. Citrus trees cannot be shipped into California except for those coming from Arizona. Regulations will vary from state to state.

Hiring Employees. Most horticulture firms employ a number of workers. The proprietor has responsibility for hiring employees. This in-volves interviewing people, establishing pay rates (in line with federal minimum pay standards), and deciding which person or persons to hire. The owner of the firm has to tell those who are not hired that they were not hired.

After employees are hired, the owner will assign jobs to the employees and give directions for accomplishing the work. A good employer will be diplomatic and use courtesy and tact in giving directions to workers. Workers do not enjoy working for someone who is mean or gives orders in an unpleasant manner. However, there will be times when the employer will have to correct a worker or encourage the worker to perform better work faster. This is often a difficult responsibility.

At times it may be necessary for a proprietor to dismiss (fire) an employee. This may be an unpleasant responsibility for the owner of a horticulture business.

Growing and Caring for Plants. In addition to managing the business affairs of a horticulture operation, Ms. Martinez must make decisions about growing and caring for plants. She must decide when to make cuttings, plant the plants, shade them, water them, transplant them, fertilize them, and apply pesticides. She needs an in-depth knowledge of plant production.

Marketing Horticultural Products. The owner of a horticulture business makes money only when the product is sold or a service is performed. In a floral shop the florist needs to know how to make various types of floral arrangements and how much they should be sold for. Landscapers have to develop landscaping plans, convince the homeowner that the plan is good, formulate a price for installing the landscape, and then install the landscape. Fruit and vegetable producers need a plan for marketing their produce. They may sell through a marketing cooperative, or directly to grocery stores, or through their own roadside market. Ms. Marti-

Figure 22-6. The proprietor of a firm has many decisions to make including the hiring of employees.

nez will need to develop a plan for marketing her ground cover plants. No matter what type of horticulture business is being operated, the owner has to develop plans for marketing the product in order to make a profit. Too often horticulturists are concerned primarily with the production of horticultural products and not their marketing. If a profit is to be made and the business is to be a success, marketing is crucial. One possible method of gaining experience in marketing is through your SOE program.

Being the owner of a horticulture business is not an easy job. The proprietor has many responsibilities and must make decisions every day. The rewards of owning a business are numerous, but so are the responsibilities.

Special seminars and classes designed to improve management skills are often offered in area vocational-technical schools, community colleges, and universities. These courses are available during the day or at night.

Manager Responsibilities

In some horticulture firms a manager is hired to operate the business. This is a common practice in large firms that are owned by a corporation. Home Lawn Care, Inc., hired a manager to operate the business. The manager assumed most of the responsibilities that the owner of a smaller business would perform: keeping records, hiring employees, and overseeing their work. The manager also makes many of the decisions concerning the day-to-day operation of the business. Big decisions about the overall operation of the business are made by the board of directors or the stockholders of the corporation.

Smaller horticulture businesses may hire managers. Often the owner of a small business acts as manager as well. As small businesses expand, it may be necessary to hire someone to help manage part of the business. The manager then

Figure 22-7. A manager needs both a knowledge of horticulture and a knowledge of how to work well with others.

assumes certain responsibilities according to the desires of the owner(s). The manager might be asked to handle the production end of the operation while the owner handles selling.

A manager should have a good knowledge of horticulture and be able to perform many horticultural skills such as operating equipment, propagating plants, or making floral arrangements, depending upon the type of horticulture

business. A good manager will also know how to work with people. Managers must be able to get employees their direction to work productively. The manager must also be able to get along with the owners of the business. Both horticultural skills and human relations skills are needed by the manager. In vocational agriculture, horticultural skills are taught through the SOE program, and human relations skills are gained through participation in FFA activities.

The responsibilities of employees are described in Chapter 23.

Managing the Horticulture Business: A Review

Owning or managing a horticulture business is enjoyable but requires hard work and decision making. Before starting a business, one should first determine whether there is a need for the business, what legal regulations must be met, and what is the best financial plan. The business can be organized as a sole proprietorship, partnership, or corporation. The manager or owner is responsible for making decisions, keeping records, maintaining a safe business, remaining aware of state and federal regulations, hiring employees, growing or caring for plants, and marketing horticultural products. A number of the skills needed by an owner or manager can be developed by participating in FFA activities and conducting an SOE program.

THINKING IT THROUGH

1. How might you determine whether a community can use a new horticulture business?
2. If you were going to start a horticulture business, how could you determine what the legal requirements were?
3. Why is a financial plan needed for a business?
4. What are the advantages and disadvantages

of owning your own horticulture business?
5. How does a firm organized as a partnership differ from a firm organized as a corporation?
6. What types of records are needed by the owner of a horticulture business?
7. What are the responsibilities of the owner of a horticulture business?

23

Getting and Keeping a Job in the Horticultural Industry

Lee and Carroll were sitting in the Dairy Delight discussing the events of the past week. They had both graduated from high school a week earlier. Lee was not very happy. All week long he had been trying to find a job but couldn't. Carroll found a job early in the week and was enjoying the job mowing fairways and watering greens at the Sycamore Canyon Golf Course.

Lee didn't know why Carroll was able to find a job so quickly. They had attended the same school and taken basically the same courses, including two years of horticulture. Their grades were about the same and both were interested in careers in horticulture.

Why do you think Carroll found a job more quickly than Lee? Was it luck? No! There were several reasons why Carroll was employed while Lee was still looking for a job. In this chapter you can learn how to locate jobs.

CHAPTER GOALS

In this chapter your goals are:

- To formulate career goals
- To develop a plan for locating jobs
- To outline a procedure for applying for jobs
- To prepare a résumé
- To practice for a job interview
- To identify the traits needed by horticulture employees

Identifying Job Opportunities

As Lee and Carroll were describing their experiences during the past week, Lee asked, "What's the secret of finding a job so quickly?"

Carroll replied, "There's really no secret. The first thing I did was to think about my career goals."

Figure 23-1. The first step in finding a job is to examine your career objectives.

Establishing Career Goals

A good starting point in locating a job is to first examine your own career goals. What type of work would you like for a career? Do you want to own a landscaping firm, manage a nursery, be a florist, be the head groundskeeper at a golf course, or teach horticulture in a high school? Would you rather work indoors or outdoors? Do you mind working evenings and on weekends? Do you want to work in the city or in a small town? Your answers to these questions will help determine the type of job you would be suited for. By asking yourself the questions listed below you can determine what your goals are. Carroll's answers give you an idea how these questions can really help. Some horticulture jobs such as landscaping, turf management, and fruit production are primarily outdoor jobs, while floral designers and greenhouse workers spend most of their time inside. Garden centers and retail nurseries are often open on Sundays. There are more opportunities for landscapers, turf specialists, and florists in larger cities.

Employment Outlook. Before establishing your career goal you should also consider the employment outlook for that career. Will the type of work you are interested in be needed 10 to 20 years from now? Are their job opportunities to pursue your career locally or will you have to move? For example, if your career goal is to be a floral designer and you live in a small town with only one florist, you will probably have to move to a larger town in order to find a job. The availability of horticulture jobs in your community should be considered in planning career goals. During the 1970s many students studied forestry, only to learn that there were very few jobs available. It would be unwise to pursue a career goal where no jobs are available.

Both personal preference and employment opportunities should be considered in establishing career goals.

Personal Survey. After a career objective is

formulated you should then ask what skills, knowledge, and experience are needed to reach this goal. Carefully examine the type of work performed in the career you desire and then prepare a plan to reach your goals. Your plan may contain several steps. Before you can manage or own a business you will need experience. This experience may include three or four jobs that will eventually lead to your career goal.

For example, sometime in the future Carroll would like to be the head groundskeeper at a country club. Before Carroll went out to look for a job he took an inventory of himself. He sat down with a piece of paper and wrote out the following questions:

1. What can I do?
2. What have I done in the past?
3. What do I know?
4. What are my strengths?
5. What are my weaknesses?

Carroll's Personal Survey

This is the way Carroll answered the personal survey questions

- What can I do?

Operate a tractor, use a push lawn mower, prune shrubs, make floral arrangements, install lawns, use a rototiller, draw landscape plans, work in a greenhouse

- What have I done in the past?

Worked selling fruits and vegetables, worked in the school greenhouse

- What are my strengths?

Strong background in greenhouse work, like to work with people, good grades in school, like mechanics

- What are my weaknesses?

Don't know alot about landscaping or turf management, not very good in English.

- What type of work would I like?

Outdoors, I like to work with equipment, would enjoy lawn or turf work. In the future I would like to be head groundskeeper of a country club.

- What types of jobs are available?

Numerous groundskeeping jobs are advertised in trade journals. I will need more experience before I would qualify for that type of job. My immediate goal is to get experience in working with turf.

After doing this, Carroll realized that he knew about basic plant growth, propagation, floristry, landscaping, plant diesease, and pest control. Carroll could operate a tractor and a standard push lawn mower. His weaknesses were in turf management and the operation of turf equipment. In horticulture class Carroll had not had the chance to operate riding mowers used on golf courses and did not spend much time studying turf management. A personal inventory will help determine what type of job you are suited for and will help you decide what type of additional experience you need.

If Carroll were to reach the goal of head groundskeeper, he would need experience in operating the turf equipment and working with turf. This information plus the fact that Carroll had a career goal was helpful when it came to finding a job. Carroll did not bother applying for work at floral shops or garden centers, because he knew that type of work would not be very helpful in reaching his career goal. Instead Carroll concentrated his job-finding efforts on golf courses, country clubs, and landscaping firms.

When the owner of the Sycamore Canyon Golf Course asked Carroll why he wanted a job mowing fairways, Carroll replied, "My career goal is to be the head groundskeeper at a golf course. Mowing fairways will give me valuable experience in reaching this goal." The owner was impressed with the answer. The fact that Carroll had a goal in life was helpful in finding a job.

Figure 23-2. Having well established career goals will help you find a job.

Lee began to realize that establishing a career goal might be a good idea. During the week of job hunting, Lee had applied for jobs at florist shops, orchards, lawn and garden centers, golf courses, a sod farm, and several landscaping firms. When asked why he wanted the job, Lee's reply had been, "I need the money." There is nothing wrong with wanting to earn money, but when this is the only answer given for wanting a job the employer may think you are interested only in money and not in working. Lee realized that the establishment of a career goal could be helpful in getting a job.

It is best to determine your career goals while you are still in school. By doing this you can plan your SOE program to include skills that will be needed in the future career. The questions Carroll answered, you should ask yourself. People who have a direction and purpose in life will generally find jobs easier than those who do not plan.

In addition to planning career goals, you may also want to think about other types of goals such as social, religious, or personal goals. What type of person do you want to be? What type of home do you want to live in? What type of friends would you like to have? Working in a career is only one part of your life. The type of career you select will have an influence on your social, religious, and personal goals. Some careers require work at night or on the weekends. Some careers pay more than others. This will influence your life outside of work.

Locating A Job

Lee was impressed with Carroll's preparation before applying for a job. Lee had never thought about conducting a personal inventory or establishing career goals. "After you did all this, what did you do next?" Lee asked.

Carroll replied, "The next step was to find where the available jobs were."

Locating jobs is not an easy task and may require effort. There are a number of ways in which a person can go about finding jobs.

Talk to your Horticulture Teacher. Often your vocational agriculture teacher will know about job openings. Sometimes local horticulture firms will call the agriculture teacher to ask for help in locating good employees. Also, the teacher will probably be familiar with most of the horticulture firms in the community and may be able to suggest places where you could apply.

Look in the Newspaper Want Ads. In the classified section of the newspaper the "Help Wanted" advertisements may list job openings in horticulture. You should read the ad carefully and relate the job requirements to your own interests and skills. You are wasting your time and the company's time by applying for jobs that you are not interested in or qualified for. Be sure to notice whether you should send a letter of application or a résumé, or call for an interview.

Figure 23-3. Jobs are often listed in the "Help Wanted" section of the newspaper.

Apply Directly to Firms You Would Like to Work for. Often people find jobs by applying directly to the business without knowing whether a job is available. This method of applying for a job can involve either personal contact or a letter. A good method is to actually go to the horticulture firm you are interested in working for and ask to see the manager or owner. Present the owner or manager with a copy of your résumé and ask to fill out an application form. When a job becomes available the person who hires new employees will look through the job applications and select several applicants to interview. The fact that you personally came and asked for a job and had a résumé could be to your advantage.

A second method of applying for a job directly is by writing a letter. In the letter you should tell what type of position you are interested in, when you will be available for employment, and why you would like to work for the company. Include your résumé.

The letter should be neatly written or typed, with the words spelled correctly.

One way to identify horticulture firms to work for is to look through the yellow pages of the phone book. By doing this you will probably identify some firms you did not know existed. Some headings to look under include:

- Crop Dusters
- Exterminating and Fumigating
- Fertilizers
- Florists
- Fruits and Vegetables
- Garden Centers
- Garden and Lawn Equipment
- Golf Courses
- Greenhouses
- Landscape Contractors
- Landscaping Equipment and Supplies
- Lawns
- Lawn Maintenance
- Nursery Stock
- Orchards
- Sod and Sodding Service
- Soil
- Tree Service

Look in Trade Journals. Employers often advertise for employees in horticulture journals. These jobs are often for top-level workers such as managers or head groundskeepers.

Talk with Friends and Relatives. Let your friends and relatives know that you are looking for a job. They can serve as extra eyes and ears. If you have friends working in horticulture firms, they can keep alert for job opportunities in that firm.

Place an Ad in the Paper or in a Trade Journal. You might want to consider placing an advertisement in the local paper or in a trade journal. In the ad you would tell what type of work you are looking for and some of your qualifications. You will have to pay a fee for running the advertisement.

Read the Business Section of the Newspaper. Look for newspaper articles that tell about expansion of horticulture businesses, branch store openings, new horticulture businesses, or a change in the ownership of a horticulture firm. These types of activity often signal a need for new employees.

Contact Employment Agencies. In many states an employment agency is operated by the state government. The agency helps locate jobs and does not charge for its services. There are also private employment agencies that will help find jobs. You have to pay a fee for this service; normally a certain percentage of your pay for a length of time is paid to the agency.

Visit Past Employers. If you have worked part-time for a horticulture firm in the past, go see the owner or manager. The firm may hire you if you had a good work record when you worked there, or they may be able to refer you to another firm that is looking for an employee.

Contact Nonhorticultural Firms. Often firms that are not horticulturally oriented need workers with horticulture training. Cemeteries, memorial gardens, and city parks often need people to care for flower beds, shrubs, and lawns. Large apartment complexes may need someone to care for the grounds. Home builders might consider installing lawns and landscapes themselves if they had someone with a horticulture background. You may be able to find a job using your horticultural skills with a firm that is not horticulturally oriented.

Check with the School Guidance or Placement Counselor. A placement service is operated in many vocational schools and in some regular high schools. Talk with the person who is in charge of the placement service. If your school does not have this service, your guidance counselor may have knowledge of job openings.

How Employers Find Employees

There are a number of ways to locate jobs. All these are good methods of finding jobs, but some are successful more often than others. In a survey of employers in Texas the question was asked, "How do you locate prospective employees?" The top five methods were:

1. Word-of-mouth
2. People applying without knowing whether an opening is available
3. Newspaper want ads
4. Public employment agencies
5. Private employment agencies

Knowing how employers find workers will give you some clues about how to go out and find a job.

Applying for a Job

After a job has been located, the next step is to apply for it. This may involve calling or writing the prospective employer, filling out an application form, and preparing a résumé.

```
                                        521 Morehouse Road
                                        Cedar Falls, Ohio  44305
                                        May 19, 19__




Mrs. J. D. Blanton
Custom Lawns, Inc.
1813 Washington Avenue
Cedar Falls, Ohio  44305

Dear Mrs. Blanton:

     Mr. Ray Parsons, my vocational agriculture teacher at Cedar Falls
High school told me that you had an opening in your firm on the lawn
installation crew.  I would like to apply for the job.

     I will graduate from Cedar Falls High School at the end of this
term.  I have completed two years of horticulture and have experience
in installing lawns.  In the horticulture class we prepared several
sites for initial lawn plantings and installation of sod.

     My future goal is to be the head groundskeeper at a country club.
I believe the type of experiences I would gain in working for your
company would be beneficial.

     May I have a personal interview?  You can contact me at 583-2526
after 4:00 p.m.




                                        Sincerely yours,

                                        Carroll Martin

                                        Carroll Martin
```

Figure 23-4. A letter of application is sent to prospective employers.

The Letter of Application

After locating a job you must let the employer know that you are interested in the job. The purpose of the letter of application is to let the employer know you are interested and that you should be considered. You are trying to sell yourself through the letter. A carefully written letter can help you land the job.

You should first write a rough draft of the letter. Ask your parents, friends, or vocational agriculture teacher to read it and offer suggestions. In the first sentence you should mention where you learned about the job. This serves as a bridge between you and the employer. For example, you might write, "Ms. Hodges at the State Employment Office advised me of an opening in your business." In the second sentence you should state that you desire to be considered for the job. You could say, "I am interested in applying for this position," or "Please consider me as an applicant for the job."

In the second paragraph of your letter of application you should describe your educational background and/or experience. You are trying to convince the employer that you have skills or experience that will qualify you for the job. You might want to mention your career goals.

In the final paragraph of your letter you should ask for a personal interview. Include your phone number in case the firm wants to call you.

After writing a rough draft of the letter and having your friends examine it, write the final copy. The letter will look best if it is typed. It should be typed neatly, with all words spelled correctly. There should be no erasure marks. If you cannot type neatly, then a handwritten letter is acceptable if it is carefully done.

Calling About a Job

At times you may apply for a job over the telephone instead of writing a letter. If you call about a job you will say basically the same thing you would have if you had written a letter of application. Before calling you should make an outline of what you want to say.

In calling about a job you should first identify yourself and tell why you are calling. An example of this would be: "Hello! My name is Sarah Barnes and I'm calling about the job you have advertised in the newspaper for a greenhouse worker." You should then briefly tell a little about yourself and ask for an interview at the employer's earliest convenience. At the conclusion of the conversation you should say thank you before you hang up.

The Application Form

Often you will be expected to fill out an application form when you apply for a job. An application form is simply a sheet of paper on which general questions about you and your background are asked. Such items as your age, where you live, where you go to school, what your goals are, and where you have worked are on the form. An application form is not difficult to complete but should be done carefully. All the blanks should be filled in as completely as possible. If a question does not apply to you, place a NA in the blank to indicate that it is *Not Applicable*. The application form should be filled out in ink or typed. All words should be spelled correctly. If you are not sure how to spell a word, try to use another word with the same meaning. Be as neat as possible in filling out the application form because this will serve as an example to the employer of how careful you are.

The Résumé

Many people who apply for jobs develop a résumé. A *résumé* could be called a personal data sheet. It is a one- or two-page document that contains the personal information about yourself that you would like a prospective em-

Resume

Carroll Martin

Name: Carroll Wayne Martin

Address: 5251 Morehouse Road
                                Cedar Falls, Ohio 44305

Telephone: (317) 583-2526

Birthdate: February 23, 1964

Height: 5'9"

Weight: 155

Health: Excellent

Hobbies: Jogging, baseball

Position desired: Turf maintenance or installation

Long term career objective: Head Groundskeeper at a Country Club

Education: Graduated 1981
                                Cedar Falls High School
                                Cedar Falls, Ohio 44305

                                Completed two years of horticulture

Horticultural skills possessed: Plant propagation
                                Floral arrangements
                                Insect and disease control
                                Pruning shrubs
                                Drawing landscape plans
                                Planting trees and shrubs
                                Can drive a Ford 5000 tractor
                                Can operate a rototiller, landscape
                                rake, and push lawn mower.

                                Have had experience in raising:

                                a. poinsettias
                                b. Easter lilies
                                c. mums
                                d. tomatoes
                                e. various bedding plants

Resume (Continued)
Carroll Wayne Martin

Work experience: Cedar Falls High School Greenhouse
                                Cedar Falls, Ohio 44305
                                720 hours of general work

                                Elder Avenue Fruit and Vegetable Market
                                Cedar Falls, Ohio 44305
                                Clerk - Summer of 1980

References: Mr. Ray Parsons
                                Vocational Agriculture Instructor
                                Cedar Falls High School
                                Cedar Falls, Ohio 44305

                                Mrs. Betty Reynolds
                                Elder Avenue Fruit and Vegetable Market
                                Cedar Falls, Ohio 44305

                                Rev. Jerry Lacey
                                Trinity United Methodist Church
                                801 Beech Grove Street
                                Cedar Falls, Ohio 44305

Figure 23-5. A sample résumé.

ployer to know. It is sent or handed to a prospective employer as an introduction, and it often contains information that may not be asked for on a job application form.

Résumés can be written in many ways; however, there is some standard information that is normally found on all résumés:

- Name
- Address
- Telephone number
- Date of Birth
- Education
- Skills
- Experience
- Honors or awards
- Student activities
- Hobbies
- References

Carroll's résumé is shown in Figure 23-5.

The résumé should be neatly typed, with all words spelled correctly. Several copies of the résumé should be made. Each time you apply for a job a copy of the résumé should be included as a part of the application. This will provide the prospective employer with information that may not be found in a job application form. It also shows that you are serious about locating a job since you took the time and effort to prepare a résumé.

Your vocational horticulture instructor, English teacher, and guidance counselor can give you tips on how to prepare a résumé. There may also be books in the library on how to prepare a résumé.

The Job Interview

An *interview* is a meeting between a prospective employer and a prospective employee. The purpose of the interview is to allow the employer to evaluate the personality and job skills of the prospective employee, and to allow the pro-

spective employee to evaluate the employer. Both parties ask questions. The employer decides whether the prospective employee has the personality and skills needed for the job, while the prospective employee decides whether he or she would like to work at the particular job.

One of the reasons why Lee was having trouble finding a job was because of the manner in which he interviewed for jobs. Earlier in the week Lee had noticed an ad in the newspaper for a floral delivery person at Southside Flowers. He called about the job and was told to come in the next day at 9 a.m. for a job interview. The following is a description of Lee's interview. As you read it, think about the ways in which Lee could have improved the interview.

Lee's Interview

Lee turned over and shut off his alarm. He awoke a half-hour later and remembered that he had an interview in one-half hour with Ms. Scott for a job as a florist delivery person.

Lee found his favorite pair of jeans on his closet floor. He pulled the jeans on. A faded sweatshirt and dirty tennis shoes completed his outfit. Lee didn't have time to wash his face, brush his teeth, or shave. He jumped into his car and raced to the interview.

Lee was 10 minutes late for the interview. He sat down to fill out the application form. He had to ask the secretary for a pen, because he had forgotten to bring one. He sped through the application and made no attempt to make his writing neat and readable.

Lee shuffled into Ms. Scott's office. He limply shook hands with Ms. Scott and said, "Sorry I am late but the traffic was terrible today."

Lee handed Ms. Scott the completed application form. Ms. Scott asked Lee to be seated. Lee slumped into the chair and stared at the floor. Lee had no idea what interviews were all about and he was not prepared for this interview.

Ms. Scott asked Lee several questions, including "What was your favorite class in high

Figure 23-6. It is important to dress neatly and to be well groomed for an interview.

Preparing for a Job Interview

Before you interview for a job, there are a number of things you can do to help prepare.

One thing that can be done is to practice interviewing. Your friends or parents can ask you questions just as if they were actually thinking about hiring you. This will give you a chance to think how you would answer certain types of questions. Interviewing for a job is a skill that should be practiced, just as budding a fruit tree or making a corsage. Some questions that are often asked during a job interview are:

1. Why do you want to work for this company?
2. What classes have you had in school?
3. Which classes did you like best? Why?
4. What type of work would you like to do?
5. What are your plans for the future?
6. Why are you qualified for this job?
7. What type of hobbies do you have?
8. What type of school activities did you participate in?
9. What type of work have you done in the past?
10. Why did you leave your former jobs?
11. What salary do you expect?

You will want to think about how you should answer these questions. You should also try to find out all you can about the company you will interview with. What type of employee are they looking for? If you know about their history, the products they sell, and how they conduct business, you may have an advantage over other people applying for the same job. If you go into the interview with some prior knowledge of the firm you are interviewing with, the interviewer will be impressed.

After practicing for the interview, the next area of concern should be your personal appearance. The first impression people have of you is based on your appearance. You will want that first impression to be good. You should dress neatly for an interview. Your clothes should be clean and pressed and appropriate

school?'' and ''Why was this class your favorite?'' Ms. Scott also asked Lee, ''What do you know about this company?'', ''Why do you think you would like this particular job?'', and ''Why do you think you are qualified for this job?'' Lee didn't realize that he would be asked questions. He got scared and panicked, so he mumbled answers to the questions. Finally the interview was over. Lee rushed away without thanking Ms. Scott for the interview.

Do you now know why Lee was having problems finding a job? He was late for the interview, he was sloppy in his personal appearance, his job application was hard to read, and he was not prepared for the interview. Even though Lee had many skills that would be useful to Ms. Scott, Lee did not get the job. He had made an unfavorable impression during the interview.

for the occasion. Remember, you are interviewing for a job, not going to a dance or a picnic. Your hair should be properly styled and your teeth should be brushed. It is a good idea to take a bath or shower on the night or morning before an interview. You will want to wear a deodorant. The importance of a clean, neat personal appearance cannot be overstressed. Poor personal appearance was the number one reason why job applicants were not hired, according to a recent survey conducted by a major university.

The 12-12-12 Rule

In preparing for a job interview you should remember the 12-12-12 rule. The first 12 stands for 12 feet. Often a prospective employer will first see you at a distance of about 12 feet, so your overall general appearance at 12 feet is important.

The second 12 stands for 12 inches. Normally you will shake hands with the prospective employer. At this distance the employer will be able to smell your breath, see how clean you are, notice dirt under your fingernails, etc. You will want to practice good personal hygiene to pass the 12-inch test.

The third 12 stands for 12 words. The first 12 words you speak will make an impression on the employer. Do you remember the first 12 words in Lee's interview with Ms. Scott? Look back and see what he said. He apologized for being late. He would have made a much better impression if he could have said in a crisp, firm voice, "Good morning Ms. Scott! I'm Lee Kane. I'm interested in the delivery person position you have advertised!" Your first 12 words should be friendly and spoken clearly. The 12-12-12 rule will help when you interview for a job.

The Actual Interview

After introducing yourself you are ready for the actual interview. You should remain standing

Figure 23-7. An interview is a meeting of a prospective employer with a prospective employee. After the interview is over be sure say "thank you."

until you are asked to sit, and then you should sit facing the interviewer. If you have a handbag, notebook, or other articles with you, they should be placed on the floor next to your chair. Do not place them on the interviewer's desk. You should sit erect. You will probably be nervous at first, but this will wear off as the interview progresses.

Some interviewers will ask you a series of questions like those that were listed earlier. You should try to answer each question with a clear, concise answer. The person who is interviewing you does not have all day to listen to long drawn-out answers. Other interviewers may ask only a few questions and expect you to provide more detail. An example of this would be for an interviewer to say, "Tell me a little about yourself." The interviewer does not want to hear your life history, but would like to know a little about your family, school, goals in life, and why you think you are qualified for the job.

Often an interviewer will ask whether you have any questions. You may want to ask some questions about the company or the job, which indicates an interest in the job. There is nothing wrong with asking how much the job pays, but this should be asked near the end of the interview. If you ask about pay at the start of the interview, the person interviewing you may think you are only concerned with money and not the job.

An interview does not take long to complete; very seldom will one last for 30 minutes. The average is from 5 to 15 minutes. When the interview is over you should get up and go. Many jobs are lost because a prospective employee fails to realize that the interviewer is through. As you leave you should say thank you. If you are not offered the job during the interview, it is all right to ask whether you will be called in a few days or whether you should call back.

Follow-Up to the Interview

After you leave you should think back on the interview. What did you do right? What did you do wrong? What should you do differently in your next interview? Do not be discouraged if you are not offered a job after your first couple of interviews. The practice you gained in the interviews will make the next interview easier.

After the interview, it is a good idea to send a thank-you note to the business. This note may arrive as the employer is deciding who to hire. If you have not heard anything from the interviewer in about a week, you could call or write to ask when a decision will be made.

You're Hired: Now What?

Finding a job can be hard at times. After finding a job some employees breathe a sigh of relief and think that they have it made. However, finding a job is only the beginning. The impor-

tant task now is to keep the job. Workers in horticulture can do several things to keep their jobs.

Responsibilities of Employees

Just as an employer has responsibilities (as discussed in Chapter 22), so do the workers. The employer pays the salary of the employees. Therefore the employer can expect employees to be responsible in several areas:

1. Dependability
2. Cooperation
3. Productive Work
4. Initiative
5. Honesty

Dependability. One of the most important traits needed by employees is *dependability*. People who are dependable are trustworthy and reliable. They can be counted on to do their work. When a task is given to a dependable person, the employeer knows that the job will be done. The following story tells about a worker who is not dependable.

Joe owns a landscaping firm. During the spring and summer when he is busy, he often hires several students to help. One student he hired recently was Phil. One morning he told Phil to drive over to the Berry home and prepare the ground for laying sod. The yard had been plowed and disced the day before, so stones had to be picked up and the soil had to be raked smooth. When Phil arrived at the Berry home, he decided he would sit in the truck and listen to the radio for a few minutes. About an hour later, Joe arrived with the rest of the crew to lay sod. Phil was still sitting in the truck listening to the radio.

Phil was not dependable. This type of behavior will soon result in the dismissal of the employee. In many types of horticulture jobs you will be expected to perform tasks without close supervision. You need to be dependable.

Being prompt and reporting to work on time is another type of dependability. An employer expects you to be at work on time. A survey of employers in Texas found that the number one reason for workers' being fired (terminated) was absence from work. Dependability is one responsibility of a worker.

Cooperation. The ability to get along with fellow workers and with the employer (*cooperation*) is another important responsibility of horticulture employees. The proprietor of the business pays your salary and has the right to expect your cooperation. You should perform the jobs assigned to you in a cheerful manner. You should not argue with the manager or owner. Employees who cooperate with the management and with other employees will soon advance.

Productive Work. An old saying is that people should put in a day's work for a day's pay. This means that an employee should work hard and earn his or her pay. An entrepreneur will not make a profit and thus will be unable to pay the employees if they do not work hard to make money for the business.

Productive work implies both quality and quantity. A worker should concentrate on doing a good job. No matter what type of horticulture job you are involved in, the emphasis should be on performing an outstanding job. In designing floral arrangements the form, scale, and color should be as nearly perfect as possible. If you are grafting fruit trees, each graft should have well-matched cambium layers. An employee in a garden shop should concentrate on learning everything possible about the products being sold. Doing a job well gives a person a sense of pride.

In addition to performing a good job, employees should be concerned about the quantity of work they do. Often horticulture jobs will involve performing the same tasks for extended periods of time, such as mowing greens, picking vegetables, transplanting starts, and pruning

Figure 23-8. A person should put in a day's work for a day's pay.

shrubs. Performing these tasks with speed is important. A person who can mow three golf greens in an hour is a more valuable employee than someone who can mow only one green in an hour.

At certain times of the year the ability to work rapidly is crucial. When fruits and vegetables are ripe they must be picked. On special occasions such as Valentine's Day and Mother's Day, florists must be able to turn out a large number of floral arrangemnts.

Employees in horticulture have the responsibility to be productive workers. They should perform a quality job as rapidly as possible.

Initiative. *Initiative* is defined as the ability to perform a task without being told. In other words, when something needs to be done, employees with initiative will do the job without waiting for someone to tell them to do it. After the worker completes the job that was assigned, she or he does not sit around waiting for the next assignment. However, employees should not attempt to perform jobs that they do not know how to perform, or that require specialized skills. Some tasks that could be performed by a horticulture employee without that employee's being told are:

1. Cleaning the work area
2. Dusting and straightening displays
3. Cleaning equipment
4. Checking plants for disease or pest damage
5. Pulling weeds
6. Straightening up the tool room
7. Cleaning and oiling tools

Technical Knowledge. Horticulture is a rapidly changing industry. New varieties of flowers, vegetables, and trees are constantly being introduced. Some pesticides may be banned from use, and new ones will be placed on the market. A good employee will strive to keep up with the continuous change and new technology in horticulture.

Employees in horticulture need a knowledge of mechanics, since many horticultural tasks are performed with the aid of mechanical equipment. The ability to service or repair the equipment is an important technical skill.

Remaining up-to-date can be accomplished in a number of ways. Some vocational agriculture teachers and county agricultural extension agents will conduct night classes for adults. Attending these night classes is an excellent way to keep up with new techniques and changing technology in horticulture. You can also exchange ideas with your fellow class members.

Subscribing to horticulture journals is another way to keep up with the latest in the field. There are trade journals for all phases of horticulture. These journals are specifically designed to help people in the industry stay current with the latest developments in horticulture.

An employee displaying initiative shows the boss that additional responsibilities can be handled, which may lead to promotions and salary raises.

Honesty. An employee has the responsibility to be *honest*. Dishonesty is bad for the employee and the employer. The employer may lose money with a dishonest employee, and the employee will eventually be discovered and lose the job. In such cases another job is frequently difficult to find.

There are several types of dishonesty. One involves stealing property from the business, in the form of tools, plants, or supplies. This is costly to the business.

A second type of dishonesty is stealing time. Time is stolen when an employee comes to work late, leaves early, or takes longer breaks than are allowed. During working hours, time belongs to the employer. An employee who does not put in the full amount is dishonest.

Employability Skills

In addition to the responsibilities just discussed, there are several skills that are desirable in workers, which are called *employability skills*. Those who work at developing these skills will almost certainly remain employed.

If you try to improve your technical skills and keep up-to-date with what is going on in horticulture, you will be a valued worker.

Figure 23-9. Subscribing to horticulture journals is one way to remain current in horticulture.

Mathematics. The ability to use mathematics is an employability skill needed in horticulture. A horticulturist is constantly using mathematics. Can you think of some ways in which mathematics is used in horticulture? A few are:

1. Calculating fertilizer rates
2. Making sales
3. Mixing pesticides
4. Setting prices
5. Billing customers
6. Developing landscape plans
7. Measuring area and distance
8. Weighing produce
9. Giving discounts
10. Regulating temperatures and light

This list could continue. A working knowledge of fractions, percents, addition, subtraction, division, and multiplication is required in many horticulture jobs. As a student you may not see much need for math, but after you get a job you will discover how important it is.

Communication. The ability to communicate is also needed in horticulture. Whether you are working with customers or fellow employees, you will need to express your ideas and thoughts clearly. Often you will have to describe a plant or floral arrangement to a customer. How would you tell a customer what a taxus yew looks like? Could you describe what a $15 floral arrange-

ment looks like to a customer over the telephone? Employees in horticulture must speak clearly and loud enough to be heard. They should use words that will adequately describe the plants in question, without using horticultural terms that the customers may not know.

If you work at improving your communication and mathematical skills, you will be a valuable asset to the horticulture business you work for.

Getting and Keeping a Job: A Review

Before looking for jobs, conduct a personal inventory of your skills and establish your career goals. Sources may include the vocational horticulture teacher, friends and relatives, want ads in the newspaper and in trade journals, and employment agencies. Often jobs are located through applications without the applicant's knowing whether jobs are available. Most companies will interview prospective employees before hiring them. In preparing for an interview practice answering questions, dress neatly and cleanly, and be on time. After the interview is over you should thank the interviewer and leave promptly. After you get a job work hard to keep it. Improving technical, math, and communication skills will help.

THINKING IT THROUGH

1. Why should people develop career goals?
2. What factors should be considered in developing career goals?
3. What is a personal inventory?
4. What is a résumé and what should be included in it?
5. If you were going to look for a job, what steps would you take?
6. How can a person prepare for a personal interview?
7. List five questions that may be asked during a personal interview. How would you answer them?
8. After you have been hired, what can you do to keep the job?

GLOSSARY

Aesthetic   The beauty found in art or nature.

Aggregate fruit   A fruit developed from a single flower that has many ovaries.

Agronomy   The study of field crops.

Amperes   The amount of electrical flow through a circuit.

Anther   The pollen-bearing part of the stamen.

Arboriculture   The study and culture of trees.

Assimilation   The incorporation of plant food into the building of the plant's cell walls and cytoplasm.

Bacteria   Microscopic one-celled living plants that cannot produce their own food.

Balance   When creating a landscape or floral display, the arranging of the plants and flowers so they appear to have the same weight on each side of a centerpoint (focal point).

Balling and burlapping   The process of digging a plant with a tight soil ball, wrapping it in burlap, and getting it ready for transport or sale.

Bark grafting   Accomplished by splitting the bark of the stock. The scion is placed in the slit and held there by nailing.

Botany   The study of the properties and life phenomena of plants.

Bud dormancy   Growth of buds is temporarily halted even though the environmental conditions required for growth are provided.

Budding   When the graft involves the implanting of a bud into the stem of another plant.

Carbohydrate   A simple sugar produced by photosynthesis.

Cash flow   A prediction of how and when money (income and expenses) will be received and spent.

Cell differentiation   A process in which the cells become more specialized and development of different tissues takes place.

Chlorophyll   The pigment substance that gives plants their green color and in the presence of sunlight converts carbon dioxide and water into carbohydrates.

Chloroplast   A cell body that contains chlorophyll and where photosynthesis takes place.

Corms   A short, fleshy, underground stem having few nodes; a solid, swollen stem base which is enclosed in dry scale-like leaves. Has small cormels which develop at the base of the parent corm and may produce new plants.

Convection tube   A long plastic tube about 18 inches in diameter that runs the length of the greenhouse below its ridge. Evenly spaced holes (2 to 3 inches in diameter) permit ventilation and heating.

Cold frame   A square or rectangular pit dug into the ground covered with glass or plastic. It is heated by the sun.

Conditioning   A process the florist uses on the flowers they receive to help flowers regain their vigor after being shipped so that they will have maximum life in a cut flower arrangement.

Crown   Fleshy lump of compressed stems, just above or below ground level.

Cutting   A part of the plant used for propagation; for example, leaf, stem, or root cuttings.

Cytoplasm   An extremely complex substance found in cells. Contains 85 to 90 percent water and 10 to 15 percent salts, carbohydrates, proteins, and fats.

Dehiscent fruits   When ripe these fruits split open to release their seeds.

Dioecious plant   Male and female flowers are on separate plants.

Disbud   To remove some of the stems of the plant so that fewer but larger flowers are produced per plant.

Disorder   Describes a plant when it is not healthy.

Division   When a rhizome is manually cut into sections.

Drupe   Has a seed such as a stone fruit.

Electrolyte   Liquid in a battery, composed of an acid and water solution.

Entrepreneur   The sole owner of a business.

Enzymes   Various substances contained in plants (cells) that cause changes without being changed themselves.

Epidermis   The outer layer of a cell.

Fertilization   The uniting of the pollen (sperm) with the ovule to produce a seed.

Fibrous root-system   The primary root develops a complex, multibranched root-system.

Financial plan   A record showing the income and expenses expected for the upcoming year. It analyzes how much money will be needed to start the business and keep it running.

Floral mechanics The materials and supplies used in making floral arrangements.

Focal point In most floral designs or landscapes, the area to which your eyes are drawn upon just looking at the design.

Forestry The study and care of the forest.

Formulation The shape or form that pesticides come in. Common formulations are liquids, dusts, granules, and powders.

Foundation A substance used to hold plant materials in place.

Fruit A ripened ovary.

Fungi Plants that lack chlorophyll, often microscopic but some forms may be seen with the human eye and may contain more than one cell.

Fungicide A substance used to control fungi.

Germination A process in which a seed changes into a developing seedling.

Grafting Inserting buds or shoots from one plant into the stems or roots of another plant.

Granules Small pellets or balls of pesticide.

Greenhouse A building characterized by the roof and walls made largely of glass or other transparent materials that provides environmental control of plant growth.

Greenhouse range A group of greenhouses at one location.

Guard cells Cells that cause the stomata to open and close depending on environmental conditions. These structures form the stomata.

Guying and staking Mechanically straightening a transplanted tree or shrub that has grown in a slanted or crooked manner.

Hardwood cuttings Cuttings taken in the winter months when the plant is in the dormant stage.

Harmony When all parts of the floral design look as if they belong together.

Head house A building attached to the greenhouse where supplies are stored and work space is available.

Herbaceous cuttings Cuttings made from plants whose stems do not turn woody.

Herbicide Material used to control undesirable plants.

Home project Applying knowledge learned at school to home situations and experiments.

Hormones Chemical substances secreted by the plant or artificially injected to regulate plant growth.

Horticulture The science and art dealing with growing fruits, flowers, vegetables, and ornamentals.

Host An organism (such as a plant) that provides a home for another organism (such as a disease).

Human relation skills The ability to get along with others, to work well with fellow workers, and to relate well with the public.

Indehiscent fruit Fruit that does not split open and release its seeds when ripe.

Inert Inactive. The term is often used to describe ingredients in a pesticide that are not toxic but are used to provide bulk or moisture to a pesticide formulation.

Insect Small animal with six legs, three body regions, and generally wings.

Insecticide Material used to control insects.

Internal combustion When fuel and air are compressed inside the cylinder of an engine and ignited.

Internode Section of the stem between the nodes.

Intracurricular activities Functions that are done *in* school.

Layerage A method of propagation where the plant develops roots on its stem while it is still attached to the mother plant.

Leadership skills The ability to take action at the appropriate time to see that action is taken when something needs to be done.

Leaf cuttings Taken on certain specific plants that have the ability to reproduce both new stems and new roots from a single leaf or a leaf and its petiole.

Maturity The process of a plant completing its natural growth and becoming fully developed. Usually indicated by flowering.

Meristem Tissue that uses plant foods to make various parts of the cells and to produce energy.

Metamorphosis The changes in the life cycle of insects and insectlike pests.

Middle lamella Layer inside of the cell wall that gives structure to the cell.

Mitochondria The powerhouse of the cell. Acts along with several enzymes to function in the oxidative process with the plant.

Monecious plant Both male and female flowers on the same plant.

Multiple fruit Fruit that has flowers that are separated but closely clustered.

Nodes Enlarged portion of the stem from which the leaf grows.

Nucleus Control center of the cell; a dense, spherical shaped body located within the cytoplasm of the cell.

Nutrients Substances plants use to produce leaves, stems, roots, fruit, and other parts to ensure growth.

Ohms The resistance that allows one ampere to flow when the electrical pressure is one volt.

Olericulture The production, processing, and marketing of vegetables.

Ornamental A part of horticulture that involves the production and sale of ornamental plants.

Osmosis The process in which water is absorbed by the root system.

Ovules Females' eggs that when fertilized will develop into seeds.

Oxidation The process of mixing oxygen with another element.

Parlimentary procedure Used to ensure that meetings are run smoothly and that business is transacted in an orderly fashion.

Pests In horticulture, any living organism that competes with or damages desirable plants.

Peninsula bench arrangement A bench arrangement in which the greenhouse benches are at a 90° angle to a main aisle in the center of the greenhouse.

Pesticide Material used to kill or repel pests.

Petals The colored portion of the plant containing the fragrance that attracts birds and insects.

Photosynthesis The process in which carbon dioxide and water, when exposed to a light source are converted to carbon-rich, energy-powered organic compounds. This is a most significant life process.

Placement on a job Gaining experience by working for a business.

Plant growing surface A special surface made of asphalt, gravel, or other similar material on which plants in containers are grown.

Plant management plan An outline of what needs to be done and when it should be done, for growing and maintaining healthy plants.

Plugging Small pieces of sod 2 in x 2 in (5 cm x 5 cm) are placed one per square foot.

Pollen Male reproductive unit that is produced when the anther dries up and releases the pollen grains.

Pollination Transfer of pollen from the anther to the stigma.

Pome A fruit with no stones but several seeds within a chamber, such as an apple.

Pomology The production, processing and marketing of fruits.

Pot bound When the plant's roots have grown to the shape of the pot, and have become tight.

Preservative A substance used when conditioning flowers to make them last longer and remain beautiful.

Prick-off The transplanting of seedlings from the seed flat to pots or other containers.

Primary root The first root that develops.

Proficiency award A form filled out by the student relating to a specific area of agriculture. Competition among FFA members can win chapter, district, state, or national recognition.

Propagation The reproduction of plants by seed, cuttings, budding, or grafting.

Proprietor The sole owner of a business.

Prune The judicious removal of any part of a plant to enhance its shape or function.

Pustules Small, round or tube-like projections which release fungus spores.

Quarantine To physically isolate a pest.

Reach Length of the threads on a spark plug.

Record book An overall management plan to keep track of expenditures and cash. It contains helpful information that can aid in obtaining a loan, a higher FFA degree, or a proficiency award.

Redressing The procedure used to restore a sharp edge to the blade.

Regeneration Plants that have the capacity to replace or grow missing parts.

Reproductive growth The maturing of the vegetative parts and the development of flowers, seeds, and fruit.

Respirator A device that fits over the mouth and nose that allows a person to breathe but filters out harmful substances.

Respiration The process of obtaining energy from organic material. It is the reverse process of photosynthesis.

Rhizomes Underground horizontal stems that are able to develop roots and shoots at any node.

Rhythm Created when one's eye can move smoothly over an arrangement or landscape by carefully spacing flowers or shrubs of different colors, sizes, and forms.

Root Part of the plant located underground that anchors it, supplies it with nutrients and water, and stores food for it.

Root cutting Cutting taken by using a 2 to 4 in (5 to 10 cm) section of the root and placing it in moist rooting media.

School laboratory Includes a greenhouse and nursery in addition to the school grounds.

Scion A piece of twig or shoot that is inserted on another plant with an established root system when grafting.

Seed A miniature plant enclosed in a protective coat.

Semi-hardwood cuttings Cuttings made from narrow-leaf and broad-leaf evergreens, taken from the latest growth on the plant after it has finished its rapid summer growth.

Sepals Green leaflike structures that enclose the bud that are located at the base of the flower and collectively called the calyx.

Separation A portion of the plant such as a runner, bulb, or corm is naturally divided.

Shadehouse A frame covered with a material such as saran or plastic screen which reduces the intensity of the sun.

Simple fruit Develops from single ovary; can be fleshy or dry.

Sodding Establishing a lawn through the use of an already established sod rather than directly from seed.

Softwood cuttings Cuttings made from new growth in the spring or early summer when that part of the plant is growing rapidly.

Soil test A test made to determine the fertilizer status of the soil.

Spores Microscopic one-celled bodies produced by fungi for reproduction purposes. They are similar to seeds produced by plants.

Sprigging Technique used mainly on warm-seasoned grasses. A sprig consists of a rhizome with the roots attached. Sprigs are planted 6 to 12 in (15 to 30 cm) apart.

Stigma Expanded tip or uppermost end of the pistil.

Stockholders People who buy shares of stock in a business.

Stolons Stems that grow horizontally above the ground. Roots and shoots form from any node that touches the ground.

Stomata An outer surface structure of the cell that is composed of two guard cells that form a pore.

Stroke The movement of the piston from one end of the cylinder to the other.

Supervised Occupational Experience Program (SOE program) A planned sequence of practical activities that enables you to put into practice what you have learned.

Symptom The way a plant looks, smells, or feels that would indicate an unhealthy condition.

Systemic A type of pesticide that is absorbed into the plant and moves throughout the entire plant. It may be an insecticide, fungicide, or herbicide.

Taproot system A system where the primary root becomes the main root of the plant.

Tendrils Stem modifications that occur on grape vines and pea plants. Plant attaches itself to objects by these structures.

Translocation The process by which plants move a substance in a liquid form from one place to another such as stem to roots or throughout the plant system. As an example, systemic pesticide.

Transpiration The process that a plant releases water through the stomata openings. It can be affected by air movement, humidity, and temperature.

Toxic Dangerous or poisonous. The more toxic a pesticide is, the more poisonous it is.

Tuber A greatly enlarged fleshy portion of an underground stem; Has nodes that produce buds.

Turgor The firmness given to the cells by the pressure of the fluids that fill them.

Vacuoles Located in the cytoplasm of a cell. These large empty spaces store cell sap which is nutrient rich and also contains salt and pigment.

Vegetative growth Development of the leaves, roots, stems or the vegetative parts of the plant.

Vegetative propagation The reproduction of plants by cuttings, budding, or grafting.

Viruses Extremely small organisms that cause several plant diseases.

Voltage An electrical term used to describe the pressure of electricity.

Watt An electrical term which describes the rate at which electricity performs work.

Weed Any plant growing where it is not desired.

Wintering house A structure made of plastic used for protecting plants in the winter.

Zoning laws Restrictions that determine the type of business that can operate in a certain location.

INDEX

355